METALLURGICAL FAILURES
IN FOSSIL FIRED BOILERS

METALLURGICAL FAILURES IN FOSSIL FIRED BOILERS

Second Edition

DAVID N. FRENCH

A Wiley-Interscience Publication
JOHN WILEY & SONS, INC.
New York / Chichester / Brisbane / Toronto / Singapore

A NOTE TO THE READER
This book has been electronically reproduced from digital information stored at John Wiley & Sons, Inc. We are pleased that the use of this new technology will enable us to keep works of enduring scholarly value in print as long as there is a reasonable demand for them. The content of this book is identical to previous printings.

This text is printed on acid-free paper.

Copyright © 1993 by John Wiley & Sons, Inc.

All rights reserved. Published simultaneously in Canada.

Reproduction or translation of any part of this work beyond that permitted by Section 107 or 108 of the 1976 United States Copyright Act without the permission of the copyright owner is unlawful. Requests for permission or further information should be addressed to the Permissions Department, John Wiley & Sons, Inc., 605 Third Avenue, New York, NY 10158-0012.

Library of Congress Cataloging in Publication Data:
French, David N.
    Metallurgical failures in fossil fired boilers / by David N. French. — 2nd ed.
      p. cm.
    "A Wiley-Interscience publication."
    Includes index.
    ISBN 0-471-55839-7 (cloth)
      1. Steam-boilers—Failures. 2. Metals—Fracture.
  3. Fractography. I. Title.
TJ288.F7 1992
621.1′83—dc20   92-8747
                                          CIP

To Louise

CONTENTS

PREFACE TO THE FIRST EDITION   xi

PREFACE TO THE SECOND EDITION   xiii

1 INTRODUCTION   1

    Failure Statistics / 2
    Types of Fuel / 6
       ʿher Fuels / 11
       .oys of Construction / 16
       echanical Properties / 16
       ailure Analysis / 17
       .eferences / 19

DESIGN CONSIDERATIONS   21

    Introduction / 21
    Boiler Operation / 22
    Stress Calculations / 25
    Heat Transfer / 31
    Fluid Flows / 41
    Effects of Fuel on Furnace Size / 47
    Case Histories / 49
    References / 74

3 METALLURGICAL PRINCIPLES: FERRITIC STEELS   75

Atoms and Crystals / 75
Solid Solutions / 77
Grain Boundaries / 77
Phase Diagrams / 83
Iron–Carbon Diagram / 85
Lever Rule / 85
Transformation in Iron–Carbon Alloys / 87
T-T-T Curves / 88
Mechanical Properties / 93
Impact Properties / 106
Fatigue / 107
Thermal Fatigue and Corrosion Fatigue / 108
Spheroidization / 110
Graphitization / 115
Equilibrium Microstructures / 120
Heat Treatment / 120
Specific Effect of Alloying Elements / 122
Sample Preparation and Metallography / 123
Definitions / 124
Microstructures / 125
Case Histories / 147
References / 240

4 STAINLESS STEELS   242

Austenitic Stainless Steels / 242
Work Hardening / 243
Heat Treatment / 245
Solution Anneal / 246
Sigma-Phase / 245
Sensitization / 249
Welding / 251
Higher-Alloy Austenitic Materials / 252
Case Histories / 254
References / 276

5 FAILURES CAUSED BY GAS–METAL REACTIONS   277

Steam-Side Reactions / 277
Exfoliation / 279

Decarburization / 284
Fire-Side Reactions / 284
Carburization / 286
Erosion / 290
Case Histories / 293
References / 332

6 CORROSION-CAUSED FAILURES 324

Basic Principles / 324
Types of Corrosion / 326
Morphology Superheater and Reheater Corrosion / 346
Control of Ash Corrosion / 350
Furnace-Wall Corrosion / 351
Oil-Ash Corrosion / 361
Corrosion in Refuse-Fired Boilers / 364
Effects of Chlorine / 366
Effects of Carbon / 367
Corrosion Morphology / 367
Additives / 367
Corrosion Prevention / 369
Effects of ID Scale / 370
Chemical-Cleaning Decisions / 372
Superheaters and Reheaters / 375
Low-Temperature Corrosion / 378
Stress-Assisted Corrosion / 378
Boiler-Feedwater Control / 381
Case Histories / 384
References / 419

7 WELD FAILURES 422

Heat-Affected Zone / 422
Heat-Affected Zone Microstructures / 429
Internal Configuration / 439
Dissimilar Welds / 439
Dissimilar-Metal Weld Failure Prevention / 447
Stress-Assisted Corrosion / 448
Microstructural Effects on Corrosion / 455
Reheat Cracking / 455

Copper Penetration / 458
References / 461

8 FAILURE PREVENTION   462

Design of Superheaters and Reheaters / 463
Design of the Furnace / 468
Operational Problems / 469
Fuel / 470
Remaining-Life Assessment / 472
Computer Simulations / 483
References / 489

APPENDIX   491
INDEX   506

PREFACE TO THE FIRST EDITION

The study of boiler failures can be beneficial to both the plant and design engineer. At the plant level, an accurate explanation of the cause of a forced outage can prevent a similar future failure. For the design engineer, a catalog of past metallurgical problems can lead to alternative designs that reduce the chance of failures. When an old unit is repaired—for example, to replace a superheater—knowledge of the metallurgical condition of the original components can help in the installation of an improved design. In each case, the result is more trouble-free operation, lower maintenance costs, and better availability and reliability.

For nearly 25 years, the Riley Stoker Metallurgical Laboratory has been conducting failure analyses of boiler-tube samples submitted by its customers. One of the problems in dealing with these failures has been the lack of suitable examples for reference. Until the last few years, examples of and explanation for metallurgical failures were seldom published. Some information is now available, but it is scattered throughout the many publications that cover metallurgy. Also, the several books on failures give only an occasional example that relates to boilers. No single text is available to cover metallurgical failures in this important field. Therefore, I have presented here, through the use of actual case histories, a broad range of the more typical failures.

The examples of boiler failures cover a wide range of topics, including corrosion, high-temperature-related phenomena, welding, fabrication defects, changes in microstructure, oxidation, exfoliation, and decarburization. Not everyone interested in failure has the same level of metallurgical training. The water chemist has a different perspective from the mechanical engineer or the welding specialist. To help understand the cause and the reasoning behind the explanation, and to place most readers on an equal footing, basic concepts in metallurgy are also given. Finally, suggestions are made for the prevention of these failures from design, operation, and maintenance viewpoints.

This book was written while I was Director, Corporate Quality Assurance for the Riley Stoker Corporation, Worcester, Massachusetts. While there, several people helped me in the development and understanding of the ideas expressed herein. I gratefully acknowledge their assistance: Mr. James J. Farrell, former president, who gave permission to use the files of the Metallurgy Department for the case histories and encouraged the writing, and Mr. Kenneth J. Heritage, president, for his continued support; of the many engineers and staff, C. W. Arcari, J. C. Finneran, R. E. Hallstrom, P. J. Hunt, F. M. Kightlinger, C. M. Kotseas, T. A. Mellor, and R. K. Mongeon deserve a special note of thanks. R. Holdridge did the line drawings; James H. Bauer ably did all of the metallography and photomicrographs. A. H. Rawdon of Riley Stoker, Dr. P. Weihrauch, Technical Director, Massachusetts Materials Research, and Prof. C. H. T. Wilkins of the University of Alabama read the manuscript and made useful and valuable suggestions. I especially thank Miss Katherine M. McManus who typed the entire text. Finally, to my wife, Louise, I give my heartfelt appreciation for her support throughout the many months of this project.

DAVID N. FRENCH

Northborough, Massachusetts
December 1982

PREFACE TO THE SECOND EDITION

In the nearly 10 years since the publication of the first edition of this book, there has been a dramatic increase in the interest and understanding of boiler-tube failure analysis. Boiler-tube failures are, by far, the primary cause of forced outages in power plants. Improved availability and reliability are most easily obtained by reduction in the frequency and number of boiler-tube failures. The Electric Power Research Institute (EPRI) has been instrumental in developing methodology and in sponsoring programs to that end. Important publications have increased our understanding of boiler-tube failure mechanisms, including EPRI-sponsored conference proceedings, an EPRI Manual on Boiler-Tube Failures, and the NALCO Guide to Boiler-Tube Failure Analysis. With the increased interest in boiler-related problems, it seems an appropriate time to revise the first edition of this volume.

In 1984, I started a metallurgical consulting firm specializing in the failure analyses of boiler components. Thus, the exposure to boilers manufactured by companies other than Riley Stoker became an important part of my experience. The current revision includes a wider variety of boiler equipment than represented in the first edition. Further, important changes in the combustion of fossil fuels to reduce sulfur- and nitrogen-oxide emissions were mandated. Low-NOx burners have led to an increase in the frequency of reducing-condition corrosion problems.

The idea that a boiler operation is a balance between heat flow from the combustion of fuel and the steam generation in the furnace, or steam superheating in the superheater or reheater, suggested incorporating heat- and fluid-flow considerations to the volume. Since an improved understanding of a failure is often related to an understanding of the stresses involved, a short section on stress analysis has been included. Additional case histories have been presented that reflect the wider variety of problems examined.

Finally, virtually all steam-generating equipment is more than 10 years old, and half is more than 20 years old. As plant equipment ages, operators become interested in the determination of the end-of-life and material-condition assessments. Thus, information and methodology on these topics have been added.

Several people have been especially helpful to this second edition. I gratefully acknowledge Mr. A. H. Rawdon and Mr. James Ciulik of Radian Corp. Mr. Rawdon was most helpful in the chapter on heat flow, stress analysis, and fluid dynamics. Finally, to my wife, Louise, the corporate factotum of David N. French, Inc., Metallurgists, who has been particularly helpful in editing and correcting grammar and spelling, and without whom the revision would have been impossible, my heartfelt gratitude.

DAVID N. FRENCH

Northborough, Massachusetts
November 1992

CHAPTER ONE

INTRODUCTION

When a power plant is forced off the line because of a single component, the cost of the lost electric-power generation can run to several hundred thousand dollars a day. When process steam is lost in a plant manufacturing synthetic textile fibers and the liquid polymers solidify within the pipe work of the plant, the cost of repair of the steam generator is minor compared with the cost of lost production and repair outside the boiler.

Understanding the cause of failures can lead to prompt corrective measures. Failures caused by the improper selection of materials for a particular application lead to changes in design standards and manufacturing processes. A catalog of weld failures by type of joint design, for example, has eliminated the use of backing-ring joints in water circuits. The crevice between the inside diameter of the tube and the backing ring led to premature failure caused by corrosion initiated at the crevice. These joints are now welded by using tungsten–inert-gas or metal–inert-gas techniques without the use of backing rings. Materials used in support systems for superheater and reheaters have, over the years, gradually changed from medium-chromium ferritic alloys to 18 chromium–8 nickel austenitic stainless steels to the higher chromium–nickel alloys similar to Inconel 601® or RA333®. Again, these changes have been made because analyses of failures indicated that greater corrosion resistance was required in these portions of the boiler.

High-temperature failures in waterwall tubes in the areas of higher heat release have been identified as having been caused by film boiling. Rather than have the steam form in discrete bubbles, the heat transfer through the tube walls is so high and the steam generation so rapid that bubbles from many nucleation sites cannot be entirely swept away from the surface, and a

2 CHAPTER 1 INTRODUCTION

film of steam forms. To improve the turbulence and decrease the chance of film boiling, rifled tubing is commonly used. The identification of the cause of these premature high-temperature failures aided greatly in the change in tubing configuration used in these high-heat-release areas of the furnace.

FAILURE STATISTICS

The following analysis reflects the experience of the Metallurgy Department of the Riley Stoker Corp. for the 25-year period ending in 1980. We list the breakdown between mechanical and corrosion failure and further classify the various kinds of failures, locations, and materials:

Mechanical	81%
Corrosion	19%

Of the mechanical failures, the various types of failures are:

High temperature (short time)	65.8%
Creep (high temperature/long time)	8.6%
Erosion	3.5%
Graphitization	4.9%
Fatigue	3.8%
Weld failures	3.1%
Swages	2.6%
Coefficient of thermal expansion	2.2%
Tube ties, lugs	2.6%
All others	2.9%

Of the corrosion failures, the breakdown by cause is:

Boiler feedwater	37.2%
Hydrogen damage	20.0%
Ash	19.2%
oil 9.2%	
coal 10.0%	
Oxygen pitting	10.8%
Stress corrosion cracking	8.1%
Caustic attack	4.5%
Others	4.0%

The corrosion failures listed above add to more than 100% because some problems have been counted in more than one category; for example, an upset in boiler-feedwater chemistry may cause hydrogen damage or caustic attack.

FAILURES BY LOCATION

Waterwalls	29.4%
Superheaters	44.8%
Reheaters	13.5%
Economizers	4.8%
Roof	1.9%
Floor	1.4%
Others	4.2%

FAILURES BY MATERIAL

Carbon steel (SA-178A, SA-210, etc.)	40.4%
Carbon + ½ Mo (SA-209 T-1)	9.5%
1¼ Cr − ½ Mo (SA-213 T-11)	18.0%
2¼ Cr − 1 Mo (SA-213 T-22)	24.0%
Stainless steel (SA-213, 304, 321H, etc.)	5.1%
Others	3.0%

There may be certain inherent distortions in the distributions of these data. Only those samples submitted are included; if a failure occurs and is not reported, there is, of course, no record. The materials reflect the Riley Stoker use and may not be representative of the rest of the industry. However, their records are believed to be typical of the industry's problems.

David N. French, Inc., Metallurgists has provided metallurgical services to the boiler industry since 1984. The following is a tabulation of the top 10 causes of failure, again, based only on samples submitted.

Top 10 Failure Causes

1. Creep (long-term overheating)		23.4%
2. Fatigue		13.9%
thermal	8.6%	
corrosion	5.3%	
3. Ash corrosion		12.0%
coal	8.1%	
oil	1.4%	
refuse	2.5%	
4. Hydrogen damage		10.6%
5. Weld failures		9.0%
dissimilar metal welds	3.4%	
6. High temperature (short-term overheating)		8.8%
7. Erosion		6.5%
8. Oxygen pitting		5.6%
9. Caustic attack		3.5%
10. Stress corrosion cracking		2.6%
		95.9%

The service environment of fossil-fired steam generators is among the most severe of any large engineered structure. Engineering alloys in these applications are expected to have a useful service life of 30 or more years in a hostile atmosphere. On the fire side there are the rigors of high temperature and corrosion by the products of combustion from a variety of fossil fuels. On the steam side, the oxidation of the boiler tube by high-temperature steam and the corrosion by contaminated boiler-feedwater can seriously reduce the tube-wall thickness and lead to premature failure. Since the life expectancy is measured in thousands of hours, and service temperatures range from 700 to 1200°F (370 to 650°C) and higher, alloy microstructures will tend toward their thermodynamic equilibrium structures (or forms). To be useful, steels need to possess good high-temperature strength and creep resistance; but these properties are usually achieved by additions of alloying elements and by the presence of microstructural constituents that may be unstable when exposed to long times at high temperatures. For example, lamellar pearlite in a plain carbon steel will be transformed progressively from platelets to spheroids of iron carbide to, ultimately, graphite and ferrite when exposed to service temperatures for long times. Also, austenitic stainless steels will develop extensive networks of carbide precipitates and/or sigma phase, a brittle intermetallic compound of iron and chromium. These microstructural changes tend to weaken the material. Furthermore, metals tend to revert to the much more stable oxide and waste away through oxidation.

Steam pressure varies from a few hundred pounds per square inch (psi) for small package boilers to 4000 psig and higher for supercritical or once-through boilers. The American Society of Mechanical Engineers (ASME) Boiler and Pressure Vessel Code[1] establishes minimum design requirements and governs the strength of pressure-part components. Hence, the stress levels are the same for a given material at a particular temperature regardless of the size of the boiler. For design purposes, the Code has established minimal allowable stresses based on the following criteria:

1. One-quarter of the minimum room-temperature tensile strength
2. One-quarter of the tensile strength at temperature
3. Two-thirds of the minimum room-temperature yield strength
4. Two-thirds of the yield strength at elevated temperature
5. In the creep range, the stress to give 1% creep in 100,000 hr, or two-thirds of the average stress to produce rupture in 100,000 hr or four-fifths of the minimum stress for rupture in 100,000 hr

However, while the Code may set design-stress levels, it is usually the oxidation limit that sets the maximum useful temperature for a given alloy. Highly corrosive fuels will dictate the use of more-corrosion-resistant materials than would otherwise be required. For example, stainless steels are substituted for medium-chromium ferritic alloys, and low-chromium ferritic alloys are substituted for the plain carbon steels to reduce oxidation.

Steam boilers vary in size from a few thousand pounds of steam generation per hour to more than 10 million pounds of steam per hour for large central-station utility units. For boiler sizes of less than 200,000 pounds of steam per hour, shop assembly of the entire unit is the preferred method of fabrication; the larger boilers are field-assembled. The sizes of large utility boilers are truly surprising. Furnace sizes of 100 ft × 50 ft × 150 ft high are not uncommon. Such steam generators would have more than 50 miles of tubing in the furnaces, and the superheater, reheater, and economizer would have more than 160 miles of tubing. Coal consumption of 10,000 tons per day is not uncommon.

For units of equal steam output, the type of fuel burned dictates the relative size of the boiler. The type of fuel also dictates the materials used to support the superheater and reheater. The selection of materials for such support is governed by the expected gas or metal temperatures and the corrosiveness of the fuel ash. The severity of corrosion attack from the products of combustion in an oil-fired boiler can vary from nearly innocuous, for a highly refined low-sulfur oil similar to home heating oil, to highly corrosive, for oil ashes containing appreciable amounts of vanadium pentoxide, sodium oxide, and sodium sulfate. Under severe conditions, a stainless-steel reheater can be corroded to the point where steam leaks in less than 18 months.

The use of coal as a fuel is now the overwhelming choice in new boiler construction. The type of coal, from lignite to eastern bituminous, will dictate the design of the boiler. The amount of ash affects the furnace-ash buildup and slagging rate; the fouling of superheater, reheater, and economizer surfaces; the quantity of particulate and gaseous emissions; and the amount of bottom ash requiring disposal. The composition of the coal and the way in which it is burned dictates the corrosiveness of the residue. The corrosiveness of coal ash is related to the formation of complex sulfates, usually sodium or potassium pyrosulfate or sodium or potassium iron sulfate. These form under deposits in the hottest portions of the superheater and reheater tube surfaces.[2]

Another major difference among oil-, gas-, and coal-fired boilers is the effect the quantity and type of ash have on the erosion of boiler surfaces. The quantity of ash is small for oil- and gas-fired boilers, but in coal-fired boilers the ash may be as much as 15–20 wt% of the coal. Gas velocities as they pass through the superheater and reheater surfaces must be lower for coal-burning boilers than for oil- or gas-fired boilers. Economizer tubes must be spaced farther apart; bare-tube economizers typically are used for coal firing, whereas finned-tube economizers may be used for oil- and gas-fired boilers.

TYPES OF FUEL

Coal

Coal is the most abundant and perhaps the most complicated and troublesome of the major fuels. Problems arise in the use of coal almost from the moment of its delivery to the coal pile at the power plant. High-sulfur coal will weather. Sulfur leaches out, forms sulfuric acid, and is extremely corrosive to the coal-pulverizing equipment. Special alloys are needed to handle the problems of corrosion and erosion and still give satisfactory performance. From the coal-pulverizing equipment, the coal travels through coal piping to the burners for uniform combustion within the furnace. The handling of coal involves problems of erosion in all of the coal-piping system, and combinations of erosion and corrosion occur at the high-temperature burners. After coal is burned, the ash continues to cause problems within the furnace. Ash may contain corrosive components that lead to corrosion and fouling within the high-temperature portions of the boiler. Since coal can contain up to 15% or more ash, provisions must be made for its handling and removal. Coal ash can be highly fouling, necessitating the use of sootblowers and other deslagging devices, and thereby increasing the erosion and wear of boiler tubes in the vicinity of these devices. Table I[3] of the appendix shows typical composition of coal mined and used in the United States, while Figs. 1.1 and 1.2 show cross-sectional views of typical coal-fired boilers.

Natural Gas

The use of natural gas as a boiler fuel has been curtailed severely in recent years. However, many large boilers still use it. Natural gas has the fewest design restrictions of all of the major fuels, since it is clean and easy to burn. If only natural gas is burned, fuel-storage facilities, ash-handling equipment, sootblowers, and dust collectors are unnecessary. Superheaters, reheaters, and economizer surfaces can be arranged for maximum heat transfer and minimum draft loss without consideration of ash deposits and erosion. The total furnace size is the smallest of all major fuels for a given steam output. Figure 1.3 shows a side elevation of a gas-fired utility boiler.

Fuel Oil

Many of the desirable features of natural gas, including ease of handling, are found in fuel oil. Since the amount of ash found in fuel oil rarely exceeds a few pounds per ton, ash hoppers, ash pits, and other ash-handling equipment may be eliminated from design considerations. It does require, however, fuel storage, heating, and pumping facilities.

TYPES OF FUEL 7

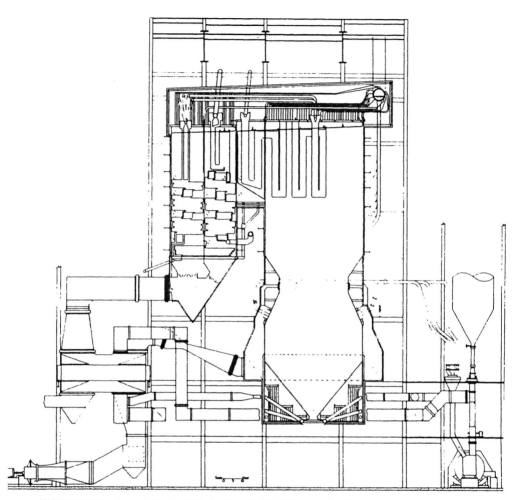

FIGURE 1.1. A coal-fired utility boiler: 3,900,000 pounds of steam per hour at 2630 psig operating at 1005°F/1005°F (540°C) superheat and reheat steam temperature. Courtesy of Riley Stoker Corp.

As will be discussed in greater detail in Chapter 6, fuel-oil ash is not without its corrosion problems. Compounds of vanadium, sodium, and sulfur may, under certain firing conditions, form low-melting-point slags that are extremely corrosive to boiler tubing. However, since the volume of ash is so small, it is possible through the use of suitable oil additives to adjust the chemical composition of these slags to reduce or eliminate their corrosive tendencies. In the appendix, Table II[4] lists the range of analyses of oil fuels, while Table III[2] shows analyses of the ash from fuel oils taken from various sources. Figure 1.4 is a schematic of a typical oil-fired utility boiler.

8 CHAPTER 1 INTRODUCTION

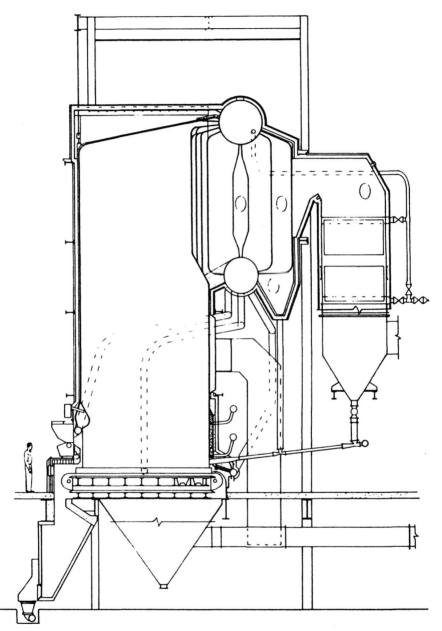

FIGURE 1.2. A stoker-fired industrial boiler: 150,000 pounds of steam per hour at 260 psig operating pressure at saturated steam temperature. Courtesy of Riley Stoker Corp.

FIGURE 1.3. A gas-fired utility boiler: 2,942,000 pounds of steam per hour at 2620 psig and 1005°F/1005°F (540°C) superheat and reheat steam temperature. Courtesy of Riley Stoker Corp.

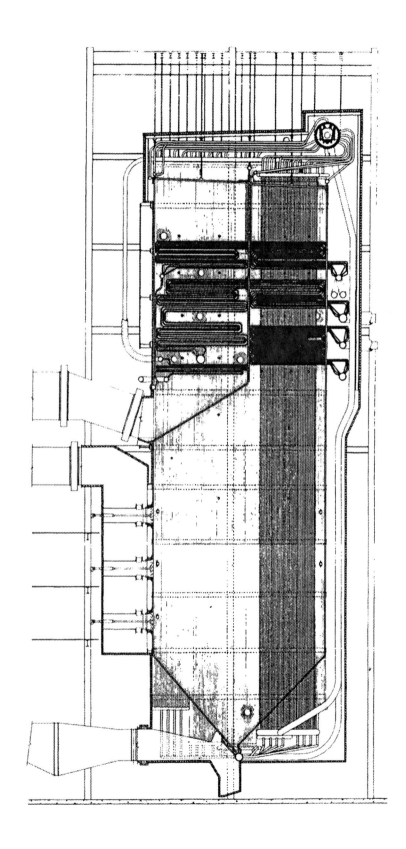

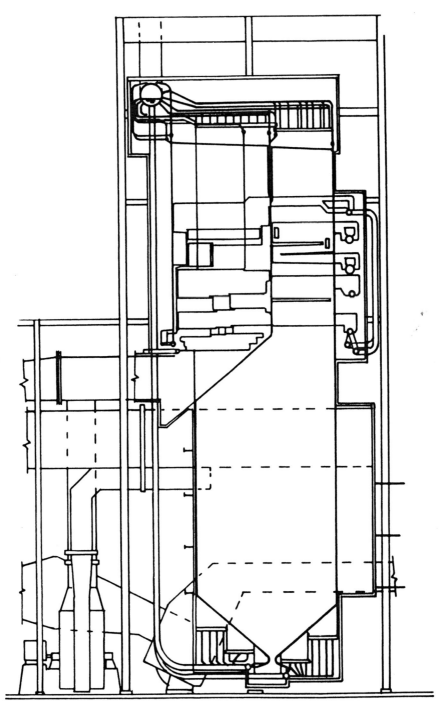

FIGURE 1.4. An oil-fired utility boiler: 3,550,000 pounds of steam per hour at 2500 psig and 1005°F/1005°F (540°C) superheat and reheat steam temperature. Courtesy of Riley Stoker Corp.

OTHER FUELS

By-product Gases

Within the petroleum and steel industries, by-product gases rich in carbon monoxide (CO) are plentiful and relatively inexpensive. Boilers have been designed specifically to use these low-Btu fuels. Since CO is a reducing gas, care must be exercised in the firing to prevent CO buildup adjacent to furnace walls: CO can reduce the protective magnetite (Fe_3O_4) film to iron powder, leading to severe metal wastage or corrosion by the reducing environment. High dust loading, especially in blast-furnace gas, can lead to plugging of burner ports and excessive fouling of heat-absorbing surfaces. Dust must be removed from the blast-furnace gas by washing or by using electrostatic precipitators before the gas is introduced into the boiler. Figure 1.5 shows a cross-sectional view of a blast-furnace gas-fired boiler.

Pitch and tar residues from the distillation of petroleum and coal are also suitable for use as boiler fuels.

Bagasse

Refuse from the milling of sugarcane is called bagasse. Bagasse consists of matted cellulose fibers with fine particles of cellulose and has a heat content on a dry basis of 4000 Btu/lb. Depending on the type of sugarcane operation, the supply of bagasse can meet the plant steam demands. Figure 1.6 shows a schematic of a bagasse-fired boiler.

Wood

Lumber, pulp and paper, furniture, plywood, and turpentine industries are typical of operations where wood-derived refuse is available in apreciable quantities and can be used as a boiler fuel. On a dry basis, all wood has essentially the same composition and heat content. The analysis of bark ash shown in the appendix, Table IV,[5] is representative of the residue from the combustion of all wood. When wood is fired alone, there is no serious slagging problem. Ashes from more than one fuel may combine to form an ash of a lower melting temperature than either alone, thus causing increased fouling or corrosion problems. Figure 1.7 shows a schematic of an oil- and wood-bark-fired boiler.

Municipal Refuse

Quite a bit of study has been done, primarily at Batelle Memorial Institute, Columbus, Ohio, on the use of municipal refuse as a boiler fuel.[6-10] In these studies, the municipal refuse has been sorted carefully to remove glass,

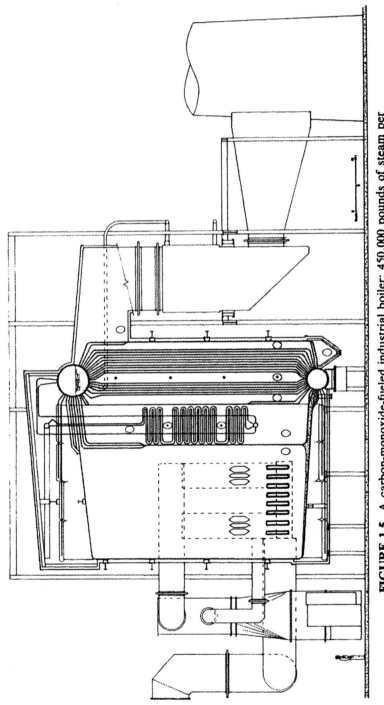

FIGURE 1.5. A carbon-monoxide-fueled industrial boiler: 450,000 pounds of steam per hour at 500 psig and 750°F (400°C) superheat steam temperature. Courtesy of Riley Stoker Corp.

OTHER FUELS 13

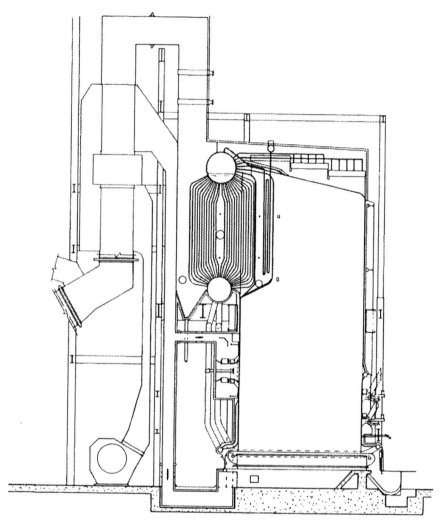

FIGURE 1.6. A bagasse-fueled industrial boiler: 240,000 pounds of steam per hour at 400 psig and 550°F (290°C) superheat steam temperature. Courtesy of Riley Stoker Corp.

metal, and other noncombustibles. Pneumatic tubes are used to blow the fuel into the furnace for ancillary combustion with oil or coal.

Stoker-fired boilers have been used to burn unclassified refuse to generate steam for industrial use. Figure 1.8 is a cross section of such a burner now in operation. There are problems with municipal-refuse-derived fuel (RDF) that come from polymer waste. Polyvinyl chloride, when burned, contributes to higher chlorine or to hydrochloric acid components within the flue gas.[7] For the most part, if tube-metal temperatures are kept below about 750°F (400°C), these corrosion problems are minimized.

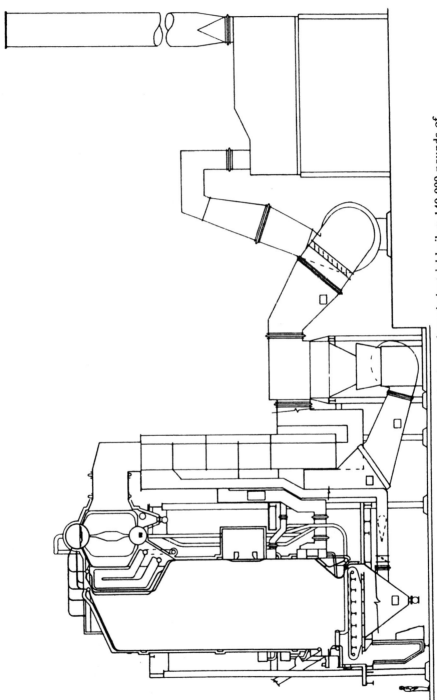

FIGURE 1.7. An oil- and wood-bark-fired two-drum industrial boiler: 140,000 pounds of steam per hour at 400 psig operating pressure and 725°F (385°C) steam temperature. Courtesy of Riley Stoker Corp.

OTHER FUELS   15

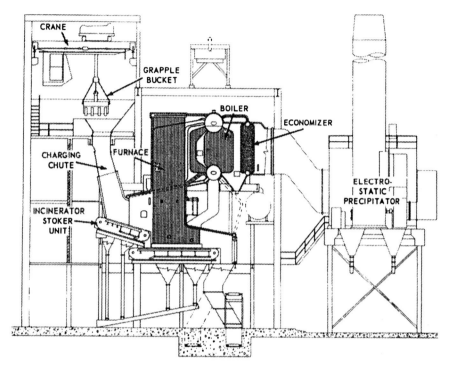

FIGURE 1.8. A municipal-refuse-fired boiler; 30,000 pounds of steam per hour at 250 psig and 406°F (208°C) saturated steam temperature. Courtesy of Riley Stoker Corp.

Work done at Battelle has shown a reduction in corrosion by chlorides if sulfur is added in sufficient quantities to the refuse.[8]

Coke

Coal heated in the absence of air has the volatile portions of the bituminous coal removed, leaving carbon behind as a residue. This carbonaceous residue, containing all of the ash and most of the sulfur of the original coal, is called coke. Coke is primarily used in the reduction of iron ore in a blast furnace and in the melting of iron in cupolas. "Coke breeze" (small particles of coke) is unsuitable for blast furnace and cupola use, and was at one time available for use as fuel in boilers. The modern trend to use lower grades of iron ore as part of the blast furnace charge requires a benefication treatment prior to smelting. Coke breeze is finely ground and mixed with iron-ore powder and clay, and the mixture is pelletized and sintered. These sintered pellets are a major part of the charge for a blast furnace, leaving little coke breeze for use now as a boiler fuel.

16 CHAPTER 1 INTRODUCTION

Miscellaneous Fuels

The food-processing industry produces numerous wastes based on vegetable products that can be used as fuel. These include nutshells, grain hulls, coffee grounds, tobacco stems, and others; but they are available in such limited quantities as to be insignificant in our total energy needs.

In the pulp and paper industry, the residue from the manufacture of paper, called "black liquor," is used as a fuel source for certain specially designed boilers. Wood chips are cooked under pressure in a solution of sodium hydroxide and sodium sulfide, which dissolves the lignin binder that holds the cellulose fibers of the wood together. The spent cooking liquor containing the lignin dissolved from the wood is the "black liquor." Combustion of such fuels produces some of the most corrosive environments of any boiler service.

ALLOYS OF CONSTRUCTION

Figure 1.1 shows the cross section of a large utility boiler. The furnace tubing and membrane walls are fabricated from plain carbon steel containing a maximum of 0.25% carbon. In the high-heat-release areas around the burners, a steel alloy containing 0.5% molybdenum or 0.5% chromium and 0.5% molybdenum may be used. In the superheater and reheater portions of the boiler, the materials of construction are carbon steel for the inlet end where steam temperatures are less than about 800°F (425°C), chromium–molybdenum steel alloys where steam temperatures are no higher than 950°F (510°C) and metal temperatures are less than about 1075°F (580°C), and austenitic stainless steels in the outlet portions where the steam temperatures are 1005°F (540°C) and metal temperatures may be as high as 1200°F (650°C). Alloys used for supports in these high-temperature portions of the boiler include the nickel–chromium high-temperature steel alloys. In the appendix, Table V[1] lists the commonly used alloys, and Table VI[11] shows the oxidation limits for tubing material.

MECHANICAL PROPERTIES

The allowable stresses used in the design of boilers are governed by the ASME Code. Table VII of the appendix lists the allowable stresses from the ASME Code, Section 1, for materials listed in Table IV.[1] Table VIII[12] lists the short-term, high-temperature, tensile-strength properties for the same alloys. Table XII presents tensile and yield strengths, elongations, and Rockwell B hardnesses of the commonly used boiler steels. These data are taken from mill test reports and, thus, reflect the measured values and give a range of values for these alloys. In most cases, the strength is more than 10%

greater than the specification requirements. These data will be useful in interpreting high-temperature failures, as will be discussed later.

FAILURE ANALYSIS

During the 1980s considerable attention was focused on preventing and understanding boiler-tube failures, which are the principal cause of forced outages in power plants. Understanding boiler-tube failures may lead to substantially improved performance, availability, and reliability of steam-generating equipment. The explanation of the root cause of the failure may lead to reduced frequency of future failures, since appropriate corrections may be made in a timely fashion.

The Electric Power Research Institute has suggested 22 failure mechanisms, which may be broadly categorized as:

1. Thermal degradation of the microstructure
2. Metal wastage by oxidation, corrosion, and erosion
3. Weld failures occurring in pressure parts and attachments
4. Fatigue, thermal fatigue, and corrosion fatigue
5. Human errors or poor overall quality

Thermal degradation of the microstructure occurs at elevated temperatures and weakens the steel. The onset of this change is temperatures above about 850°F for carbon steels and above about 1000°F for chromium–molybdenum steels. Microstructural changes precede creep failures. Occasional short-term, high-temperature excursions hasten the process. Excursions to very high metal temperatures will, by themselves, cause instantaneous failure. Examination of the microstructure will indicate its temperature at the moment of rupture.

Metal wastage reduces the thickness of the pressure part and increases the operational stress that hastens the onset of failures. Erosion by fly ash, sootblowers, or coal particles leads to metal loss. Usually erosion patterns are well known and may be repaired as a part of an annual outage. Fire-side corrosion comes from liquid–ash attack in the superheater and reheater at temperatures above 1000°F (540°C), on furnace walls at temperatures below 750°F (400°C), and dew-point corrosion by condensing sulfuric acid occurs at temperatures below 300°F (150°C). Ordinary high-temperature oxidation consumes steel also. Metal loss from oxidation is 1–2 mils per operating year, while liquid–ash corrosion rates may be as high as 15–25 mils per operating year, although they usually are 2–4 mils per operating year. Water- or steam-side corrosion comes from impurities, most often oxygen, or excursions in boiler-feedwater chemistry that are outside of the normal control range.

Welds are a continuing source of trouble, especially in high-temperature components. When a superheater and reheater support or alignment clip breaks, oxidizes, or corrodes, or slip spacers do not slip because of fly-ash pluggage, welds break. These failures cause bundles to fall out of alignment, and improper flue-gas distribution through the pendant sections results. Poor flue-gas distribution then puts added thermal stress on these high-temperature components. Weld backing rings are a source of corrosion, hideout, and fatigue cracks. Chemical cleaning may lead to preferential corrosion of the heat-affected zone of welds in drums, headers, and downcomers. Rolled tube ends are also subject to more rapid corrosion during chemical cleaning. Dissimilar-metal welds between ferritic and austenitic alloys have local high-temperature stresses that lead to premature failure.

Fatigue, thermal fatigue, and corrosion fatigue are causes of failures that are difficult to identify before actual failure occurs. Thermal-fatigue cracks can form on the insides of headers and drums that suffer temperature excursions. Superheater and reheater outlet headers may have large tube-to-tube differentials in steam temperature that accentuate and promote thermal-fatigue or creep-fatigue failures. Waterwall tubes suffer corrosion fatigue, especially at welded attachments or buck stays, when there is insufficient allowance for differential thermal expansion.

Failures caused by human error or poor overall quality can occur in any manufactured item. However, within a boiler, human error can be operator error (e.g., a low-water upset) design error, or poor choice of material. Material defects, laps, seams, or poor-quality welds are amenable to prevention by appropriate inspection techniques. For the most part, these kinds of mistakes have been well identified and are not a principal focus here.

Since the first edition was published in 1983, there has been considerable interest in boiler-tube failure analysis. EPRI has held several conferences, and numerous volumes on the subject have been published. The more useful texts are listed at the end of the chapter.

The book is divided along two general lines: engineering and metallurgical principles useful in understanding the "why" of a failure, and a catalog of case histories that illustrate the majority of boiler-tube failures. Chapter 2 presents stress, heat-flow, and fluid-flow analyses to help determine conditions that led to failure. Chapters 3 and 4 give metallurgical principles of ferritic steels and stainless steels, respectively. In Chapter 5 the effects of oxide formation from steam with steel, carburization, decarburization, and other gas–metal reactions are discussed. Corrosion, both fire-side and steam-side, the interrelation between steam-side scale and fire-side corrosion, and principles of boiler-water control are outlined in Chapter 6. Chapter 7 covers weld-related problems. Chapter 8 highlights approaches to failure prevention and the difficulties of predicting the end of useful service life of a high-temperature component. Data tables are included in the appendix.

REFERENCES

1. *The ASME Boiler and Pressure Vessel Code Section I, Power Boilers*, ASME, New York, 1980.
2. W. T. Reid, *External Corrosion and Deposits Boilers and Gas Turbines*, Elsevier, New York, 1971.
3. H. E. Burbach and A. Bogot, "Design Consideration for Coal Fired Steam Generators," Presented at Annual Conference Association of Rural Electric Generating Cooperation, Wichita, Kansas, June 13–16, 1976.
4. *Steam/Its Generation and Use*, 38th ed., Babcock and Wilcox, New York, 1972, pp. 5–19.
5. R. Schwegler, "Power from Wood," *Power*, Feb. 1980, pp. 51–53.
6. H. H. Krause, D. A. Vaughn, and P. D. Miller, "Corrosion and Deposits from Combustion of Solid Waste," *Trans. ASME*, Jan. 1973, pp. 45–52.
7. H. H. Krause, D. A. Vaughn, and P. D. Miller, "Corrosion and Deposits from Combustion of Solid Waste, Part II—Chloride Effects on Boiler Tube Metals," *Trans. ASME*, July 1974, pp. 216–222.
8. H. H. Krause, D. A. Vaughn, and W. K. Boyd, "Corrosion and Deposits from Combustion of Solid Waste, Part III—Effects of Sulfur on Boiler Tube Metals," *Trans. ASME*, July 1975, pp. 448–452.
9. H. H. Krause, D. A. Vaughn, and W. K. Boyd, "Corrosion and Deposits from Combustion of Solid Waste, Part IV—Combined Firing of Refuse and Coal," *Trans. ASME*, July 1976, pp. 369–374.
10. H. H. Krause, D. A. Vaughn, P. W. Cover, W. K. Boyd, and R. A. Olexsey, "Corrosion and Deposits from Combustion of Solid Waste, Part IV—Processed Refuse as a Supplementary Fuel in a Stoker-Fired Boiler," *Trans. ASME*, Oct. 1979, pp. 592–597.
11. Riley Stoker Corp., Metallurgy Department Standard.
12. *Steels for Elevated Temperatures Service*, U.S. Steel Corp., Pittsburg, Pennsylvania, 1972.

Useful References on Boiler Design and Operation

J. G. Singer, ed., *Combustion, Fossil Power Systems*, Combustion Engineering, Windsor, Connecticut, 1981.

Steam/Its Generation and Use, 38th ed., Babcock and Wilcox, New York, 1972.

David Gunn and Robert Horton, *Industrial Boilers*, Longman/Wiley, New York, 1989.

Useful References for Boiler-tube Failure Analysis

G. A. Lamping and R. M. Arrowood, *Manual for Investigation and Correction of Boiler Tube Failures*, EPRI Final Report, Project 1890-1 EPRI CS-3945, 1985.

R. D. Port and H. M. Herro, *The NALCO Guide to Boiler Failure Analysis*, McGraw-Hill, New York, 1991.

H. Thielsch, *Defects and Failures in Pressure Vessels and Piping,* Van Nostrand Reinhold, New York, 1965.

ASM Metals Handbook, 9th ed., Vol. 11, "Failure Analysis and Prevention," ASM International, Metals Park, Ohio, 1986.

ASM Metals Handbook, 9th ed., Vol. 13, "Corrosion," ASM International, Metals Park, Ohio, 1987.

I. M. Rehn, *Corrosion Problems in Coal-Fired Boiler Superheater and Reheater Tubes,* EPRI Final Report, Project 644-1 EPRI CS-1653, 1980.

International Conference on Fossil Power Plant Rehabilitation Proceedings, ASM International, Metals Park, Ohio, 1989.

R. Viswanathan, *Damage Mechanism and Life Assessment of High Temperature Components,* ASM International, Metals Park, Ohio, 1989.

R. W. Bryers, ed., *Incinerating Municipal and Industrial Waste,* Hemisphere, New York, 1991.

M. F. Rothman, ed., *High Temperature Corrosion in Energy Systems,* The Metallurgical Society of AIME, Warrendale, Pennsylvania, 1985.

I. G. Wright, ed., *Proceedings of the Symposium on Corrosion in Fossil Fuel Systems,* The Electrochemical Society, Pennington, New Jersey, 1983.

M. O. Speidel and A. Atrens, eds., *Corrosion in Power Generating Equipment,* Plenum Press, New York, 1984.

Conference on Boiler Tube Failures in Fossil Plants, Atlanta, Georgia, Nov. 1987, EPRI Research Project 1890, 1987.

J. P. Dimmer, G. A. Lamping, and O. Jonas, *Boiler Tube Failure: Correction, Prevention, and Control,* EPRI Final Report Project 1890-7 EPRI GS-6467, 1989.

R. H. Richman and T. W. Rettig, eds., *Failures and Inspections of Fossil-Fired Boiler Tubes: 1983 Conference and Workshop,* EPRI CS-3272, 1983.

CHAPTER TWO

DESIGN CONSIDERATIONS

INTRODUCTION

The two types of boilers are fire tube and water tube. In a fire-tube boiler, the hot gases are on the inside of the tubes and water is on the outside, and the boiler is usually used without superheat. In a water-tube boiler, the water is on the inside of the tubes and the hot gases are on the outside. Only water-tube boilers are used in large installations, usually with superheaters and, often, reheaters; hence, they are the focus of the following discussion. Depending on pressure and temperature, the design of water-tube boilers varies from small, shop-assembled, two- or three-drum boilers of 20,000 lb steam per hour to field-erected, single-drum units of several million pounds of steam per hour.

Any combustible material may be used for fuel: the main fuels are natural gas, oil, and coal. Examples of other fuels are organic refuse from coffee grounds, wood chips, and waste gas of carbon monoxide. Solid fuel may be ground to fine powder and burned, as pulverized coal, or fired on a stoker in larger pieces, as are refuse or wood chips.

The conventional steam cycle for central-station power plants is the Rankine cycle. The Rankine cycle consists of compression of liquid water by the boiler feed pump, heating to the saturation temperature in an economizer, evaporation in the furnace, expansion work of the steam in the turbine, and condensation of the exhaust steam in a condenser. Steam-cycle efficiency can be improved by adding superheaters to heat the steam above its saturation temperature. Reheat cycles further improve plant efficiency. Superheated steam is partially expanded and cooled, and then removed from the turbine and returned to the boiler to be reheated. In a reheat superheater (reheater), more heat is added to the steam, which is then returned to the turbine. Regenerative cycles use some partially expanded steam to preheat the boiler feedwater. Double reheater units in supercritical 3600-psi designs have been built to further improve efficiency. Thus, a current plant design for subcritical operation may be 2600-psi drum pressure, a 2450-psi superheater outlet at 1005°F, with a 600-psi reheater at 1005°F. A supercritical design may be a 3800-psi superheater outlet at 1005°F with two reheat stages: 1600 psi at 1005°F, and 600 psi at 1005°F.

BOILER OPERATION

The essential elements of a boiler are shown in Fig. 2.1.[1] Heat is added to the risers. As heat is added, steam forms at the saturation temperature, the equilibrium point between liquid and gas. For natural-circulation designs,

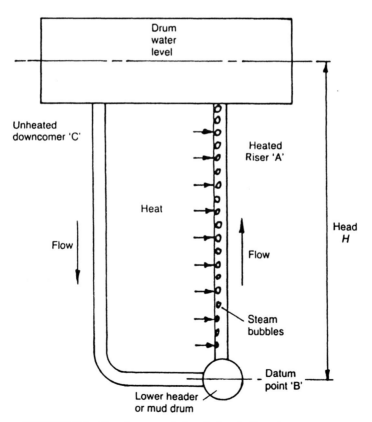

FIGURE 2.1. Simple natural-circulation boiler circuit (Ref. 1).

the fluid temperature in the risers remains constant. The water–steam mixture thus formed is of lower density than that of water alone in the downcomer. The driving force for a natural-circulation boiler is this difference in density of the fluid between the heated risers and unheated downcomers. At the top of the unit is a steam drum for separating steam from water. The steam may then be used at its saturation temperature for useful heat elsewhere, as in a paper mill or chemical plant, or piped to a superheater to be further heated, as in a plant for generating electricity. Condensed steam, condensate, is returned to the drum and thence to the furnace through unheated downcomers to start the cycle again. Mechanical pumps may be added to assist in forcing the water through the various circuits in a forced-circulation boiler. The circulation ratio is the total amount of water entering the riser divided by the amount of steam produced in the riser.

The typical steam cycle has the following elements:

1. It condenses steam output from the turbine in a condenser.
2. Condensate is circulated through appropriate purification equipment to ensure correct levels of dissolved solids and oxygen. The pH is controlled in the 8.5–10 range. These steps are necessary to prevent corrosion of the boiler.
3. The condensate is pressurized with the boiler feed pump.
4. The condensate is preheated in a feedwater heater or heaters. Steam for this step is taken from the turbine at various points. The temperature leaving the feedwater heaters may be about 450°F.
5. From the feedwater heaters, condensate travels to the economizer, which further raises the temperature to saturation temperature minus about 5°F.
6. The economizer outlet feeds the drum, and the drum feeds the lower waterwall headers through the downcomers.
7. Steam is generated in the furnace at constant temperature and pressure.
8. Steam and water are separated in a steam drum. Water returns to the downcomer, and steam is fed to the superheater.
9. The superheater raises the steam temperature from saturation to, for example, 1005°F.
10. Expansion work is performed in a turbine.
11. Steam is removed at lower pressure and temperature and returned to the reheater. Some steam is extracted at this point and fed to the feedwater heaters.
12. Reheated steam is returned to the turbine to perform more expansion work.
13. The steam exhausts from the turbine to the condenser, where the cycle repeats.

For pressures below 3208.2 psia, a steam drum is necessary to provide space to separate steam from water. The separated water, together with makeup feedwater from the economizer, is returned to the downcomers. At pressures above 3208.2 psia, there is no distinction between water and steam or between liquid and gas. Supercritical boilers—units that operate above 3208.2 psia—require mechanical pumps to help force the fluid through the heat-absorbing circuits. Since there is no steam separation, no steam drum is required.

Natural circulation is most effective when there are large differences in density between vapor and liquid. At pressures above about 2900 psig, a natural-circulation system becomes too large to be cost effective. Hence, natural-circulation boilers are limited to a design pressure of 2900 psig and an operating pressure of about 2600 psig. Forced circulation, however, may be used over all pressure ranges, including supercritical and subcritical.

In the design of all boilers with natural circulation, the primary objective is to maintain sufficient fluid flow through the heated risers for adequate control of tube-metal temperatures. The expected heat-transfer regime on the fluid side is nucleate boiling for all operating conditions. For nucleate boiling to be maintained, the heat flux from the combustion of fuel must be balanced by appropriate fluid flow. Steam bubbles form at discrete locations on the inside surface of the heated risers. When they reach a critical size, each bubble is swept away by the moving liquid. So long as single bubbles of steam form and are removed, nucleate boiling is said to occur. When the volume of steam or steam quality becomes too great, individual bubbles cease to exist; they are no longer swept away, but instead form a continuous film. This is referred to as *departure from nucleate boiling* (DNB). The DNB point depends on many variables, including pressure, heat flux, and fluid-mass velocity.

Sloping tubes, particularly those heated from above in which gravity aids in steam separation and holds the steam against the heated side of the tube, are especially likely to form steam blankets. Screen tubes—tubes heated around the whole perimeter—also require fluid velocities substantially higher than vertical risers. One way to keep the fluid—the steam–water emulsion—well mixed is to use rifled or internally ribbed tubing in these regions of high heat flux or low fluid velocity. Imparting a swirl to the fluid motion keeps the steam bubbles well mixed and prevents steam blankets (DNB), and a substantial improvement in the safe-operating heat-flux region may be obtained. Such tubes are useful in the burner zones, locations of highest heat input, sloped tubes, and other areas in the furnace where steam separation and DNB are likely.

The rate of heat absorption is rarely uniform along the entire length of the riser path. The maximum is at, or near, the burner position.

In any boiler, the design is a dynamic balance between heat transfer, from the fuel combustion, and fluid flow. Heat is transferred through a thin steel pressure boundary from hot flue gas or flame to the cooler steam or water. In "equilibrium," the temperature of the steel is within the design limit, the stress is low enough to meet design requirements, and the boiler will last for several decades. Failures occur because the stress or temperature is too high. Erosion, corrosion (either fire side from fuel ash or water side from impure boiler water), or oxidation reduce the wall thickness until the pressure boundary is too weak to contain the steam pressure. Finally the tube-metal temperature may be raised above safe design levels by conditions on both the gas (fire) side and steam side. The following sections present elementary stress analysis, heat transfer, and fluid flow. These concepts will be useful in the understanding of some boiler-tube failure problems.

STRESS CALCULATIONS

Since virtually all of the examples deal with pressurized cylinders, and the hoop stress is the principal stress in a tubular cross section, the following stress analysis deals with this stress only. In its simplest form, a boiler tube fails when the hoop stress, due to the internal steam pressure, equals the strength of the material. Table VIII in the appendix lists elevated-temperature tensile strengths for the common boiler steels. Thus, it is sometimes desirable to calculate the hoop stress in a tube to help estimate metal temperature at the instant of rupture.

The ASME Boiler and Pressure Vessel Code governs the design of the operating stress of steam-generating equipment. The equation of interest for boiler tubes is found in paragraph PG27.2:[2]

$$W_{min} = \frac{PD}{2S + P} + 0.005D + e \qquad (2.1)$$

where W_{min} = minimum tube-wall thickness, in
P = design pressure, psi
D = outside tube diameter, in
S = B & PV Code allowable stress from Table 1A, psi
e = rolling allowance, zero, for welded construction

Equation (2.1) gives the specified minimum wall thickness as required under the rules of the ASME Boiler and Pressure Vessel Code. There is no allowance made in Eq. (2.1) for metal loss due to oxidation, corrosion, or erosion. This equation is a little more conservative than the formula for hoop stress in a thin-walled cylinder, which can be found in texts on stress calculations:[3]

$$S = \frac{PD_M}{2W} \qquad (2.2)$$

where D_M is the mean tube diameter in inches and is equal to

$$D_M = D - W \qquad (2.3)$$

Rearranging Eq. (2.2) and (2.3) gives

$$W = \frac{PD}{2S + P} \qquad (2.4)$$

For welded construction, the B & PV Code requires a thicker wall by $0.005D$; compare Eq. (2.1) with (2.4).

Occasionally, the stress in a tube must be estimated just prior to failure. Two equations may be used. One is the equation for thick-walled cylinders with internal pressure,[3]

$$S = P \left(\frac{r_o^2 + r_i^2}{r_o^2 - r_i^2} \right) \qquad (2.5)$$

where $r_o = D/2$, the outside radius, in
 $r_i = r_o - W$, the inside radius, in

which is used if the thickness is more than one-tenth the radius. However, for most boiler-tube applications involving failure analysis, the most convenient equation, and it is sufficiently accurate for all stress calculations, is a rearrangement of Eq. (2.4):

$$S = \frac{P(D - W)}{2W} \qquad (2.6)$$

Equation (2.6) gives, within about 10%, the same calculated values of S for $D/W > 4$ as the more mathematically correct Eq. (2.5). Virtually all boiler tubes fall into the $D/W > 4$ range (2.00 in OD × 0.500 in minimum wall thickness (MWT), for example). In general, except where metal wastage has occurred due to corrosion, the calculated stress is lower than the allowable stress used in Eq. (2.1). The design pressure is about 5% greater than the actual operating pressure (e.g., 2730 psi versus 2600 psi drum pressure and 2450 psi superheater-outlet pressure). In seamless tubing, the actual wall thickness is 8–10% greater than the specified minimum wall. Thus, tube for tube, the actual operating stress is some 15+% lower than that used in Eq. (2.1) at the time the boiler was designed.

A comparison of Eqs. (2.5) and (2.6) for several conditions shows that Eq. (2.6) is more than satisfactory for purposes of failure analysis.

STRESS CALCULATIONS

OD	W	OD/W	S (Eq. 2.5)	S (Eq. 2.6)
2.00	0.200	10.0	4.556P	4.500P
2.00	0.250	8.0	3.571P	3.500P
2.00	0.333	6.0	2.597P	2.500P
2.00	0.400	5.0	2.125P	2.000P
2.00	0.500	4.0	1.667P	1.500P

Thus, for all practical purposes, Eq. (2.6) yields approximately the same results as the more awkward Eq. (2.5).

Equation (2.6) gives the hoop stress in the wall of a pressurized cylinder; the axial stress is half the hoop stress. Thus, when boiler tubes fail, the usual circumstance is a longitudinal rupture that parallels the cylinder axis. When circumferential cracks form, a secondary stress has been applied to the axial-pressure stress to make the axial stress greater than the hoop stress. Such secondary stresses can be system loads, bending moments, attachments for alignment, slip spacers that become "locked," differential expansion, thermal shock from sootblowers, or fluid-flow instabilities, etc. The one certainty is that the cracks or failure path is normal to the biggest stress.

Torus

The ASME B & PV Code makes no mention of the stresses in a torus, but it is occasionally useful to calculate the hoop stress in a tube bend. A torus is a doughnut shape; and a tube bend is half a torus (see Fig. 2.2) that is frequently found in SRE (superheater, reheater, economizer) elements. For thin-walled tubes, $D > 5W$, the hoop stress is given by

$$\text{hoop stress} = \frac{Pr}{2W}\left[\frac{2r_{bend} + r}{r_{bend} + r}\right] \quad \text{(extrados)} \quad (2.7)$$

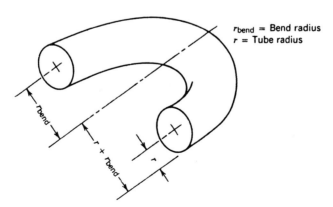

FIGURE 2.2. Sketch of half a torus, representative of a tube bend.

$$\text{hoop stress} = \frac{Pr}{2W}\left[\frac{2r_{\text{bend}} - r}{r_{\text{bend}} - r}\right] \quad \text{(intrados)} \quad (2.8)$$

Note that the maximum hoop stress is on the inside (intrados) of the bend. However, the metal is usually thicker here because of the metal flow during fabrication of the bend.

Figure 2.3 shows the cross section through a return bend. The extrados is smooth, and the intrados is thicker. The wall thickness at the intrados is about 60% thicker than the wall thickness at the extrados. Thus, even though the calculated stress, Eqs. (2.7) and (2.8), shows a higher value at the intrados, the assumption is that the wall is uniform throughout the bend. Since real bends appear more like Fig. 2.3, failures are seldom seen on the intrados. Those failures that do occur on the intrados are usually circumferential cracks from a force that tends to open the spacing between the legs. Circumferential hoop-stress cracks are very rare in this thickest part of the bend. A comparison of Eqs. (2.7) and (2.6) shows that the hoop stress in a cylinder is higher than that for a torus. Thus, tubes fail in the cylindrical section rather than in the bend, even though the extrados usually has a slightly reduced metal thickness.

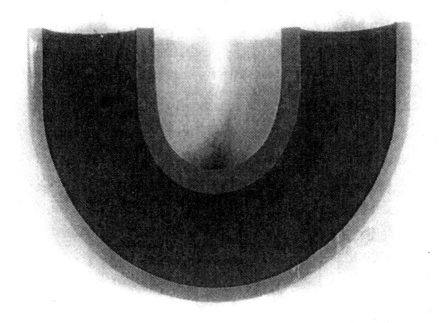

FIGURE 2.3. Cross section through a return bend. Note the wall-thickness increase at the intrados, about 60% thicker than at the extrados ($\frac{1}{2}\times$).

For a thin-wall cylinder, $D = D_M$, and Eq. (2.6) becomes

$$S = \frac{PD}{2W} = \frac{Pr}{W} \quad (2.9)$$

When Eq. (2.9) is set equal to Eq. (2.7), an estimate may be obtained for the wall thickness required to yield equal hoop stress in the extrados of a bend and in the wall of a cylinder. Thus,

$$\left(\frac{Pr}{W}\right)_C = \frac{Pr}{2W}\left[\frac{2r_{bend} + r}{r_{bend} + r}\right]_T \quad (2.10)$$

where the subscript C refers to a cylinder and T refers to the torus. Since the internal pressure P is the same for both, it drops out of the equation. The bend radius r_{bend} may be expressed as a multiple of the tube radius r. Equation (2.10) is solved for W_T/W_C for various r_{bend}. The following list shows the ratio of wall thickness in a torus to wall thickness in a cylinder for various values of r_{bend}:

W_T/W_C	r_{bend}
0.75	r (tube bent back on itself, legs tangent)
0.80	$1\frac{1}{2}r$
0.83	$2r$
0.88	$3r$
0.92	$5r$
0.96	$10r$

The thinning on the extrados determines the manufacturing process used. For close-radius bends ($r_{bend} \leq 2r$), hot bends are preferred. The intrados is heated to force maximum metal flow to the inside. The cold extrados is stronger and thus will thin less. Well-made hot bends thin less than 10% of the original wall. Computer-controlled hydraulic boosters also reduce wall thinning by pushing the tube through the bend.

In seamless tubes, the actual wall thickness is usually about 8–10% thicker than specified minimum wall thickness. To keep the stress in the extrados of the bend low enough to prevent failure but still allow some wall thinning, a wall thickness reduction of MWT minus 10% is usually used. Thus, a 2.0 in OD × 0.250 in MWT tube is acceptable if W_T is 0.225 in.

Tube Wastage and Wall Thinning

During normal boiler operations, a certain amount of tube wastage or wall thinning occurs. Such wastage may be due to erosion by fly ash, coal parti-

30 CHAPTER 2 DESIGN CONSIDERATIONS

cles, or sootblowers, or to corrosion by fuel ash or oxidation. In most cases the metal loss cannot be prevented. Guidelines are needed for tube replacement or repair—criteria for the extent of wall thinning that can be accepted without failures and forced outages. Figure 2.4 presents an example of severe wastage, a wall loss of two-thirds of the original thickness, without failure.

When tube-metal wastage occurs from the outside of the tube, both the diameter and wall thickness change, and hoop stress is redistributed within the wall. Since the stress that causes failure is in the thinnest section at an unknown diameter, it is more convenient to use the inside radius R_I and mean radius R_M. Equations (2.2) and (2.3) are altered to reflect this change from diameters to radii:

$$S = \frac{PR_M}{W} \quad \text{since } D_M = 2R_M \tag{2.11}$$

and

$$R_M = R_I + \frac{W}{2} \tag{2.12}$$

The inside radius R_I is

$$R_I = \frac{D - 2W}{2} \tag{2.13}$$

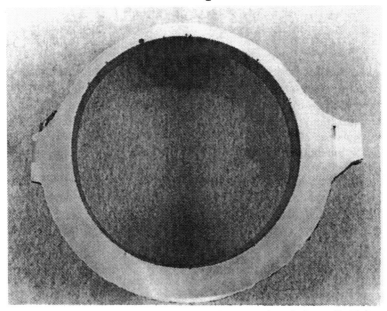

FIGURE 2.4. Molten ash erosion of floor tube has reduced the wall thickness by nearly 70% with no failure.

Since failures are not intended to occur, R_I remains essentially constant as the wall thickness and outside diameter are reduced.

Equations (2.11) and (2.12) can be combined to give

$$W = \frac{PR_I}{S - P/2} \tag{2.14}$$

Equation (2.14) is solved for the minimum wall thickness necessary for replacement.

To calculate W in Eq. (2.14), values for the pressure P and the safe stress S must be assigned. The Code uses design pressure for all stress-related calculations. However, boilers actually operate at a lower pressure. Therefore it is appropriate to use operating drum pressure (or superheater-outlet pressure) about 5% less than design pressure for P. The Code-allowable stress below the creep range is the tensile strength at a given temperature divided by a safety factor of 4. For waterwalls and economizers at metal temperatures below about 750°F, the actual tensile strength at 750°F is close to the Code-specified minimum tensile strength at room temperature. For SA-210 A-1, this specified minimum strength is 60,000 psi. Boiler tubes show considerable thinning before failures occur. In the example in Fig. 2.4, wall thickness was reduced by two-thirds. It seems reasonable to use a safety factor of 2½ rather than 4, therefore. In our example, OD = 2.75 in, MWT = 0.290 in:

$$R_I = \frac{2.75 - 2(0.290)}{2} = 1.085 \text{ in}$$

$$P = 2600 \text{ psig}$$

$$S = 60,000/2.5 = 24,000 \text{ psi}$$

Solving Eq. (2.14) gives

$$W = \frac{(2600)(1.085)}{24,000 - 2600/2} = 0.124 \text{ in}$$

The wall thickness, when replacement is necessary, is 43% of the original specified minimum thickness. No failure occurred in the sample shown in Fig. 2.4 at one-third of the original wall thickness. Thus, for boiler tubes operating below the creep range, it is quite safe to allow tube wastage to reduce the wall to about 40% of the original.

HEAT TRANSFER

Plane Areas

There are three modes of heat transfer: conduction, convection, and radiation. The operation of a boiler transfers the heat energy from the combustion

32   CHAPTER 2 DESIGN CONSIDERATIONS

of a suitable fuel through a steel tube to generate steam or to superheat the steam; for all practical purposes, all three modes occur within a boiler. Under normal service conditions, heat transfer reaches steady-state conditions. Thus, steady-state heat transfer only is considered, and is given by[4]

$$\frac{Q}{A_0} = U_0 \Delta T = \frac{\Delta T}{R_0} \tag{2.15}$$

where Q/A_0 = heat flux per unit of outside area, Btu/hr · ft²
$\quad\quad\quad U_0$ = overall heat-transfer coefficient based on outside area, Btu/hr · ft² · °F
$\quad\quad\quad \Delta T$ = temperature difference that drives heat flow, °F
$\quad\quad\quad R_0 = 1/U_0$ = combined resistance to heat flow based on outside area, ft² · hr · °F/Btu

The coefficient U_0 may be thought to include contributions from conduction, radiation, and convection; but it is usually more convenient to think in terms of thermal resistance R_0, the reciprocal of U_0. For steady-state conduction through a flat, planar wall, the steady-state heat-flow equation is

$$\frac{Q}{A_0} = \frac{k \Delta T}{\lambda} = \frac{k(T_1 - T_2)}{\lambda} \tag{2.16}$$

where k = thermal conductivity, Btu/ft² · hr · °F/ft
$\quad\quad\quad \lambda$ = plate thickness, ft
$\quad\quad\quad T_1$ = hot-side temperature, °F
$\quad\quad\quad T_2$ = cold-side temperature, °F
$\quad\quad\quad \lambda/k$ = thermal resistance
$\quad\quad\quad k/\lambda$ = thermal conductance

If two or more plates are in a series heat-flow path, the resistances are additive. For example, in Fig. 2.5,

$$\frac{Q}{A_0} = \frac{\Delta T}{R_{12} + R_{23} + R_{34}} \tag{2.17}$$

where $\Delta T = T_1 - T_4$
$\quad\quad\quad R_{12} = \lambda_{12}/k_{12}$
$\quad\quad\quad R_{23} = \lambda_{23}/k_{23}$
$\quad\quad\quad R_{34} = \lambda_{34}/k_{34}$

Cylindrical Areas (Tubes)

For circular cross sections, λ is replaced by an equivalent λ_e:

$$\lambda_e = r_o \ln\left(\frac{r_o}{r_i}\right)$$

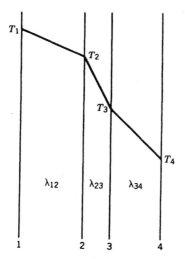

FIGURE 2.5. Schematic presentation of the temperature profile through three planes of different thermal conductivities.

where r_o = outside radius, ft
r_i = inside radius, ft

The overall heat transfer in tubes can be expressed similarly to Eq. (2.17):

$$\frac{Q}{A_0} = \frac{\Sigma \Delta T}{\Sigma R} = U_0(\Sigma \Delta T) \qquad (2.18)$$

where $\Sigma \Delta T$ = sum of ΔT across individual layers
ΣR = sum of all resistances

The outside coefficient h_o is the sum of the convective and radiative coefficients (both of which may include outside fouling factors built into their respective values):

$$h_o = h_c + h_r$$

where h_c = convective coefficient based on outside tube area, Btu/ft² · hr · °F
h_r = radiative coefficient based on outside tube area, Btu/ft² · hr · °F

In an operating superheater or reheater, heat is transferred from hot flue gas through the steel tube to the cooler steam. After a short service time, an oxide scale forms on the inside of the tube as a result of the reaction of steam

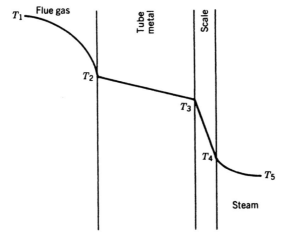

FIGURE 2.6. Schematic presentation of the temperature profile for a superheater or reheater tube with an oxide layer on the inside.

with steel. There is an additional resistance across the laminar film between the turbulent steam flow and the inside surface of the tube or oxide. Figure 2.6 presents a schematic of the temperature across these four thermal resistances. The thermal resistance through each portion of the heat flow path may be expressed by:

Gas side $\quad \dfrac{1}{h_o} = R_{12}$

Tube wall $\quad \dfrac{r_o \ln(r_o/r_i)}{k_{METAL}} = R_{23}$

ID scale $\quad \dfrac{r_o \ln(r_i/r_s)}{k_{SCALE}} = R_{34}$

Steam side $\quad \dfrac{r_o}{r_s h_s} = R_{45}$

Combining these resistances gives an expression for the overall heat-transfer coefficient:

$$U_0 = \left(\frac{1}{h_o} + \frac{r_o \ln(r_o/r_i)}{k_{METAL}} + \frac{r_o \ln(r_i/r_s)}{k_{SCALE}} + \frac{r_o}{r_s h_s} \right)^{-1} \quad (2.20)$$

Using Eqs. (2.18) and (2.20) gives

$$\frac{Q}{A_0} = \frac{T_1 - T_5}{1/h_o + r_o \ln(r_o/r_i)/k_{METAL} + r_o \ln(r_i/r_s)/k_{SCALE} + r_o/r_s h_s} \quad (2.21)$$

To calculate the temperature drop through each section, separate Eq. (2.21) as follows:

$$T_1 - T_2 = \frac{Q}{A_0}\left(\frac{1}{h_o}\right) \quad \text{gas-side temperature drop} \quad (2.21a)$$

$$T_2 - T_3 = \frac{Q}{A_0}\left(\frac{r_o \ln(r_o/r_i)}{k_{\text{METAL}}}\right) \quad \text{tube-wall temperature drop} \quad (2.21b)$$

$$T_3 - T_4 = \frac{Q}{A_0}\left(\frac{r_o \ln(r_i/r_s)}{k_{\text{SCALE}}}\right) \quad \text{ID scale temperature drop} \quad (2.21c)$$

$$T_4 - T_5 = \frac{Q}{A_0}\left(\frac{r_o}{r_s h_s}\right) \quad \text{steam-side temperature drop} \quad (2.21d)$$

Convection (outside tube)

A typical convective heat-transfer coefficient equation is[4]

$$h_c = 0.28\, F_a \left(\frac{k}{\mu^{0.6}}\right)\left(\frac{G^{0.6}}{D^{0.4}}\right)$$

where F_a = tube arrangement factor (dimensionless)
k = thermal conductivity of flue gas, Btu/ft² · hr · °F/ft
μ = viscosity, lb/hr · ft
D = tube OD, ft
G = mass flow, lb/ft² · hr

Radiation

$$\frac{Q}{A_0} = \sigma \varepsilon T^4 \quad (2.23)$$

where σ = Stefan–Boltzmann constant,[4] 1.71×10^{-9} Btu/hr · ft² · °F
T = absolute temperature, °F + 460
ε = emissivity

For two blackbodies, $\varepsilon = 1$, one emits all radiation and the other absorbs all radiation. The net rate of heat transfer is

$$\frac{Q}{A_0} = \sigma(T_1^4 - T_2^4) \quad (2.24)$$

As a practical matter, the radiation contribution in a boiler, while substantial, is usually "lumped" into the overall heat-transfer coefficient and becomes part of h_o, the gas-side heat-transfer coefficient in Eq. (2.21). It also includes the effects of external scale, slag, or ash deposits. Thus, the overall heat transfer in a "seasoned" boiler is different—poorer—than in a new or "unseasoned" unit.

Inside Film Coefficient

For nonboiling heat transfer inside circular tubes with fully developed turbulent flow, similar to steam in a superheater or reheater, the following gives a reasonable correlation:[4]

$$\frac{h_s D}{k} = 0.023 \left(\frac{DG}{\mu}\right)^{0.8} \left(\frac{C_p \mu}{k}\right)^{0.4} \tag{2.25}$$

where C_p is the specific heat (Btu/lb · °F). The following conditions apply to Eq. (2.25):

1. Fluid properties are evaluated at the arithmetic-mean bulk temperature.
2. Re > 10,000.
3. 0.7 < Pr < 100.
4. L/D > 60.

where Nu = $h_s D/k$ is the Nusselt number
Re = DG/μ is the Reynolds number
Pr = $C_p \mu/k$ is the Prandtl number

To a first approximation, the steam-side heat-transfer coefficient h_s is proportional to the mass flow rate G raised to the 0.8 power. Thus, as G decreases, so does h_s; and the net effect is a slightly higher tube-metal temperature.

Heat Flux

The use of the equation for the temperature drop through the tube wall in a rearranged form leads to the calculation of heat flux. Chordal thermocouples implanted in the OD and ID surfaces of a tube, a known distance apart, provide data on the temperature gradient (see Fig. 2.7). Since both spacing and thermal conductivity are known, a measurement of temperature gradient gives the heat flux. Thus,

$$\frac{Q}{A_0} = \frac{k \Delta T}{r_1 \ln(r_1/r_2)} = \frac{k(T_2 - T_3)}{r_1 \ln(r_1/r_2)} \tag{2.26}$$

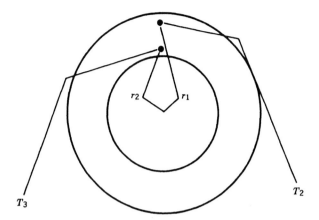

FIGURE 2.7 Schematic drawing of a chordal thermocouple arrangement. Sometimes there is a third thermocouple to measure the fluid temperature as well as the fireside temperature gradient.

Figure 2.8[5] plots the thermal conductivity for various common boiler steels as a function of temperature.

The rate of heat absorption is rarely uniform along the entire length of the riser path. Figure 2.9[6] gives the heat-absorption rates as a function of elevation in a typical furnace. The maximum is at or near the burner position.

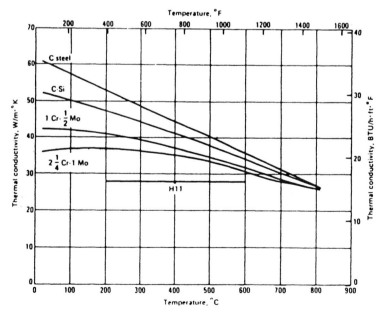

FIGURE 2.8. Thermal conductivity of carbon and low-alloy steels at the indicated temperatures (Ref. 5).

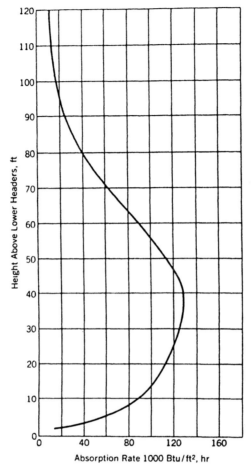

FIGURE 2.9. Heat absorption rates at different elevations in a furnace at maximum output (Ref. 6).

Since the temperature gradient through a waterwall tube is a function of heat flux (Eq. 2.21b), the highest metal temperatures correspond to the highest heat-flux zones of the furnace. Thus, any corrosion or creep problem related to steam-side scale or deposit will be exacerbated by high temperature and will occur first at the burner elevation.

Rifled tubing can be used in furnace regions where high heat fluxes may, under some operating conditions, lead to DNB. Figure 2.10 shows the configuration of rifled tubing. Imparting a swirl to the fluid motion keeps the steam bubbles well mixed and prevents steam blankets. Figure 2.11[6] plots the safe and unsafe regions for smooth-bore and ribbed tubing on a fluid-velocity–steam-quality diagram. Substantial improvement in safe-operating regions may be observed. Such tubes are useful in the burner zone of highest

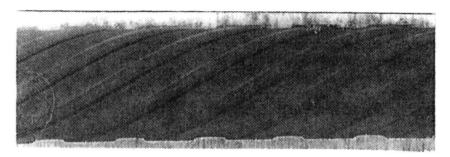

FIGURE 2.10. Photograph of one form of rifled tubing 0.5×.

heat input, sloped tubes, and other areas in the furnace where steam separation is likely.

Chordal thermocouples can also be used to determine the need for chemical cleaning. The effect of Eq. (2.21c), an added thermal resistance to the inside of a boiler tube, is to raise the tube-metal temperature. To a first

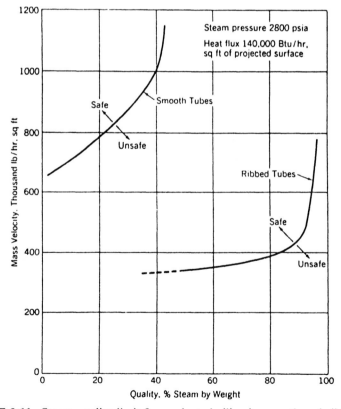

FIGURE 2.11. Steam-quality limit for nucleate boiling in smooth and ribbed tubes as a function of mass velocity (Ref. 6).

approximation, the temperature drop through ID scale is numerically equal to the tube-metal temperature increase. Thus, as water-side deposits form, T_2 and T_3 increase; and when $T_2 = 800°F$, the boiler should be chemically cleaned to remove the deposits.

Design of Superheater and Reheater

For understanding the connection between metallurgical failures and the design of these convection-pass components, a superheater or reheater can be viewed as a large heat exchanger. A simple heat balance is sufficient. The enthalpy loss in the flue gas is balanced by the enthalpy increase in the steam, thus

$$\Delta H_{STEAM} = \Delta H_{FLUE\ GAS}$$
$$\Delta H = [WC_p \Delta T]_{STEAM} = [WC_p \Delta T]_{FLUE\ GAS}$$

where ΔH = Btu/hr
W = flow, lb/hr
C_p = specific heat, Btu/lb · °F
ΔT = temperature rise (steam) or fall (flue gas), °F

Based on heat flow, the size (number of square feet of tubing surface) depends on the overall heat-transfer coefficient; see Eq. (2.15). To improve efficiency and allow higher steam temperatures, the location of the steam inlet corresponds to the location of the flue-gas outlet; in effect, the superheater or reheater is a counterflow heat exchanger. With this arrangement, it is possible to have outlet steam temperatures higher than the flue-gas temperature leaving the superheater or reheater.

The heat transfer from flue gas to steam and the enthalpy gain by the steam raise the tube-metal temperature. Figure 2.12 is a schematic presentation of the metal temperature and steam temperature from inlet to outlet. Changes in materials from carbon steel to stainless steel and increases in wall thickness for the same alloy are required as tube-metal temperatures increase. The ASME Boiler and Pressure Vessel Code gives allowable stresses as a function of metal temperature for all alloys used in boiler construction. Table VII of the appendix is a shortened version of such data.

The Code specifies how the minimum wall thickness is to be calculated. Equation (2.1) incorporates the allowable stress. As the design temperature increases, the allowable stress for a given material decreases; thus, the wall thickness increases as the pressure changes only slightly from inlet to outlet. When the safe oxidation limits are reached, changes in alloy are required. Thus, in Fig. 2.12, the kinds of alloys found in superheater and reheater are indicated.

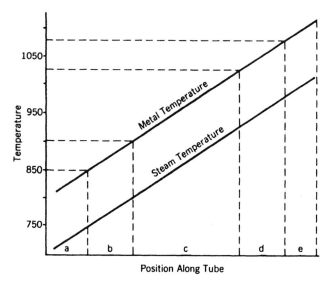

a. Carbon steel, SA-178, SA-192, SA-210
b. Carbon + $\frac{1}{2}$ Mo. SA-209
c. $1\frac{1}{4}$ Cr-$\frac{1}{2}$ Mo. SA-213 T-11
d. $2\frac{1}{4}$ Cr-1 Mo. SA-213 T-22
e. Stainless steel, SA-213 TP-304, 321, 347

FIGURE 2.12. Schematic presentation of the change in temperature and materials in a superheater or reheater from inlet to outlet.

Problems arise when deviations from design occur. Figure 2.13[7] plots steam temperature (actually the metal temperature in the stub tube at the outlet header in the penthouse) across a unit. Note the very wide variations. Thus, failures would be expected in all tubes at all locations where metal temperatures exceed design conditions.

FLUID FLOW

The flow of liquid or gas through a boiler tube is governed by the laws of fluid flow. These laws apply the laws of conservation of mass, energy, and momentum. In simplest form, the law assumes an incompressible fluid in a frictionless tube and gives Bernoulli's equation.[8] For pressure drops less than $0.15P$, incompressible flow is used without serious error:

$$P_1 V_1 + Z_1 \left(\frac{g}{g_c}\right) + \frac{v_1^2}{2g_c} = P_2 V_2 + Z_2 \left(\frac{g}{g_c}\right) + \frac{v_1^2}{2g_c}$$

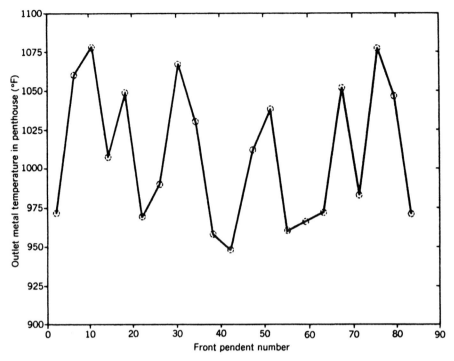

FIGURE 2.13. Final superheater, outlet temperature. Copyright © 1985. Electric Power Research Institute. EPRI CS-4252. *Dissimilar-Weld Failure Analysis and Development Program.* Reprinted with permission.

where P = pressure
  V = specific volume
  Z = elevation
  v = velocity
  g = gravitational acceleration
  g_c = conversion constant

all in consistent units.

The total stored mechanical energy in a moving fluid consists of pressure energy, gravitational energy, and kinetic energy.

In a boiler, each tube in a waterwall or in a superheater or reheater is but one of many in parallel connecting the inlet and outlet headers. The understanding of factors that upset the balance between heat flow and fluid flow on the fluid side may be improved by examination of the resistance to flow by the tube walls.

The flow of a fluid, whether gas or liquid, through a closed pipe or duct is resisted by frictional forces between the fluid and the conduit walls. Internal frictional forces, called viscosity, also exist. At every bend or turn, momentum and energy losses change. They convert mechanical energy into heat. To keep the fluid moving takes energy to overcome these resistances. Energy may be added by a pump, fan, or heat, or for example, removed by a turbine, or changed from potential to kinetic; but the sum total of all energy can be accounted for.

The following list outlines the energy balance from point 1 to point 2 within a fluid-flow circuit.

Energy Balances

1. *Kinds of Energy*
    Heat content
    Energy content
    Potential
    Kinetic
    Pressure
2. *Energy Exchanges*
    Heat transfer
    Mechanical work added by a fan or pump
    Mechanical work subtracted by a turbine
    Mechanical work converted to heat by friction
3. *Energy unit is pound force (lbf)*. Since there is also pound-mass (lbm),

$$g_c = 32.2 \frac{\text{lbm} \cdot \text{ft}}{\text{lbf} \cdot \text{sec}^2}$$

4. *Potential Energy.* Energy realized by allowing a fluid to fall to earth under force of gravity, numerically equal to height in feet Z (actually $Z \cdot g/g_c$ since g varies from point to point on the surface of the earth, but for our purposes $g = g_c$). Increase in potential energy is

$$Z_2 - Z_1 \frac{\text{ft} \cdot \text{lbf}}{\text{lbm}}$$

5. *Pressure Energy.* Fluid under pressure is similar to a coiled spring. The change in going from P_2 to P_1 is

$$\frac{P_2}{\rho_2} = \frac{P_1}{\rho_1}$$

where P is in lbf/ft² and the density ρ is in lbm/ft³.

44 CHAPTER 2 DESIGN CONSIDERATIONS

6. *Kinetic Energy.* Energy contained in fluid by virtue of motion,

$$\frac{v_2^2 - v_1^2}{2g_c}$$

where the velocity v is in ft/sec and g_c is defined above.

7. *Heat Content or Enthalpy*

$$H = E + PV$$

where H = heat content or enthalpy
E = energy content
PV = pressure and volume

$$\Delta H = Cp \, \Delta T$$

where Cp = specific heat at constant pressure
ΔT = temperature change

8. *Energy Content*

$$E_2 - E_1 = Q - W$$

where Q = heat absorbed by system
W = work done by system

9. *Self-expansion*

$$\int_1^2 P \, d\left(\frac{1}{\rho}\right) = 0 \quad \text{if density is constant}$$

For boiler systems, density changes, so

$$\int_1^2 P \, d\left(\frac{1}{\rho}\right) \neq 0$$

10. *Friction.* Between moving fluid and walls or within fluid, converts mechanical energy into heat.
11. *Mechanical Energy.* Add to fluid by means of a pump or fan (increase pressure or velocity or both). Take energy out of system as in a turbine in a power plant.
12. *Total Energy Balance.* When the principle of conservation of energy is applied to any fluid-flow system, the concept of a total energy balance follows.

Total energy output = Total energy input

$$Z_2 + \frac{P_2}{\rho_2} + \frac{v_2^2}{2g_c} + E_2 = Z_1 + \frac{P_1}{\rho_1} + \frac{v_1^2}{2g_c} + E_1 + Q$$

Friction factor f and friction F for fluid of constant density flowing in a horizontal pipe of constant cross section is

$$\frac{\Delta P}{\rho} = -F$$

$$F = f \frac{v^2}{2g_c} \frac{L}{D}$$

where L = pipe length
D = pipe diameter

$$-\Delta P = f \frac{\rho v^2}{2g_c} \frac{L}{D}$$

where v = average velocity over whole cross section
$v\rho = G$, mass flow velocity, lbm/ft² · sec

Note, for fluid flow in pipe of constant cross section, both v and ρ vary with temperature, but G does not.

Basic Pressure Drop

In the design of any fluid circuit consisting of several tubes in parallel, it is essential to maintain equal flow resistance in all tubes. Friction forces are usually given in terms of the pressure drops ΔP needed to overcome them. Friction losses are the sum of the resistance to flow from the inside surface of the tube plus all of the resistances through bends, swages, and fittings and is expressed by

Friction loss + turn loss = Pressure drop

$$\frac{fL}{144D} \frac{G^2}{2g_c\rho} + \frac{\Sigma K}{144} \frac{G^2}{2g_c\rho} = \Delta P \qquad (2.27)$$

where ΔP = pressure drop, psi
f = friction factor
L = length of tube
D = tube ID, ft
G = mass flow, lb/sec · ft²
g = gravity constant, ft/sec²
ρ = density, lb/ft³
K = turn or fitting constant

The constant K is usually expressed in terms of the equivalent L/D; thus, $\Delta P \propto L/D$. Rearranging (2.27) for a superheater or reheater, where the ΔP between inlet and outlet headers is fixed, gives

$$\frac{G^2}{2g_c\rho}\left[\Sigma f \frac{L}{D}\right] = \Delta P \qquad (2.28)$$

The term in brackets is the summation of all frictional resistances from inlet to outlet through each circuit. Note, then, that

$$G^2 \propto \frac{\Delta P}{L/D} \qquad (2.29)$$

or

$$G \propto \left[\frac{\Delta P}{L/D}\right]^{1/2} \qquad (2.30)$$

For a given ΔP, as L/D increases, G decreases.

This little exercise helps us better understand where to look for premature failures in superheater and reheater elements. In Eq. (2.25), the steam-side heat transfer is proportional to the steam mass flow rate to the 0.8 power:

$$h_s \propto G^{0.8} \qquad (2.31)$$

In Eq. (2.30), the mass flow rate is inversely proportional to the square root of the flow resistance expressed as L/D:

$$G \propto \left(\frac{L}{D}\right)^{-1/2} \qquad (2.32)$$

Combining (2.31) and (2.32) gives

$$h_s \propto \left(\frac{L}{D}\right)^{-0.4} \qquad (2.33)$$

Finally, Eq. (2.21d) indicates that the temperature gradient through the laminar film is inversely proportional to the steam-side heat-transfer coefficient:

$$\Delta T \propto \frac{1}{h_s} \qquad (2.34)$$

or

$$\Delta T \propto \left(\frac{L}{D}\right)^{0.4} \qquad (2.35)$$

If the assumption is made that larger temperature gradients through the steam-side laminar-film layer translate into higher tube-metal temperatures, then those tubes with the largest flow resistances will have the highest temperatures and will fail first. In practice, this is the case. Wrapper tubes—tubes bent out of the plane of a platen and surrounding it to ensure proper alignment—do indeed experience creep failures to a greater extent than do tubes within a platen.

EFFECTS OF FUEL ON FURNACE SIZE

Table 2.1 presents heating values for various fossil fuels and clearly illustrates the effects of fuel on furnace size. Since the same amount of steam generated has the same heat content regardless of the fuel burned, for a given boiler temperature, pressure, and efficiency; it takes $3\frac{1}{2}$ times as many

TABLE 2.1. Heat Values of Fossil Fuels

Fuel	Heat Content, Btu/lb
Coal	
Eastern bituminous	13,240
Midwest bituminous	10,500
Subbituminous-C	8,125
Texas lignite	7,590
North Dakota lignite	6,520
Oil	
No. 1	19,810
No. 2	19,430
No. 4	18,860
No. 5	18,760
No. 6	18,300
Gas	
Pennsylvania	23,170
South Carolina	22,904
Ohio	22,077
Louisiana	21,825
Oklahoma	20,160

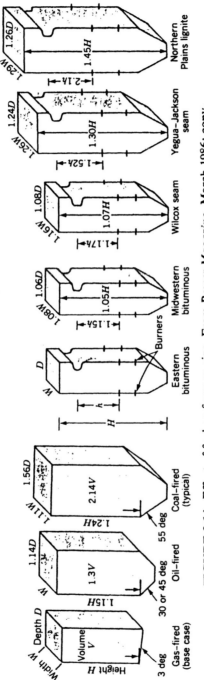

FIGURE 2.14. Effect of fuel on furnace size. From *Power Magazine*, March 1986; copyright McGraw-Hill, Inc., 1986.

pounds of North Dakota lignite as it does Pennsylvania gas. The required heat absorption Q is determined from the equation

$$Q = B^{\mathrm{I}}(H_2^{\mathrm{I}} - H_1^{\mathrm{I}}) + B^{\mathrm{II}}(H_2^{\mathrm{II}} - H_1^{\mathrm{II}}) \qquad (2.36)$$

where Q = rate of heat absorption, Btu/hr
B^{I} = primary steam or feedwater flow, lb/hr
H_1^{I} = enthalpy of feedwater entering, Btu/lb
H_2^{I} = enthalpy of primary steam leaving SH, Btu/lb
H_1^{II} = enthalpy of steam entering RH, Btu/lb
H_2^{II} = enthalpy of steam leaving RH, Btu/lb
B^{II} = reheat steam flow, lb/hr

and SH = superheater and RH = reheater. There is no consideration of the type or heat content of the fuel.

Gas is virtually ash-free, oil contains but a few pounds of ash per ton, while coal may contain several hundred pounds of ash per ton. Thus, the amount and composition of the coal ash dictate the size and spacing of the convective heat-transfer surfaces.

Figure 2.14[9] shows the relative sizes of coal, oil, and gas furnaces and the effect of coal rank on furnace size. Coals are ranked as either high fouling or low fouling. There are various methods for measuring the stickiness of coals, usually a measure of ash-fusion temperature. The ash composition, often the sodium and potassium content, defines the melting or softening point. The main idea is to have the coal-ash leaving the furnace at a low-enough temperature to prevent its sticking to convective surfaces. As a practical matter, high-fouling coals may have such a low fusion point that convective regions have to be very large to prevent excessive ash buildup. All of these factors dictate the bundle spacing of the superheater and reheater, and the number, location, and frequency of use of sootblowers. The higher the fouling value of the coal, the wider the superheater and reheater spacing, and thus, the larger the furnace, need to be. Compare, in Fig. 2.14, the relative sizes for an eastern bituminous, a low-fouling coal, with a Northern Plains lignite, a very high-fouling coal. The one is nearly $2\frac{1}{2}$ times the size of the other.

CASE HISTORIES

The concepts of heat flow, stress analysis, and fluid dynamics presented in this chapter can be applied to several problems that plague boiler operations. While the various analyses represent approximate solutions, the results are useful in understanding some general microstructural features and failure conditions; they also suggest useful schemes for prevention.

50 CHAPTER 2 DESIGN CONSIDERATIONS

Case History 2.1

Design Considerations

The ASME Boiler and Pressure Vessel Code[2] provides rules for calculating minimum wall thicknesses in pressurized cylinders. Paragraph PG27.2 of Section I of the Code gives the governing equation for welded construction:

$$W = \frac{PD}{2S + P} + 0.005\, D \qquad (2.1.1)$$

where W = specified minimum wall thickness, in
 P = design pressure, psi
 D = tube OD, in
 S = maximum allowable stress value at the operating temperature of the metal, psi

Equation (2.1.1) is used for all calculations and contains no allowance in the thickness for corrosion or oxidation wastage during service.

Sometimes manufacturing considerations overrule the specified minimum wall thickness calculated from Eq. (2.1.1). Low-pressure boilers or reheaters come quickly to mind. Two calculations will suffice:

A. *Reheater Tube in a Utility Boiler.* Conditions: 500 psi, 2.0 in OD, SA-213 T-22 material, design metal temperature of 1050°F. The allowable stress for T-22 at 1050°F is 5800 psi. Thus,

$$W = \frac{(500)(2.0)}{2(5800) + 500} + 0.005(2)$$

$$= 0.092 \text{ in}$$

B. *Waterwall Tube in an Industrial Boiler.* Conditions: 750 psi, 3 in OD, SA-178 A material, design metal temperature 700°F. The allowable stress for SA-178 A at 700°F is 11,500 psi. Thus,

$$W = \frac{(750)(3)}{2(11,500) + 750} + 0.005(3)$$

$$= 0.110 \text{ in}$$

Manufacturing a reheater or an all-welded waterwall from tubes of wall thickness less than 0.125 in would be difficult, if not impossible. The tubes would be hard to bend without considerable distortion, and they would be hard to weld without excessive burn-through. Fin and attachment welds would be especially hard to make. Thus, for these manufacturing reasons, waterwall tubes are seldom less than 0.180 in, and reheaters are usually at least 0.150 in, regardless of material specified.

The ramifications of using thicker tubes than required by Code are lower operating hoop stress or, conversely, higher metal temperatures before premature creep failures are likely. For a waterwall tube in a low-pressure boiler, thicker scale or deposits can form before a chemical cleaning is needed. The microstructures that develop also reflect either the lower hoop stress or higher temperatures. Reheaters are particularly abused during start-up; metal temperatures in excess of 1300°F are common before steam flow and proper cooling are established.

Case History 2.2

Long-Term, High-Temperature Failures of Reheater Tubes

BOILER STATISTICS

Size	1,600,000 lb steam/hr (725,000 kg/hr)
Steam pressure	1005°F/1005°F (540°C/540°C)
Steam pressure	1975 psig (140 kg/cm^2)
Fuel	Pulverized coal

The heat-transfer concepts presented in Chapter 2 may be used to calculate the effects of steam-side scale on tube-metal temperatures. The following case history on the high-temperature failure of a reheater tube is used to discuss these concepts.

This investigation covers several reheater tube samples, including a tube that failed. Figures 2.2.1 and 2.2.2 show the condition of the as-received samples. The tubes are 1¾ in OD × 0.150-in-thick wall, grade SA-213 T-22 alloy steel.

The visual examination of these tubes revealed the following:

A. The rupture was a wide-open burst with the tube wall drawn to a knife-edge at the failure.

B. The OD scale measured 0.011 in thick, and the deposit from the products of combustion was 0.04 in thick.

C. The internal scale measured from 0.015 to 0.026 in thick.

D. The tube-wall thickness 12 in away from the failure and in the same plane measured 0.127 in. The tube-wall thickness 180° from this point measured 0.171 in.

E. The remaining reheater-tube sections showed ID scale varying from 0.018 to 0.021 in thick, and OD scale and deposit up to 0.040 in thick.

F. Wall-thickness measurements on these unfailed tubes varied from 0.170 to 0.127 in.

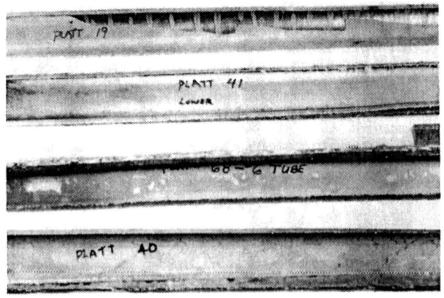

FIGURE 2.2.1. The as-received reheater tubes cut longitudinally to reveal the excessive internal magnetite scale.

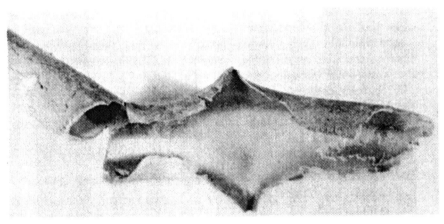

FIGURE 2.2.2. Failed reheater tube; note the thick ID scale visible on the right-hand edge.

Microstructural analysis of the samples, including the failure, showed complete spheroidization of the carbides, indicating that exposure to elevated temperatures was for an extended period, as shown in Fig. 2.2.3.

The reaction of steam and iron to form magnetite leaves the tube with an insulating oxide layer on the ID, which will cause the metal temperature to increase.[9] A calculation may be done to estimate the peak tube-

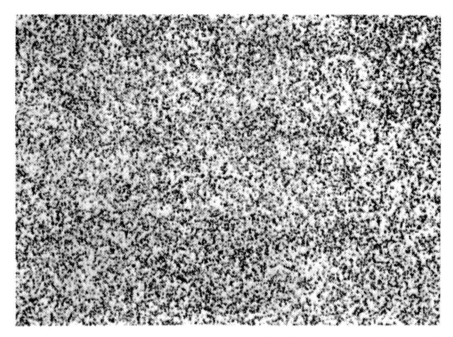

FIGURE 2.2.3. The microstructure typical of all reheater-tube samples received. The microstructure shows complete spheroidization of the carbide (Magnification 500×, nital etch).

metal temperature resulting from the formation of the oxide scale on the inside of the tube. The scheme follows the heat-flow analysis of Krieth[10] for the steady-state condition.

Nomenclature:

Q = heat flow, Btu/hr
U_0 = overall heat transfer coefficient, Btu/hr · ft² · °F
A_0 = area of the outside tube surface, ft²
T_0 = flue-gas temperature, °F
T_3 = tube-metal temperature on the OD surface, °F
T_2 = temperature on the OD surface, °F
T_1 = temperature at the ID of the tube, °F
T_s = bulk steam temperature, °F
r_3 = radius of tube OD, ft
r_2 = radius of tube metal ID, ft
r_1 = radius of tube ID, ft
h_i = steam-side heat-transfer coefficient, Btu/hr · ft² · °F
h_0 = gas-side heat-transfer coefficient, Btu/hr · ft² · °F
k_D = thermal conductivity of the scale, Btu/hr · ft · °F
k_T = thermal conductivity of the tube metal, Btu/hr · ft · °F

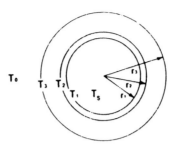

For a unit length of tube, the flow of heat Q is in a radial direction and is given by

$$Q = U_0 A_0 (T_0 - T_s) \qquad (2.2.1)$$

Schematically, the temperature profile from flue gas T_0 to bulk steam temperature T is as follows:

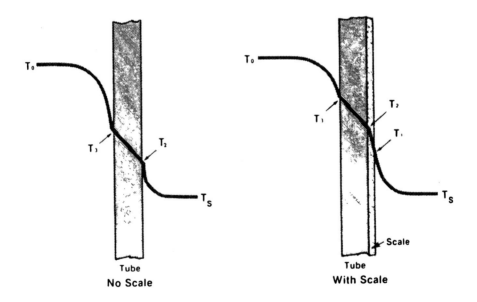

$$U_0 = \left(\frac{r_3}{r_1 h_i} + \frac{r_3 \ln(r_2/r_1)}{k_D} + \frac{r_3 \ln(r_3/r_2)}{k_T} + \frac{1}{h_0} \right)^{-1} \qquad (2.2.2)$$

The denominator of Eq. (2.2.2) has four terms, one for each of the thermal components:

$\dfrac{r_3}{r_1 h_i}$ = steam-side film resistance

$\dfrac{r_3 \ln(r_2/r_1)}{k_D}$ = scale resistance

$\dfrac{r_3 \ln(r_3/r_2)}{k_T}$ = tube-metal resistance

$\dfrac{1}{h_0}$ = gas-side film resistance

Equation (2.2.1) may be rewritten as

$$\frac{Q}{A_0} = \frac{T_0 - T_s}{r_3/r_1 h_i + [r_3 \ln(r_2/r_1)]/k_D + [r_3 \ln(r_3/r_2)]/k_T + 1/h_0} \quad (2.2.3)$$

Equation (2.2.3) may be separated into its components since the quantity of heat that flows through each is the same:

$$\frac{Q}{A_0} = \frac{T_1 - T_s}{r_3/r_1 h_i}, \quad \text{temperature drop through the steam film} \quad (2.2.4)$$

$$\frac{Q}{A_0} = \frac{T_2 - T_1}{[r_3 \ln(r_2/r_1)]/k_D}, \quad \text{temperature drop through the scale} \quad (2.2.5)$$

$$\frac{Q}{A_0} = \frac{T_3 - T_2}{[r_3 \ln(r_3/r_2)]/k_T}, \quad \text{temperature drop through the tube} \quad (2.2.6)$$

$$\frac{Q}{A_0} = \frac{T_0 - T_3}{1/h_0}, \quad \text{temperature drop on the gas side.} \quad (2.2.7)$$

To calculate the surface temperature T_3 as a result of scale formation on the tube ID, first calculate h_0 from Eq. (2.2.2) given the design parameters U_0, r_3, r_1, h_i, and k_2 for the base tube condition. The scale resistance is zero, and $r_1 = r_2$ for the condition of no scale. With this value of h_0, the design value of Q/A_0, and the design value of T_0, T_3 is calculated. In practice, during the initial design stage, T_3 is set at the practical limit minus a small margin; for grade T-22 steel, T_3 is 1075°F (580°C).

The addition of scale alters the heat flow; to calculate the effect of the scale, the scale term is added as in Eq. (2.2.3).

The final calculation of T_3 is a two-step procedure. Using h_0 just calculated and appropriate values for the other terms in the denominator of Eq. (2.2.3), assuming h_i to be constant for this calculation, and obtaining T_0 and T_s from design parameters, one finds a new, smaller value of Q/A_0. From Eq. (2.2.7), T_3 is calculated from this smaller Q/A_0.

In this case history, the conditions from the initial heat-transfer design calculation are as follows:

$U_0 = 15$ Btu/hr · ft² (includes the effects of OD scale and deposits)
$Q/A_0 = 7300$ Btu/hr · ft²; see Table 2.2.1
$T_0 = 1485°F$ (810°C)
$T_s = 1000°F$ (540°C)
$h_i = 300$ Btu/hr · ft²; see Table 2.2.1

The tubing is 1¾ in OD × 0.150-in-thick wall, T-22 alloy steel:

$$k_1 = 0.342 \text{ Btu/hr} \cdot \text{ft} \cdot °F$$
$$k_2 = 16.7 \text{ Btu/hr} \cdot \text{ft} \cdot °F$$

From these data and Eq. (2.2.2), h_0 is found to be 16.2 Btu/hr · ft². The formation of scale is assumed to occur half at the expense of the tube and half in decreasing the radius r_1 of the tubing; thus,

$$r_3 = 0.875 \text{ in } (0.0729 \text{ ft})$$

The scale thickness is 0.019 in.

$$r_2 = 0.735 \text{ in } (0.0613 \text{ ft})$$
$$r_1 = 0.716 \text{ in } (0.0597 \text{ ft})$$

From Eq. (2.2.3) the new Q/A_0 is found to be 6700 Btu/hr · ft². The higher T_3, from Eq. (2.2.7), is 1072°F (578°C), or about 40°F (22°C) higher.

TABLE 2.2.1. Range of Values for Heat-Transfer Coefficients (Btu/hr · ft²)

Economizer	$1000 h_i$	$5-7 U_0$
Waterwalls	$4000-8000 h_i$	$20-22 U_0$
Reheaters	$300-400 h_i$	$11-17 U_0$
Superheaters	$200-500 h_i$	$11-18 U_0$

Thermal Conductivity [1100°F (590°C)], Btu/hr · ft · °F

Carbon steel (SA-210 A-1)	23.0
SA-209 T-1	21.7
SA-213 T-11	17.5
SA-213 T-22	16.7
SA-213 TP-321	14.0
Iron oxide (Fe_3O_4) scale	0.342

It is instructive to calculate the various contributions to the overall flue-gas-to-steam temperature each component makes:

	With Scale of 0.019 in	Without Scale
Steam film	27°F (15°C)	29°F (16°C)
Scale	38°F (21°C)	0
Tube	7°F (4°C)	8°F (4°C)
Gas side	413°F (229°C)	449°F (249°C)
Q/A_0	6700 Btu/hr · ft²	7300 Btu/hr · ft²

From this comparison it is evident that aside from the metallurgical implications of the higher tube-metal temperatures, there is a slight loss of overall thermal efficiency of the boiler. The calculation is not meant to be precise, but to show the effects of scale on tube temperature. It is safe to say that if the original design limit was 1075°F (580°C), the present condition is some 40°F (22°C) hotter, or in the neighborhood of 1115°F (600°C).

Scale affects the tube-metal temperature; Fig. 2.2.4 plots the calculated temperature as a function of scale thickness for the present reheater tube study. Figure 2.2.5 shows the stress for 1% creep in 10,000 and 100,000 hr

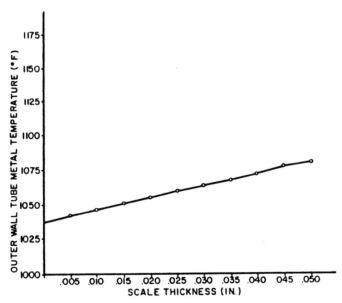

FIGURE 2.2.4. A plot of outer-wall tube-metal temperature versus scale thickness for the conditions of this reheater study.

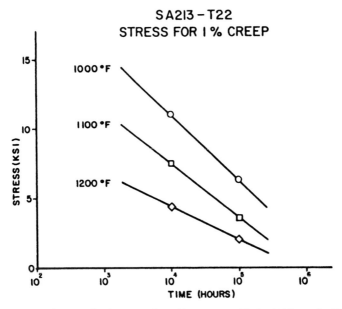

FIGURE 2.2.5. The required stress for 1% creep at 1000, 1100, and 1200°F (538, 593, and 649°C) for grade SA-213 T-22 steel.

for grade T-22 steel.[11] Two things are evident from the samples, calculations, and deductions:

1. The ID scale raises the crown or peak temperature of the tube by about 40°F (20°C).
2. The wall thickness has been reduced by oxidation or corrosion by 15% or so, and the hoop stress in the tube wall is correspondingly increased.

Since the design stress for grade T-22 steel per the ASME Code is 5800 psi at 1050°F (565°C), the actual operating conditions would have been over 1100°F (590°C) with a stress of 6700 psi.[12] As can be estimated from Fig. 2.2.5, the time required to give a 1% creep deformation decreases from about 100,000 hr for the design condition, to about 10,000 hr for 1100°F (593°C) and 6700 psi.

There are two unknowns in Eq. (2.2.5): the heat flux and the thermal conductivity of the steam-side scale. The preceding analysis used an average value of the heat flux for the entire reheater. However, there are substantial variations in the local heat flux from the average. For example, depending on the position, the variation can be a factor of 2½ to 3 from the average, that is, a range of 2500 to 22,000 Btu/hr · ft² rather than the average of 7300 Btu/hr · ft² used. Failures occur first where the heat flux is the highest, since tube-metal temperatures are highest at that location.

TABLE 2.2.2. Thermal Conductivities of Various Scales and Other Materials

Material	Thermal Conductivity, Btu/ft^2 · hr · °F · in
Analcite	8.8
Calcium phosphate	25
Calcium sulfate	16
Magnesium phosphate	15
Magnetic iron oxide	20
'Silicate scale' (porous)	0.6
Boiler steel	310
Firebrick	7
Insulating brick	0.7

Used by permission of the Drew Industrial Division, Ashland Chemical, Inc. © 1979, Ashland Chemical, Inc.

The thermal conductivity of the steam-side scale also is quite variable. Table 2.2.2 lists the thermal conductivity of various water-side scales and other materials.[10] Note the value given here for magnetic iron oxide is 1.7 Btu/hr · ft · °F (20 Btu/hr · ft^2 · °F/in).

Published work by Combustion Engineering shows that the thermal conductivity of the steam-side scale in a superheater tube varies from 1.2 to 3.6 Btu/hr · ft · °F.[11] The actual thermal conductivity depends on the porosity or, conversely, the density of the scale in a particular case. In this example we used 7300 Btu/hr · ft^2 for the heat flux and 0.342 Btu/hr · ft · °F for the thermal conductivity of the iron oxide. It appears that both values are wrong by about the same factor, nearly 3. More recent values of the thermal conductivity of the steam-side scale and the larger value of the heat flux at failure locations would, however, generate approximately the same tube-metal temperature increase. Thus, it appears that the calculations of the effects of steam-side scale are reasonably accurate, even if the precise values of thermal conductivity of the steam-side scale and the heat flux are not.

The inescapable conclusion is that the cause of the failure of this reheat tube was the combined effects of higher temperature, caused by the internal scale, and severe wall thinning, caused by either oxidation or corrosion. This combination accelerated the creep deformation of the tube until the ultimate failure by high-temperature stress rupture. Examination of the unfailed tubes shows that unless the reheater is chemically cleaned to reduce temperatures, an increasing number of failures will occur. However, since the wall thickness has already been reduced by 15% or so, it would appear that the entire high-temperature portion of the reheater should be replaced.

Case History 2.3

Waterwall Tube Blistering[12]

The operation of steam-generating tubes is a dynamic balance between the heat flow from the combustion of a suitable fuel and the formation of steam. The overall temperature difference between the flame and the steam–water emulsion can be 2500°F. When the balance is maintained, tubes heated by hot flue gas and radiation from the flame are cooled by fluid flow and steam generation. The tube-metal temperatures are kept within the design parameters of the saturation steam temperature plus about 100°F. However, when the balance is upset, the temperatures can increase to the creep range, and failures occur.

The formation of water-side deposits is one type of upset in the operational balance and leads to the formation of waterwall blisters. Blisters occur in both low- and high-pressure boilers, but more frequently in lower-pressure industrial units. The thermal conductivity of water-side oxides and deposits is low, only about 5% that of steel (see Table 2.2.2). In effect these deposits act as an insulation barrier to the cooling of the steel tube by steam formation. The net effect is to raise tube-metal temperatures. The tube-metal temperature increase depends not only on the thermal conductivity and thickness of the scale but also on the heat flux and, thus, on the location within the furnace. The heat flux varies with furnace elevation and is usually highest just above the stoker or at the burners. Thus blisters will form at these highest heat-flux regions.

Observations

The appearances of blistered furnace-wall tubes can differ markedly. Figure 2.3.1 shows one such example, and Fig. 2.3.2 displays examples of ring sections. The ID surface, Fig. 2.3.3, shows a thick, flaky deposit except where the tube expansion at a blister has removed the scale. A representative cross section is shown in Fig. 2.3.4. At this location, the ID scale thickness is approximately 25 mils (0.025 in). Note also that the scale is tightly bound to the steel tube surface.

The microstructural changes vary with the particular tube-metal composition and boiler. Waterwall tubes are usually plain-carbon steel, SA-178 A or SA-192 in industrial boilers, or SA-210 in utility units. A low-carbon steel microstructure unaffected by elevated temperature is given in Fig. 2.3.5. The structure is equiaxed ferrite and lamellar pearlite, a characteristic microstructure of SA-178 A or SA-192 material. Higher-carbon steels, SA-210, have a similar structure but more pearlite; see Fig. 2.3.6.

Several changes occur to the microstructure as these plain-carbon steels operate at temperatures above approximately 800°F; see page 110.

FIGURE 2.3.1. Typical waterwall blisters along the fire side of a waterwall tube (Magnification 0.3×). Reprinted with permission from the National Association of Corrosion Engineers. Copyright NACE.

FIGURE 2.3.2. Ring sections removed from four positions along the waterwall tube show the nonuniform expansion (Magnification 0.25×).

FIGURE 2.3.3. ID surface of the fire side displays the expected water-side deposits. Some of the scale has spalled off at the blister (Magnification 6×). Reprinted with permission from the National Association of Corrosion Engineers. Copyright NACE.

FIGURE 2.3.4. Cross section through a blister from an industrial boiler. Failure has not yet occurred, but creep expansion has cracked the ID scale and initiated OD longitudinal cracks. ID scale is 0.025 in thick (Magnification $18\frac{3}{4}×$, etched).

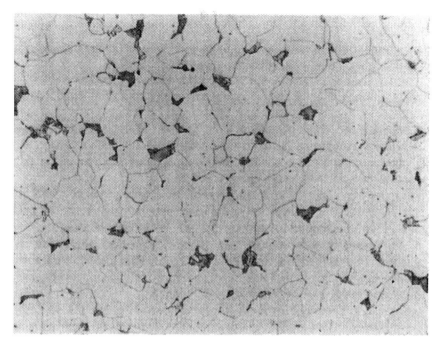

FIGURE 2.3.5. Normal ferrite and pearlite in the microstructure of a low-carbon steel similar to SA-178 A or SA-192 (Magnification 500×, etched).

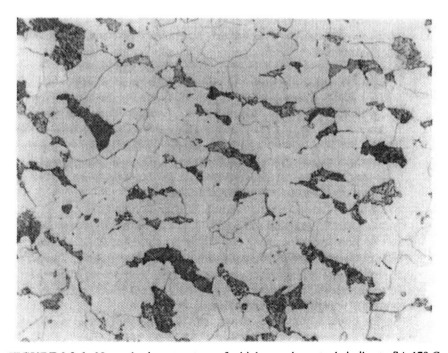

FIGURE 2.3.6. Normal microstructure of a higher-carbon steel similar to SA-178 C or SA-210 A-1. The higher carbon gives more pearlite (Magnification 500×, etched).

Blistered waterwall tubes can show all variations of microstructures that form at metal temperatures above normal.

The sequence of events that leads to the formation or failure of a bulge is as follows:

1. Steam-side scale or deposits form in sufficient thickness to impede heat flow and raise the tube-metal temperature.
2. At some time, the scale thickness is great enough to raise the tube-metal temperature into the creep range.
3. In the creep range, at a temperature in excess of about 850°F, the microstructure changes from lamellar pearlite and ferrite to spheroidized carbides and ferrite or perhaps ferrite and graphite.
4. The tube begins to expand by a creep-deformation mechanism, and creep damage appears within the microstructure. Graphitization may also occur.
5. At this point, one of two things will happen.
   A. The expanding tube metal has greater ductility than the water-side deposits and therefore expands from the deposit, leaving a gap between the steel and the scale. If the scale remains intact, the gap fills with steam. The poor thermal conductivity of steam effectively insulates the tube from the cooling flow of the fluid on the inside of the tube. The tube-metal temperature rapidly rises until failure occurs. The temperature at which failure occurs will leave within the microstructure evidence of the peak temperature at the instant of failure.
   B. As creep deformation occurs, the scale on the inside of the tube ruptures and spalls from the tube wall. Bare metal, in contact with the steam–water emulsion, is quickly cooled to the normal operating temperature and blisters are left behind. They are similar to those in Fig. 2.3.1. The blisters are intact with no fissure or rupture, but considerable wall thinning does result. Under these conditions, the boiler continues to operate, and there is no rupture that would lead to a forced shutdown. The microstructure within the blister also reflects its thermal history.

What follows is a catalog of the kinds of microstructures that develop within the fire side of a blistered boiler tube. Figure 2.3.7 shows the midwall microstructure at a blister that contained no failure. The structure is ferrite and spheroidized carbides, which indicates operating temperatures above 800°F. The black spots within the microstructure may be the very early stage of graphitization.

When a blistered tube has failed, the microstructure can sometimes show peak temperatures well above 1340°F at the moment of failure. Figure 2.3.8 shows the microstructure in the vicinity of a failed blister. Here the microstructure is ferrite, bainite, and creep voids, the black

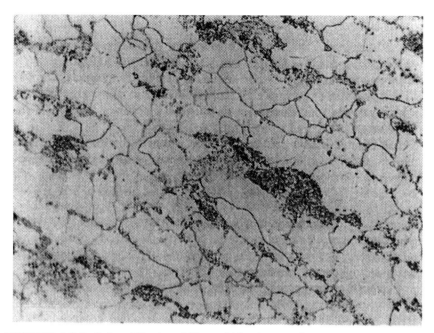

FIGURE 2.3.7. Spheroidized carbides within the pearlite indicate an operating metal temperature above about 800°F for some length of time (Magnification 500×, etched).

FIGURE 2.3.8. Microstructure through a ruptured blister is ferrite and bainite. The small black spots are creep voids (Magnification 500×, etched).

cracklike features within the microstructure. The peak temperature at the time of failure is in the neighborhood of 1400–1450°F, but extensive creep likely occurred at much lower temperatures. At temperatures of 1400–1450°F, the microstructure is a mixture of ferrite and austenite. When the rupture occurred, the austenite was quenched to bainite, leaving the creep voids intact.

The microstructure between blisters, where little or no tube deformation has occurred, can vary somewhat, depending on the scale thickness along the water side. The microstructures can vary from normal ferrite and pearlite to spheroidized carbides and ferrite, similar to Fig. 2.3.7. On rare occasions the temperature at the moment of failure can be slightly above 1340°F, not high enough to have failed but certainly high enough to transform the ferrite and pearlite to ferrite and austenite. When the rupture occurs, the austenite is quenched to bainite, but there is less bainite and more ferrite, as shown in Fig. 2.3.9.

Depending on the time and temperature, the microstructure at the blister can be graphitized to some degree and may display considerable creep damage. Figures 2.3.10 and 2.3.11 are graphitized, and Fig. 2.3.12 contains extensive creep damage. Graphitized and creep-damaged microstructures are more common in lower-pressure units. The hoop stress is lower (see Case History 2.1), and thus tubes operate at higher temperatures or longer times before failure occurs.

The appearance of the scale–metal interface is critical to the failure of a blister. Figure 2.3.13 shows the scale–metal interface of a portion of the tube that did not fail. Here the scale is tightly bound to the tube metal and has as good heat-transfer characteristics as can exist in this scale–metal configuration. In Fig. 2.3.14, the scale is reasonably intact, but the tube has expanded from the scale, leaving a gap about 0.3 mil (0.0003 in) wide. When the gap is filled with steam, heat transfer across the gap is very poor, and the metal temperature rises rapidly until failure occurs or the scale spalls.

With these observations in mind, the heat-flow analysis in Chapter 2 can be used to quantify the observations. For calculation purposes, tubes are 3.0 in OD × 0.200 in actual wall thickness. A semiquantitative approach may be understood from steady-state, heat-transfer conditions within the furnace. For a drum pressure of 700 psi, the saturation steam temperature is approximately 500°F (503°F). For steady-state heat transfer, the heat flux is equal to the temperature gradient that drives the heat flow, divided by a thermal resistance, Thus,

$$\frac{Q}{A_0} = \frac{\Delta T}{R} \qquad (2.3.1)$$

If we start at the fluid temperature, a constant 500°F, there is a thermal resistance along the interface between the inside of a clean, scale-free

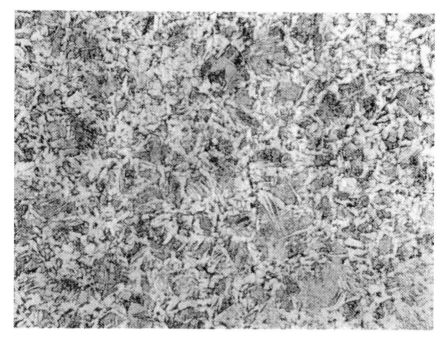

FIGURE 2.3.9. Microstructure near a ruptured blister is also ferrite and bainite. At this location the temperature was not quite high enough to fail in a short time, but was above 1340°F (Magnification 500×, etched).

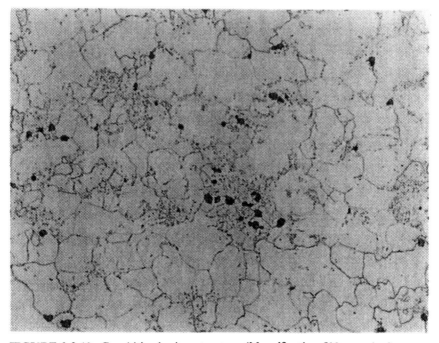

FIGURE 2.3.10. Graphitized microstructure (Magnification 500×, etched).

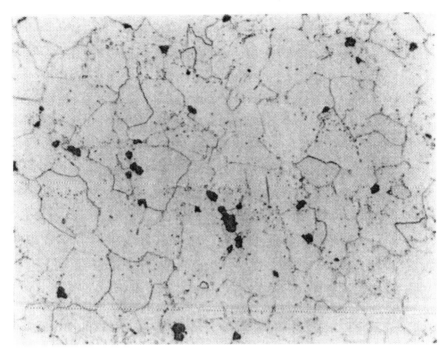

FIGURE 2.3.11. Graphitized microstructure (Magnification 500×, etched). The extent of graphitization depends on the time and temperature. Such structures are more common in lower-pressure units than in utility boilers.

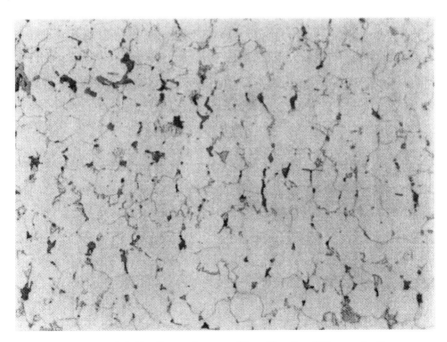

FIGURE 2.3.12. Creep damage (Magnification 500×, etched).

FIGURE 2.3.13. Tightly bound ID scale will have as good heat-transfer characteristics as possible (Magnification 25×, etched).

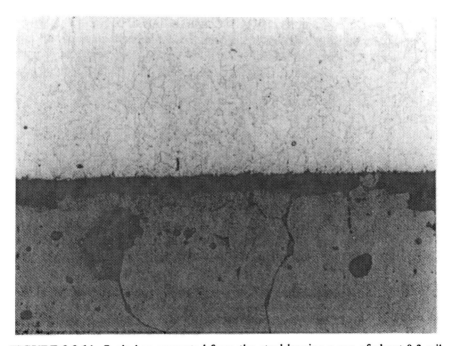

FIGURE 2.3.14. Scale has separated from the steel leaving a gap of about 0.3 mil (0.0003) (Magnification 500×, etched).

70 CHAPTER 2 DESIGN CONSIDERATIONS

tube and the moving water–steam emulsion. This resistance is

$$R = \frac{r_3}{r_2 h_i} \tag{2.3.2}$$

where r_3 = tube OD/2, or 1.50 in
r_2 = tube ID/2, or 1.30 in
h_i = fluid-side, heat-transfer coefficient, Btu/hr · ft² · °F

For nucleate boiling, h_i varies from 4000 to 8000 Btu/hr · ft² · °F. For purposes of calculation, 5000 will be used. The temperature gradient then gives the ID metal temperature:

$$\left(\frac{Q}{A}\right)\frac{r_3}{r_2 h_i} = \Delta T_i \tag{2.3.3}$$

Q/A, Btu/hr · ft	ΔT, °F	ID Metal Temp., °F
25,000	6	506
50,000	11	511
75,000	17	517
100,000	23	523
125,000	29	529

The temperature gradient through the laminar-film boundary is added to the bulk fluid temperature to get the tube-metal temperature along the ID.

Another thermal resistance to heat flow through the tube wall, R_T, is given by

$$R_T = \frac{r_3 \ln(r_3/r_2)}{k_T} \tag{2.3.4}$$

where r_3 = tube OD/2, or 1.50 in
r_2 = tube ID/2, or 1.30 in
k_T = thermal conductivity of the steel tube, 30 Btu/hr · ft · °F

The temperature gradient through the tube wall is

$$\left(\frac{Q}{A}\right)\frac{r_3 \ln(r_3/r_2)}{k_T} = \Delta T_T \tag{2.3.5}$$

Q/A, Btu/hr · ft²	ΔT_T, °F	OD Metal Temp., °F
25,000	15	521
50,000	30	541
75,000	45	562
100,000	60	583
125,000	75	609

The third column gives the fire-side tube temperature in the absence of *any* water-side deposits. It is the sum of the saturation steam temperature plus the laminar-film temperature gradient plus the steel-tube gradient.

As oxides and deposits form along the ID surface of the waterwall tube, a third thermal resistance is added to the heat-flow path, and the actual inside radius gets smaller by the thickness of the deposit. The water-side scale is taken to be 0.040 in. To a first approximation, the temperature gradient through the deposit is equal to the temperature increase in the tube. The thermal resistance of the deposit R_D is

$$R_D = \frac{r_3 \ln(r_2/r_1)}{k_D} \qquad (2.3.6)$$

where r_3 = tube OD/2, or 1.50 in
  r_2 = tube ID/2, or 1.30 in
  r_1 = ID/2 of the deposit, or 1.26 in
   ($r_2 - r_1$ = deposit thickness, 0.040)
  k_D = thermal conductivity of the deposit, Btu/hr · ft · °F

There is a small effect on the temperature gradient through the film calculated from Eq. (2.3.3) when $r_1 = 1.26$ in is used instead of $r_1 = 1.30$ in, but that is ignored here.

The thermal conductivity of deposits can vary greatly;[10] see page 59. The amount of porosity within the deposit, the bond between the scale deposit and the tube, and the exact composition all affect thermal conductivity. For purposes of calculation, assume k_D is 12 Btu/ft² · hr · °F/in:

$$\left(\frac{Q}{A}\right) \frac{r_3 \ln(r_2/r_1)}{k_D} \qquad (2.3.7)$$

Q/A, Btu/hr · ft²	ΔT, °F	OD Metal Temp., °F
25,000	98	619
50,000	195	736
75,000	293	855
100,000	391	974
125,000	489	1093

Thus, depending on heat flux and thermal conductivity of the deposit, tube-metal temperatures may easily reach the creep range for a considerable time. For utility boilers with saturation temperatures of 650°F or higher, the effects of steam-side scale are greater. Thinner scale thicknesses will raise tube-metal temperatures into the creep range, and other problems may occur; under-deposit corrosion and hydrogen damage may lead to failure before blisters can form.

The calculations for a high-pressure utility boiler with an operating drum pressure of 2500 psi are similar, but the results are quite different. A

3.0-in OD SA-210 A-1 waterwall tube in a 2500-psi unit has an actual wall thickness of about 0.300 in. The saturation temperature is nearly 670°F (668°F, actually). Equation (2.3.3) gives

Q/A, Btu/hr · ft²	ΔT, °F	ID Metal Temp., °F
25,000	6	676
50,000	13	683
75,000	19	689
100,000	25	695
125,000	31	701

Likewise, Eq. (2.3.5) yields a larger temperature gradient through a clean tube:

Q/A, Btu/hr · ft²	ΔT, °F	ID Metal Temp., °F
25,000	23	699
50,000	46	729
75,000	70	759
100,000	93	788
125,000	116	817

In a utility or high-pressure boiler, there is less "margin" between the clean-metal temperature and the onset of graphitization and creep, usually taken as 850°F for plain-carbon steels. Much thinner deposit layers raise tube-metal temperatures into the graphitization temperature range. Equation (2.3.7) with a deposit thickness of 0.005 in gives

Q/A, Btu/hr · ft²	ΔT, °F	ID Metal Temp., °F
25,000	13	712
50,000	26	755
75,000	39	798
100,000	52	840
125,000	65	882

Figure 2.9 indicates that a peak heat flux of 125,000 Btu/hr · ft² is expected at the burner elevations. At a deposit thickness of 0.005 in, graphitization occurs. This simple heat-flow calculation also suggests the reason why some chromium–molybdenum steels are often used in the highest-heat-release regions of a furnace. These low-alloy steels can better tolerate slightly higher temperatures and will not graphitize. When a gap opens between scale and steel, as in Fig. 2.3.14, the overall thermal resistance increases and the temperature gradient across the gap may grow to several hundred degrees. The gap forms as the hot tube expands from the deposit by creep deformation. When the deposit remains intact, the gap is filled with steam at the same pressure as the fluid within the

tube. Thus, there may be little pressure differential across the fragile deposit that would rupture the scale. If the scale fractures, the "bulge" is quickly cooled with little gross distortion to the tube.

To estimate the effects of this vapor barrier to heat transfer, assume the gap is 0.002 in;

$$R_v = \frac{r_3 \ln(r_2/r_v)}{k_v} \tag{2.3.8}$$

where $r_3 = 1.50$ in
$r_2 = 1.30$ in
$r_v = 1.298$ in ($r_2 - r_v = 0.002$ in)
$k_v = 0.030$ Btu/hr · ft · °F, thermal conductivity of steam

The thermal conductivity of steam varies with temperature and pressure.[13] At 500 psi, k_v is 0.0260 Btu/hr · ft · °F at 600°F and rises to 0.0352 Btu/hr · ft · °F at 1000°F. At 1000 psi, k_v is 0.0301 Btu/hr · ft · °F at 600°F and is 0.0357 Btu/hr · ft · °F at 1000°F. For purposes of calculation, the value of 0.030 Btu/hr · ft · °F is reasonable to demonstrate the effect of a thin gap between scale and steel on heat transfer and tube-metal temperature.

Again, assume the calculated temperature gradient across the tube-to-deposit gap to be equal to the temperature increase in the tube-metal temperature. Hence,

$$\left(\frac{Q}{A}\right) \frac{r_3 \ln(r_2/r_v)}{k_v} = \Delta T_v \tag{2.3.9}$$

Q/A, Btu/hr · ft²	ΔT, °F	ID Metal Temp., °F
25,000	160	712
50,000	320	1056
75,000	481	1336
100,000	641	1615
125,000	801	1894

The calculations are for illustration only; precise results are not intended. What they do show is that steam-side deposits impede heat flow; the net effect does raise tube-metal temperature. With sufficient deposit thickness, the local metal temperature rises to the creep range. Once that occurs, creep deformation expands the metal from the deposit, and a gap forms. If the deposit layer ruptures, no tube failure occurs, only a blister. If the deposit layer remains intact, tube failure does occur. The simple heat-flow analysis correctly accounts for the microstructures found in waterwall tubes that contain blisters or failures. Observations of the microstructures are similar across many analyses performed on these waterwall tubes and failures.

REFERENCES

1. David Gunn and Robert Horton, *Industrial Boilers*, Longman/Wiley, New York 1989, p. 102.
2. *ASME Boiler and Pressure Vessel Code, Section I Power Boilers*, Paragraph PG27.2, 1989.
3. Raymond J. Roark, *Formulas for Stress and Strain*, 4th ed., McGraw-Hill, New York, 1965.
4. W. H. McAdams, *Heat Transmission*, 3rd ed., McGraw-Hill, New York, 1954.
5. *ASM Metals Handbook*, 9th ed., Vol. 11, "Failure Analysis and Prevention," ASM International, Metals Park, Ohio, 1986, p. 613.
6. *Steam/Its Generation and Use*, 38th ed., Babcock and Wilcox, New York, 1972.
7. Copyright © 1985. Electric Power Research Institute. EPRI CS-4252. *Dissimilar-Weld Failure Analysis and Development Program*. Reprinted with permission.
8. James G. Knudsen and Donald L. Katz, *Fluid Dynamics and Heat Transfer*, New York, 1958.
9. *Power Magazine*, March 1986, Copyright, McGraw-Hill, New York, 1986.
10. *Principles of Industrial Water Treatment*, Drew Chemical Co., Boonton, New Jersey, 1979.
11. D. E. Gelbar and T. E. Dorazio, "Boiler Tubing Evaluation and Life Prediction Techniques," Presented at Conference on Boiler Tube Failures in Fossil Plants, Atlanta, Georgia, Nov. 10–12, 1987, EPRI Research Project 1980.
12. D. N. French, "Waterwall Tube Blistering in Industrial Boilers," NACE Corrosion '90, Paper #190, April 1990.
13. J. H. Keenan and F. G. Keyes, *Thermodynamic Properties of Steam*, New York, 1952.

CHAPTER THREE

METALLURGICAL PRINCIPLES: FERRITIC STEELS

In order to understand later discussions concerning the microstructural changes that occur in boiler steels owing to exposure to high temperature or the effects of corrosive atmospheres, we need to take a brief look at some basic metallurgical principles.[1-4]

ATOMS AND CRYSTALS

Metals, like all matter, are made up of atoms. For our purpose it is sufficient to state that the atoms within metallic crystals or grains are regularly arranged and that the arrangement of these atoms determines the crystal structure of the metal. The smallest repetitive arrangements of these atoms are called unit cells and are the basic building blocks of the crystals. Of the many possible arrangements of atoms that occur in metals, only the two commonly associated with steel will be described. The room-temperature form of iron and most steels is the body-centered-cubic structure. In this structure the unit cell of atoms is arranged as a cube with atoms at each of the eight corners and one in the center of the cube, as illustrated in Fig. 3.1. The high-temperature form of iron and most steels, and the room-temperature form of some stainless steels, is the face-centered-cubic structure. In this structure, the unit cell is a cube with atoms at the eight corners and another atom in the center of each of the six cube faces. Figure 3.2 schematically shows this structure.

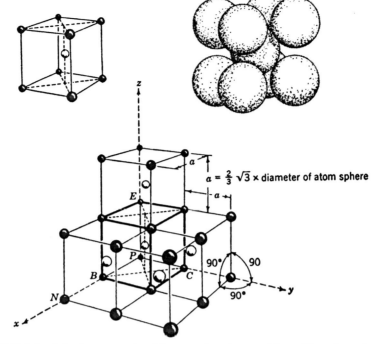

FIGURE 3.1. A body-centered-cubic crystal structure. The white sphere is at the center of each unit cell. The figure at the upper right shows how atoms fill space in the unit cell (Ref. 1).

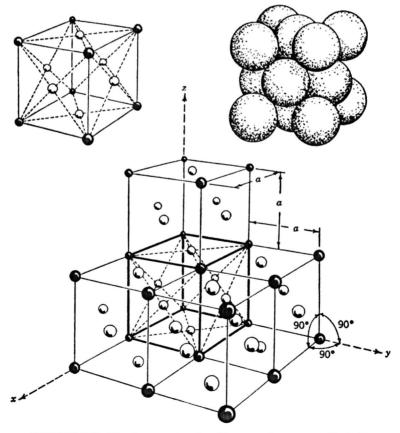

FIGURE 3.2. The face-centered-cubic crystal structure (Ref. 1).

SOLID SOLUTIONS

The two crystallographic structures previously described were for essentially pure iron. However, commercially useful metals are rarely pure elements. Stainless steel for example is an alloy of iron and chromium, with at least 11 wt% chromium. Such an alloy will consist of a random mixture of iron and chromium atoms. Since both iron and chromium atoms have approximately the same size, each will occupy any of the positions or sites within the lattice without preference. This mixture of atoms in a crystal, in which the atoms of the second element, chromium, in this example, are distributed uniformly within the parent crystallographic structure, is known as a solid solution. Since the chromium atoms are found on iron-atom sites, the solid solution is known as substitutional, with chromium atoms substituting for some of the iron atoms in the crystal. Other common alloying elements that form substitutional solid solutions with iron are manganese, nickel, molybdenum, and silicon.

When carbon is added to iron, the carbon atoms distribute themselves in the spaces between the iron atoms, because the carbon atom is considerably smaller than the iron atom. Atoms in this configuration are said to have formed an interstitial solid solution. The solution of carbon in body-centered-cubic iron, then, is an interstitial solid solution. Other elements that form interstitial solid solutions with iron are boron, hydrogen, nitrogen, and oxygen. Interstitial solid solutions are characterized by limited solid solubility.

GRAIN BOUNDARIES

The previous descriptions are of the two crystallographic structures of individual crystals of pure iron and solid solutions. Useful shapes of alloys contain many crystals that can be seen readily with a microscope. Body-centered-cubic iron containing alloying elements in solid solution is called ferrite. The ferrite microstructure shown in Fig. 3.3 is essentially pure iron containing a very small amount of other alloying elements. Face-centered-cubic iron containing alloying elements is called austenite, and an austenite microstructure is presented in Fig. 3.4. What we see in these figures is a single-phase material—that is, a material made up of many individual crystals or grains. All have the same crystal structure and similar chemical compositions; only differences in orientation lead to what are called crystal boundaries or, more commonly, grain boundaries between crystals, or simply grain boundaries. The atomic arrangement within the grain boundary is rather ill-defined compared with the regular array within the body of the grains. The photomicrograph is a two-dimensional view in which the grain boundaries appear as a series of lines. In the three-dimensional solid, the grain boundaries actually are surfaces between crystals.

78 CHAPTER 3 METALLURGICAL PRINCIPLES: FERRITIC STEELS

FIGURE 3.3. The ferrite microstructure of essentially pure iron (Magnification 100×, nital etch.)

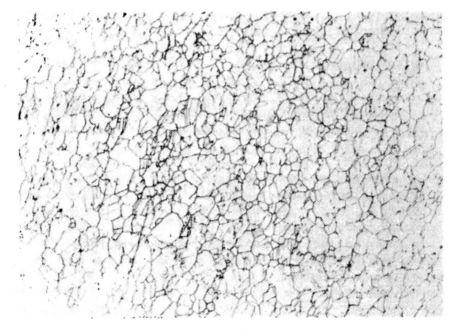

FIGURE 3.4. The microstructure of austenite. The structure taken from a solution-annealed 321H stainless steel. Note the similarities between Figs. 3.3 and 3.4 (Magnification 100×, electrolytic etch).

GRAIN BOUNDARIES

For our purposes, atoms within a metallic crystal are regularly arranged over great distances—distances that are huge when compared with atomic dimensions. For example, the iron atom has a diameter of about 2.48 Å,[1] 2.48 × 10^{-8} cm or 10^{-8} in. Across an individual grain diameter there may be 10^5 atoms. There is, then, long-range atomic order within individual crystals. Where adjacent crystals join is a grain boundary. The grain boundary is 2–10 atoms thick, so this region contains short-range atomic disorder. The grain boundaries may be viewed as a transition region in which the individual atoms are not precisely aligned with either grain. These grain boundaries determine in no small way the useful properties of engineering materials.

Grain size can vary greatly, depending on the alloy and heat treatment. A typical grain size is about 1 mil (0.001 in). Thus, there may be a billion (10^9) grains per cubic inch of alloy. The ASTM grain-size number, N, is one standard for determining the average grain size. It is defined by[1]

$$n = 2^{N-1}$$

where n is the number of grains per square inch when viewed at 100×. The usual range is from 1 to 10; note that as the grains get smaller, the grain-size number gets larger. The specification requirements for SA-213 TP-321H include a grain size of 7 or coarser, or 0.00125 in in diameter or larger.

The region of atomic disarray at the grain boundary is associated with greater reactivity. For example, the movement of atoms along the grain boundary is usually much more rapid than the movement of the same atoms through the crystal lattice. In etching of metallographic samples, the grain boundaries are attacked more vigorously by the dilute acid used, which leaves a shallow groove at the grain boundary. Under the reflected light of the microscope, since the grain boundaries are no longer planar, they show up as dark lines. The grain boundaries, although only a small fraction of the total solid, are frequently the critical factor in the failure of a given metal component.

Grain boundaries are regions of atomic mismatch and less dense atomic packing. Less density on an atomic scale implies bigger atom-sized holes through which atoms can more easily move. Such atomic mobility is called diffusion. Thus, grain boundaries oxidize or corrode more rapidly, which is usually referred to as *grain-boundary penetration*. Oxygen diffuses along the grain boundaries, reacts with the steel, and forms iron oxide within the grain boundaries. See Fig. 3.5 and page 285. In austenitic stainless steels, diffusion of atoms along the grain boundaries leads to the formation of chromium carbides. As carbides form, they deplete the region immediately adjacent to the grain boundary of chromium. As the chromium content decreases, the grain-boundary region becomes less corrosion-resistant, a condition referred to as *sensitization* (see page 289). Once these steels have been sensitized, they are subject to an intergranular corrosion attack; see Fig. 3.6 and page 249.

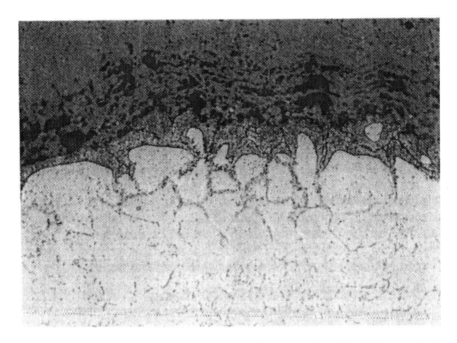

FIGURE 3.5. Oxide penetration along the grain boundaries of a T-22 superheater tube (Magnification 500×, etched).

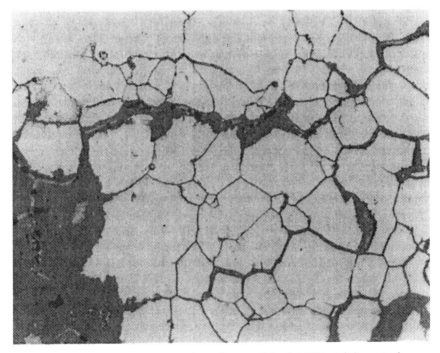

FIGURE 3.6. Intergranular corrosion of a sensitized 347H stainless-steel superheater tube (Magnification 200×, not etched).

Hydrogen damage is another corrosion process where grain boundaries play an important role. Water-side corrosion of a waterwall tube produces hydrogen as one of the corrosion products. The hydrogen atom diffuses into the steel, where it reacts with iron carbide to form methane. The methane collects at grain boundaries and ultimately forms intergranular cracks; see Fig. 3.7 and page 335.

At elevated temperatures, the strength of an individual grain is greater than the strength of the grain boundary, and creep deformation occurs by grain-boundary sliding (see page 99). Voids form where several grains join. Individual voids can then link up to form grain-boundary cracks; see Fig. 3.8 and page 101.

Under some corrosion circumstances, intergranular corrosion is nearly universal. A scanning-electron microscope photograph gives a good two-dimensional presentation of the three-dimensional shape of individual grains in a polycrystalline material; see Figure 3.9.

Under microscopic examination, the microstructure of one solid solution cannot be distinguished from another, except perhaps by differences in color. For example, compare the microstructure of the ferritic iron shown in Fig. 3.3 with the microstructure of an austenitic stainless steel, which is a solid solution of chromium and nickel in iron, shown in Fig. 3.4. Note the similarities in these structures.

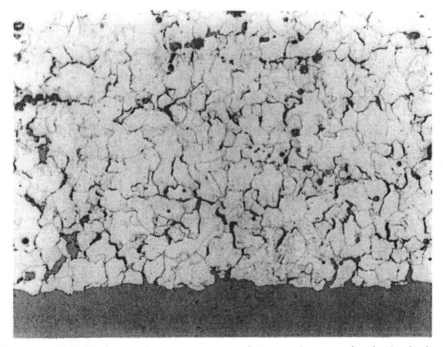

FIGURE 3.7. Hydrogen damage appears as intergranular corrosion in the ferrite and decarburization. From a waterwall tube of SA-210 A-1 (Magnification 500×, etched).

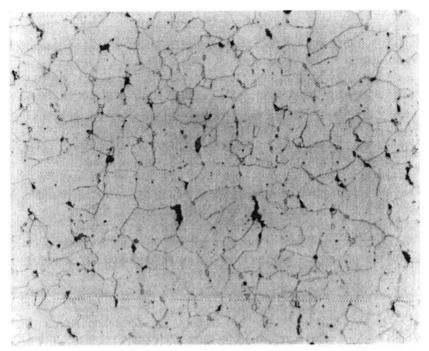

FIGURE 3.8. Creep damage starts as small voids where three grains meet and grow until a long crack appears. From an SA-192 superheater tube (Magnification 500×, etched).

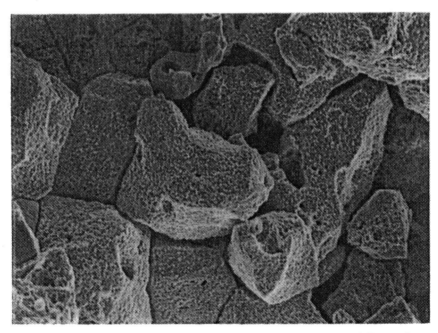

FIGURE 3.9. SEM photograph of a stainless-steel pump shaft that failed by intergranular corrosion (Magnification 500×).

FIGURE 3.10. A two-phase microstructure of spherical particles of iron carbide in ferrite (Magnification 250×, nital etch).

The ferrite in Fig. 3.3 and the austenite in Fig. 3.4 are single-phase materials. That is, nothing but ferrite in the one and austenite in the other can be seen within the microstructure. However, most useful engineering alloys contain more than one phase; in other words, they are multiphase materials. In these materials, the constituent phases are visible under microscopic examination and can be distinguished by color and shape. This has led to the description of phases as those portions of the microstructure that can be ditinguished visually or "mechanically" separated from one another. An example of a two-phase microstructure is shown in Fig. 3.10, where spheroidal particles of iron carbide are seen to be dispersed in a matrix of ferrite (the continuous white background).

PHASE DIAGRAMS

A phase diagram is a "road map" that depicts the various phases that will be in equilibrium at a given temperature and a given composition. It will show melting points and phase-transformation regions; it can predict the chemical composition of phases and can be used to predict the amount of phase present in a two-phase region.

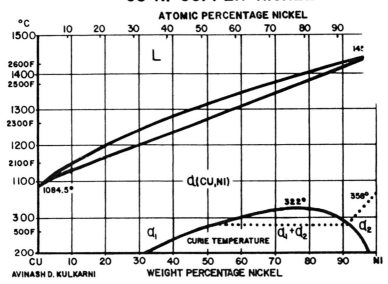

FIGURE 3.11. The copper–nickel phase diagram. The system copper–nickel is typical of solid solutions that show complete substitutional or mutual solid solubility. The field labeled L indicates liquid. The solid solution of copper and nickel, the two-phase region $\alpha_1 + \alpha_2$, the magnetic transition, Curie temperature, is also shown. From Kulkarni, Avinash D., Cu–Ni diagram, Hawkins, Donald T. and Hultgren, Ralph, "Constitution of Binary Alloys," *Metals Handbook,* 8th ed., Vol. 8, Editor, Lyman, Taylor, American Society for Metals, 1973, page 294.

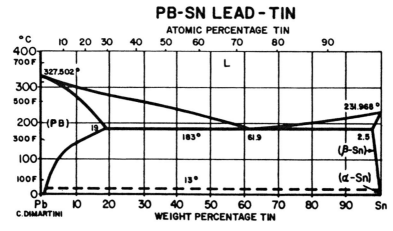

FIGURE 3.12. The lead–tin phase diagram. The system lead–tin is an example of simple eutectic alloys. There is limited solid solubility of tin-in-lead (Pb) and lead-in-tin (β-Sn). 61.9 wt% tin is the minimum-melting-point liquid, the eutectic point. Below 361°F (183°C) is the two-phase region, at 55°F (13°C) β-Sn transforms to α-Sn. From DiMartini, C., Pb–Sn diagram, Hawkins, Donald T. and Hultgren, Ralph, "Constitution of Binary Alloys," *Metals Handbook,* 8th ed., Vol. 8, Editor, Lyman, Taylor, American Society for Metals, 1973, page 330.

The simplest phase diagram shows substitutional mutual solubility in the liquid and solid states. For example, in the copper–nickel system, Fig. 3.11,[5] a face-centered-cubic solid solution is formed over the entire range of composition. These copper–nickel solid-solution alloys are ductile, easy to form, and reasonably corrosion-resistant. Condenser tubing of 70% copper and 30% nickel composition is in wide use.

The most common type of phase diagram shows a simple eutectic, similar to the lead–tin phase diagram, Fig. 3.12.[5] Lead has only limited solid solubility for tin, and tin only limited solid solubility for lead. The terminal solid solutions form the boundaries for the region showing two phases of tin-rich and lead-rich solid solutions. Over a wide range of composition, a portion of the solidification of the alloy occurs at a single fixed temperature—the eutectic temperature. The basic reaction on cooling is the eutectic liquid solidifying as a two-phase mixture of alpha solid solution and beta solid solution at a temperature of about 375°F (191°C).

IRON–CARBON DIAGRAM

Figure 3.13 shows the iron–carbon diagram.[5] Pure iron at temperatures below 1674°F (912°C) has the body-centered-cubic crystal structure. Body-centered-cubic iron can dissolve only very small amounts of carbon, a maximum of about 0.0218% at 1341°F (727°C). This interstitial solid solution of carbon in body-centered-cubic iron is known as ferrite. At temperatures between 1674°F (912°C) and 2541°F (1394°C), the iron atoms arrange themselves in the face-centered-cubic lattice. Face-centered-cubic iron has a much greater solubility of carbon, the maximum being approximately 2.11% at a temperature of 2098°F (1148°C). The addition of carbon to face-centered-cubic iron also depresses the temperature at which face-centered-cubic iron is transformed to the body-centered-cubic structure from 1674°F (912°C) to 1341°F (727°C) at 0.77% carbon. Between 2541°F (1394°C) and the melting temperature, 2800°F (1538°C), the iron reverts to the body-centered-cubic arrangement, called delta-iron in the pure state and delta-ferrite as a solid-solution alloy. For practical purposes in boiler usage, we can ignore this high-temperature transformation.

Just as carbon expands the temperature range over which austenite is stable, other alloying elements will do the same, most notably nickel. The addition of about 8% nickel to an 18% chromium stainless steel forms an alloy that has the face-centered-cubic structure all the way to room temperature. These alloys are known as austenitic stainless steels.

LEVER RULE

Using the phase diagram, we can easily discern the composition of alloy phases at a particular temperature; from this information, the relative

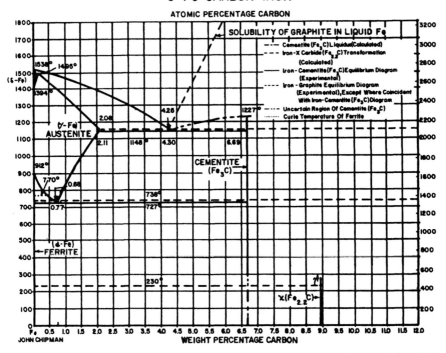

FIGURE 3.13. The iron–carbon phase diagram. Note that the phase relationships between iron and iron carbide (cementite) and between iron and graphite are superimposed on the diagram. Cementite is a metastable phase that will revert to the stable phase, graphite, on prolonged heating at temperatures above about 850°F (455°C). For steel metallurgy of alloys used in boiler construction, the carbon content is less than 0.35%. The transformation from austenite (γ) to ferrite (α) and cementite are of prime consideration. From Chipman, John, C-Fe diagram, Hawkins, Donald T. and Hultgren, Ralph, "Constitution of Binary Alloys," *Metals Handbook,* 8th ed., Vol. 8, Editor, Lyman, Taylor, American Society for Metals, 1973, page 275.

amounts of each phase can be calculated easily. For example, Fig. 3.14 shows a portion of a binary-phase diagram. As an alloy of composition OP is cooled to temperature T_1, it just reaches the phase boundary separating the one-phase gamma region from the two-phase alpha plus gamma region. At this temperature, alpha of composition A begins to form. On further cooling to temperature T_2, the composition of the alpha phase changes from point A to point B, and the composition of the gamma phase changes from point D to point E, while the relative amounts of alpha (increasing) and gamma (decreasing) are such that the overall composition of the alloy remains constant. At temperature T_2, the amount of alpha present is equal to the length of the line QE divided by the total length of the line BE. Similarly, the amount of gamma present is the length of the line BQ divided by the length of the line BE.

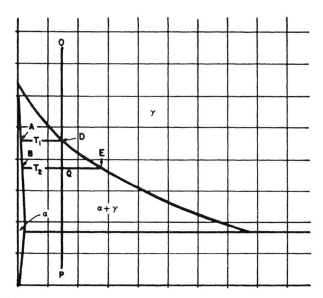

FIGURE 3.14. A schematic portion of a phase diagram showing the elements of the Lever Rule. See text for an explanation.

The Lever Rule was used above to calculate the relative amounts of the phases present at a particular temperature. Later it will be applied in the determination of temperature when the relative amounts of the phases are known from microstructural analysis.

TRANSFORMATION IN IRON–CARBON ALLOYS

A look at the iron–carbon phase diagram of Fig. 3.13[5] shows that there are really two superimposed phase diagrams. One is the true equilibrium diagram for iron and graphite, and the other is a metastable equilibrium diagram between iron and iron carbide, Fe_3C. Iron carbide, also called cementite, is a metastable compound relative to its decomposition into ferrite and graphite. Thus, one set of transformations to be discussed will be the change in iron carbide into graphite.

The eutectoid transformation of austenite at 1341°F (727°C) is the basis of the heat treatment of steel. If austenite is cooled very slowly, several hours to cool to room temperature, the structure would be one of ferrite and pearlite for carbon contents less than 0.77%. Pearlite is a mixture of iron carbide and ferrite. The morphology of the mixture depends on the cooling rate. If austenite is cooled extremely rapidly (a few seconds to cool to room temperature), the transformation to iron carbide is suppressed and the austenite transformation product is martensite, which is extremely hard.

T-T-T CURVES

Reaction rates for the phase changes in steel are shown by time–temperature–transformation (T-T-T) curves, such as shown in Fig. 3.15.[6] These are also called S curves and C curves. When the cooling rate is fast enough to miss the "nose" of the curve where pearlite transformation starts in the 1000°F (538°C) region, austenite is maintained to a low temperature. Since time is inadequate for the diffusion of carbon to form iron carbide and ferrite, the retained austenite is unstable and a different sort of transformation occurs. The transformation product is a hard material called martensite and is formed by a diffusionless shear transformation. The face-centered-cubic crystal structure of austenite transforms to the body-centered-tetragonal structure of martensite.

By the use of suitable alloying elements, nickel, chromium, and molybdenum, for example, the transformation from austenite to pearlite may be

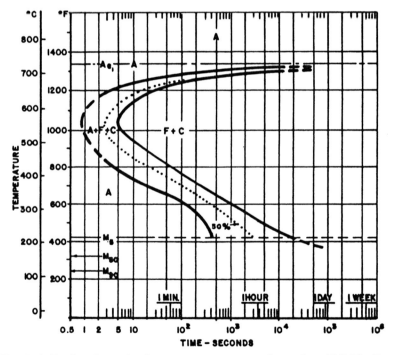

FIGURE 3.15. A schematic time–temperature–transformation (T-T-T) diagram. Since the normal transformation of austenite to ferrite and cementite may be suppressed by rapid cooling, T-T-T diagrams show the phase relationships with time. For slow cooling, austenite will transform to ferrite (F) and cementite (C). More rapid cooling will retain some austenite along with ferrite and cementite. For very rapid cooling, no austenite will transform until the M_s (martensite start) temperature, about 410°F (210°C). On further cooling, 90% of the austenite will transform to martensite at 240°F (116°C). Courtesy of U.S. Steel Corp. (Ref. 9).

retarded. These cause the nose of the pearlite transformation curve to be shifted to the right so that slower and slower cooling rates will suppress the transformation of austenite to lower temperatures. Certain tool steels contain sufficient alloy content so that cooling in still air is sufficiently rapid to miss the pearlite nose and to allow martensite to form. Figure 3.16 shows the martensitic structure of a medium-carbon tool steel.

Cooling rates that are slow enough to form pearlite are more typical of the manufacturing process for ferritic boiler steels. When a 0.15% carbon alloy, for example, SA-178 A, is cooled from the all-austenite region at a temperature of 1700°F (927°C), nothing happens to the microstructure until a temperature of 1550°F (843°C) is reached. At this temperature the austenite begins to transform to ferrite; and, as cooling continues, the amount of ferrite increases while the amount of austenite decreases. Since ferrite contains very little carbon, the carbon content of the remaining austenite must continuously increase in order to maintain the initial composition of the alloy. The carbon content in the austenite increases from 0.15% at 1550°F (843°C) to a maximum of 0.77% at the eutectoid temperature of 1341°F (727°C). At the eutectoid temperature, austenite of 0.77% carbon is in equilibrium with ferrite of 0.022% carbon and iron carbide (Fe_3C) of 6.67% carbon; refer to Fig. 3.13.

FIGURE 3.16. The microstructure of a medium carbon–chromium tool steel. The structure is composed of needles of martensite, the dark-etching constituent, and retained austenite, the white (Magnification 500×, nital etch).

The amounts of ferrite and austenite that are present at any temperature in the two-phase ferrite–austenite region can be determined by means of the Lever Rule and the iron–carbon diagram (Fig. 3.13). For example, just above the eutectoid transformation temperature of 1341°F (727°C), the 0.15% carbon alloy is determined to contain 17% austenite and 83% ferrite by means of the following calculation:

$$\text{Amount of austenite} = \frac{0.15 - 0.022}{0.77 - 0.022} \times 100 = 17\%$$

On cooling below 1341°F (727°C), any remaining austenite transforms to ferrite and iron carbide. In this process, the carbon that is rejected from the austenite and cannot be dissolved in the ferrite forms platelets of iron carbide. The microstructure resulting from this eitectoid transformation is called pearlite, and is a lamellar mixture of ferrite and iron-carbide platelets. At the eutectoid composition of 0.77% carbon, the all-austenite microstructure would transform to 100% pearlite. However, in the 0.15% carbon alloy, only the 17% austenite that remains at the eutectoid transformation temperature would transform to pearlite; the remaining 83% would be the ferrite that formed above the eutectoid temperature. The lamellar appearance of pearlite can be seen in Fig. 3.17, along with large light-colored areas of the proeutectoid ferrite, defined on page 93.

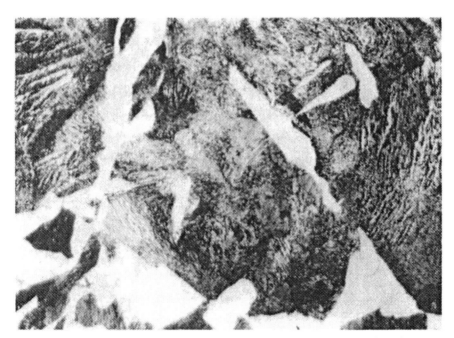

FIGURE 3.17. The microstructure of a 0.35% carbon steel, SA-515-70. The structure is pearlite and ferrite. Note that the pearlite is composed of platelets of the dark-etching constituent, cementite (Magnification 500×, nital etch).

Within limits, as the cooling rate increases, the time for carbon diffusion decreases; and the thickness of the lamellar platelets of iron carbide and ferrite decrease. The rate of cooling that produces the "normal" microstructure of pearlite is called a normalizing heat treatment. The normalized structure of a low-carbon steel, then, is one of ferrite and pearlite.

A review of the T-T-T curve of Fig. 3.15 shows that if the cooling rate is fast enough to prevent the formation of ferrite and pearlite, the temperature has to drop below about 400°F (200°C) before martensite will form. The M_s (martensite start) temperature is affected by the composition of the steel. For plain-carbon steels the M_s temperature drops from around 900°F (482°C) for 0.05% carbon to 425°F (218°C) for a 0.9% carbon.

When steel is cooled rapidly enough to miss the pearlite nose and suppress the formation of pearlite, but is not cooled below the M_s temperature, a third transformation product called bainite forms (see Fig. 3.18). Bainitic structures vary from the appearance of very fine pearlite for upper bainite to a martensitic-like structure for lower bainite. Upper and lower in this case

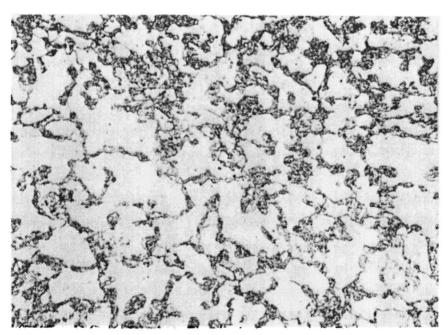

FIGURE 3.18. The microstructure of a 0.25% carbon steel similar to SA-210 A-1. The structure is composed of ferrite, the white-etching constituent, and bainite, the dark-etching constituent. The structure is formed by cooling sufficiently rapidly to miss the pearlite nose, suppressing the formation of pearlite, and is allowed to transform at a temperature above the martensite transformation temperature. The austenitic transformation product in this case is bainite. The temperature from which it was quenched is in the neighborhood of 1450°F (788°C) (Magnification 500×, nital etch).

92 CHAPTER 3 METALLURGICAL PRINCIPLES: FERRITIC STEELS

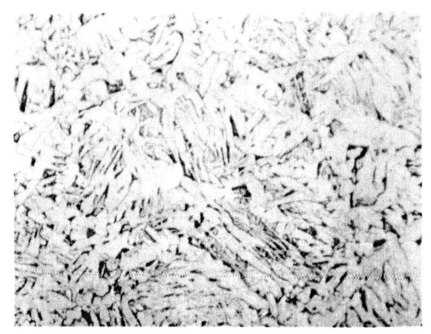

FIGURE 3.19. A Widmanstätten structure (Magnification 500×, nital etch).

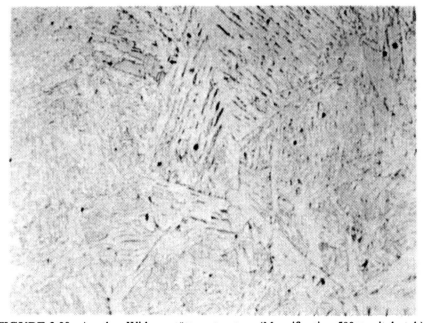

FIGURE 3.20. Another Widmanstätten structure (Magnification 500×, nital etch).

refer to the temperatures at which the bainite forms, and range from about 1000°F (538°C) to 425°F (218°C) for the curve of Fig. 3.15.

Not only is the cooling rate below 1341°F (727°C) important in determining the structure and morphology of the austenite decomposition products, but so is the cooling rate through the two-phase austenite and ferrite region. Since only the austenite can transform at temperatures below 1341°F (727°C), the morphology of the ferrite formed in cooling through the two-phase region dictates the final structure. Ferrite that forms above the eutectoid temperature is known as proeutectoid ferrite. The cooling rate through the two-phase region establishes the nucleation sites of proeutectoid ferrite. Ferrite will form not only at the austenite grain boundaries, but along certain crystallographic planes of the austenite. When the eutectoid temperature is reached, austenite remains in layers between ferrite plates; and, thus, the final structure appears to be layers of pearlite or bainite and ferrite, called a Widmanstätten structure. Figure 3.19 shows such a structure, and Fig. 3.20 shows another variation on the Widmanstätten structure.

MECHANICAL PROPERTIES

The usefulness of engineering alloys depends primarily upon their mechanical properties—such as, strength, hardness, ductility, high-temperature strength, and creep resistance. Quantitative data on these and other properties are given in handbooks. Indeed, the whole design of boilers depends upon standard test data tabulated in the ASME Boiler and Pressure Vessel Code. For our purposes, we shall limit our discussions to the more general concepts of elastic and plastic behavior of metals.

Tensile Strength

In a simple tensile test, a specimen is stretched axially until fracture occurs. Careful measurements of the force involved and change in length and/or diameter provide information for the stress–strain diagram, Fig. 3.21.

The stress, plotted on the ordinate, is computed by dividing the instantaneous load by the initial cross-sectional area of the specimen. Strain is plotted along the abscissa and is the change in gauge length divided by the original gauge length; it is a dimensionless value, but is generally reported as inches per inch.

The engineering stress–strain curve, Fig. 3.21,[1] reflects elastic and plastic behavior of plain-carbon steel similar to SA-178 A or C and SA-210 A-1. The initial portion of the curve is a straight line and is Hooke's law, where stress S is proportional to strain ε. The proportionality constant E is Young's modulus of elasticity; $S = E\varepsilon$. As long as the load remains below the proportional elastic limit, the specimen will return to its original length when the load is removed. Beyond the elastic limit, the specimen undergoes perma-

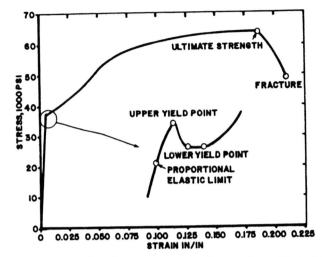

FIGURE 3.21. An engineering stress–strain diagram for mild steel (Ref. 1).

nent deformation; that is, above this stress, the material no longer obeys Hooke's law. For low-carbon steels, the elastic limit is about 30,000 psi. As load is increased in our simple tensile test, a stress is reached at which the steel will elongate without an increase in load. This stress is called the yield strength. Most deformation beyond the yield point is plastic; that is, it remains in the metal after the load is removed. The yield point for materials that do not have a well defined yield point, as does mild steel (e.g., austenitic stainless steel), is defined as that stress at which a permanent set of 0.2% or a strain of 0.002 in/in occurs.

As shown in Fig. 3.21, a peculiar behavior of mild steel leads to upper and lower yield points. However, we need not concern ourselves with this phenomenon.

As the test specimen is stressed beyond the yield point, the stress increases toward the maximum, known as the ultimate tensile strength of the material. Brittle materials will break at this point, but ductile materials begin to decrease rapidly in diameter at some localized area, forming a well-defined neck. Since the engineering stress–strain curve is based on a calculation using initial area rather than instantaneous area, the engineering stress–strain diagram for ductile metals shows a decrease in stress from the maximum to the final stress at fracture.

For ductile metals, like austenitic stainless steel or mild steel, values for the percentage elongation or percentage reduction in area are reasonably good indices of ductility.

Table XIII of the appendix presents some mechanical-property data of most of the common alloys used in boiler construction. These data are taken from mill certificates and represent actual values. Note that in every case the tensile strength is about 10% or more greater than the ASME Code-specified minimum.

Deformation

Microstructures often reflect the type of atomic process involved in the plastic deformation. Deformation occurs by the movement of one plane of atoms over another within individual crystals, sometimes referred to as *slip*. There are particular crystallographic planes and particular crystallographic directions in which atomic movement is easiest. The individual planes and directions are different, depending on the atomic arrangement—that is, different in face-centered-cubic and body-centered-cubic structures. Another deformation mechanism more common in face-centered-cubic materials is twinning, which consists of a shearing movement of the atomic planes past one another. The final orientation on either side of the twin boundary is a mirror image of the orientation in the crystal. Twinning is an important process in face-centered-cubic alloys similar to the austenitic stainless steels. Ferrite will plastically deform by twinning at room temperature under impact conditions.

The microstructural appearance of these deformation processes is somewhat varied. At large amounts of deformation, the microstructure shows entire crystals deformed and elongated; see Fig. 3.22. Up to about 15%

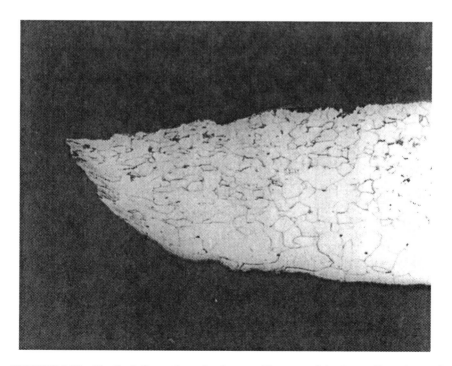

FIGURE 3.22. Plastic deformation of polycrystalline materials shows distortion and elongation to the microstructure. From the failure lip of an SA-178 A waterwall tube (Magnification 200×, etched).

deformation there is little change in the microstructure at magnifications of, say, 500×. From 15% to 20%, slip or twin lines appear within individual crystals; see, for example, Fig. 3.23. At deformation greater than about 20%, individual grains begin to show elongation. Plastic deformation in a forging, for example, displays flow lines as the metal fills the die. Figure 3.24 presents the deformation in an electric-resistance-weld line.

Under impact conditions, ferrite also shows deformation twins; see Fig. 3.25. During plastic deformation the hardness of the material increases, a process known as *work hardening*.

Hardness

Hardness can most easily be defined as the resistance of the material to indentation. Most hardness tests measure the resistance of metal to plastic deformation. Usually a spherical, pyramidal, or conical indenter is pressed into the surface of the sample, and the size of the impression is measured.

In a Brinell hardness test, the diameter of the impression made in the material by a hardened steel ball under a load of 500 or 3000 kg is measured. The ratio of the load to the area of the indentation is called the Brinell hardness number (BHN). Metals with a high tensile strength have a high

FIGURE 3.23. Deformation twins in a 304 stainless-steel bend, 18% elongation to the material at the outside or extrados of the bend (Magnification 200×, etched).

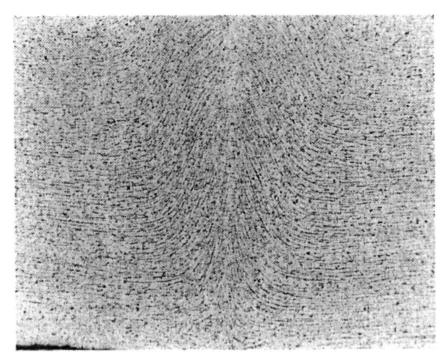

FIGURE 3.24. Flow lines in an electric-resistance weld indicate the plastic deformation of the microstructure during the weld. From an SA-178 A tube (Magnification 30×, etched).

FIGURE 3.25. Impact conditions will form deformation twins in ferrite. From an SA-209 T-1 superheater tube near a dent formed by a shotgun pellet (Magnification 500×, etched).

hardness. For carbon and low-alloy steels, the BHN times 500 is approximately equal to the tensile strength of the material in pounds per square inch.

The Rockwell test uses a hardened steel ball or diamond cone. Various loads are employed for different materials. The depth of the indentation is a measure of the hardness. Two scales are usually used: Rockwell B scale for softer materials and Rockwell C scale for harder steels. The B scale uses a $\frac{1}{16}$-in-diameter steel ball and a load of 100 kg; the C scale uses a diamond cone with a load of 150 kg. Rockwell B 100 is approximately equal to Rockwell C 20.

High-Temperature Tensile Strength

By means of a tensile-strength test performed at different temperatures, useful data are developed for the mechanical properties, tensile strength, yield strength, and elongation at temperatures extending over the entire useful range for these materials. Typically, as temperature goes up, the strength goes down; and the amount of elongation increases; see Table VIII in the appendix.

Stress-Rupture Strength

In stress-rupture tests, various loads are applied to specimens at elevated temperatures, and the time to fracture the specimen is measured. From results at high temperatures and high stresses, predictions are made of the probable stress-rupture life at lower stresses and temperatures that might be encountered in service. A commonly used method is to take the high-temperature tensile-strength data and the 100- and 1000-hr rupture-strength data and plot Larson–Miller master curves; see Fig. 3.26.[7] In the Larson–Miller method,[8] time and temperature are related by the following equation:

$$T(C + \text{Log } t) = P$$

where P is the Larson–Miller parameter, T is the absolute temperature in degrees Rankine (°F + 460), t is the rupture time in hours, and C is a constant equal to 20 for SA-213 T-22. Thus, if the rupture stress is plotted as a function of the parameter $T(20 + \text{Log } t)$, all of the data points should fall on a single curve. This is a useful parameter for judging the changes in expected life when operating conditions are altered.

The Larson–Miller plots are easy to use. For example, at a stress of 8000 psi, from Fig. 3.26 for SA-213 T-22 steel, the value of P is found to be 38,000. Thus, any combination of temperature and time that gives this value of P will cause rupture. At 1050°F (566°C) rupture should occur in about 146,000 hr, but at 1150°F (621°C) rupture occurs in only 4000 hr. Those calculations indicate life expectancy at constant temperature and stress and do not take into consideration start-up and shutdown stresses and temperature excur-

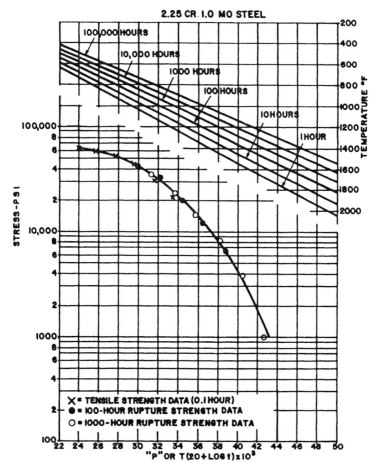

FIGURE 3.26. Larson–Miller master curves for SA-213 T-22 material. Copyright, American Society for Testing and Materials, 1916 Race Street, Philadelphia, PA, 19103. Reprinted with permission.

sions. As we shall see, this simple use of the Larson–Miller parameter is helpful in analyzing the life expectancy of a superheater tube that has a heavy scale in the bore of the tube and is thus operating at a higher temperature than desired.

Creep

At elevated temperatures and stresses much less than the high-temperature yield stress, metals undergo permanent plastic deformation, called creep.[9] Figure 3.27 shows a schematic creep curve for a constant load. Four por-

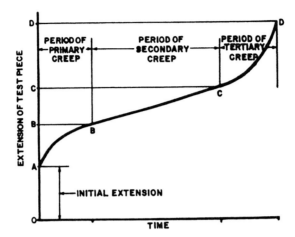

FIGURE 3.27. A schematic presentation of a creep curve. Courtesy of Babcock and Wilcox.

tions of the curve are of interest. There is an initial steep rate that consists, at least partly, of elastic origin; this is followed by a region in which the rate decreases with time, the so-called transient creep (primary creep). The next portion of the creep curve is the area of engineering interest, where the rate of creep is nearly constant (secondary creep). For boiler applications the stress to cause a 1% elongation in 10,000 or 100,000 hr is important. The allowable stresses given in the ASME Boiler Code are based on these creep values for temperatures above about 800°F for carbon steel and 1000°F for austenitic stainless steel. The fourth portion of the creep curve, beyond the constant-creep-rate portion, shows a rapidly increasing creep rate (tertiary creep), which culminates in failure.

The creep rate is a grain-size-sensitive property since much of the deformation is by grain-boundary sliding. In many alloys, a larger grain size improves creep resistance; hence, for austenitic stainless steels, the ASME Boiler Code requires an ASTM grain size of 7 or larger. Precipitation of carbides or of sigma-phase in stainless steel and the decomposition of pearlite into spheroidized carbides also influence the creep strength.

Creep may be defined as time-dependent strain or deformation at constant stress at elevated temperatures. High-temperature components of superheater and reheater usually fail by a creep or stress-rupture mechanism. There are subtle changes within an individual tube that precede any obvious deformation or microstructural damage. The first sign of creep damage is often the appearance of longitudinal cracks within the steam-side scale. The ductility of iron oxide is essentially nil compared with the ductility of steel. Thus, as the tube expands by a creep-deformation process, the brittle steam-side oxide cannot follow the expansion, and longitudinal cracks appear. Once the longitudinal cracks form, the path between steam and steel is shortened; and with continued expansion, the ID scale re-forms at the crack. Locally a cusp forms at the scale–metal interface, and the sharp point of the cusp acts as a stress raiser and intensifies the hoop stress. This stress may be also slightly higher because the wall thickness is slightly less at this cusp location.

Depending on the circumstances on the fire side, similar longitudinal cracks may appear in the fire-side scale. These OD cracks are more noticeable in gas-fired units because fuel-ash corrosion may obscure these creep effects in coal- or oil-fired units. Within the metal, creep cracks often first form in the decarburized layer, either on the OD or ID. These cracks may not penetrate through this weaker layer, see Fig. 3.28. In the metal, creep damage first appears as tiny voids or holes at triple points where three grains come together; refer to Fig. 3.29. The sequence of changes in the steam-side scale is shown in Figs 3.30–3.36.

Often the first signs of creep damage within the steel microstructure are creep voids at the cusps; see Fig. 3.34. Less than 0.1 in away from this damage, the microstructure shows no visible signs of distress at modest (500×) magnifications. As the creep deformation continues, these OD and ID cracks and the creep damage from them fan out toward each other; see Figs. 3.35 and 3.36. Creep failures, then, may initiate at the ID because the

FIGURE 3.28. Creep cracks may form in the decarburized layer but do not extend into the stronger material where the carbon content is higher (Magnification 50×, etched).

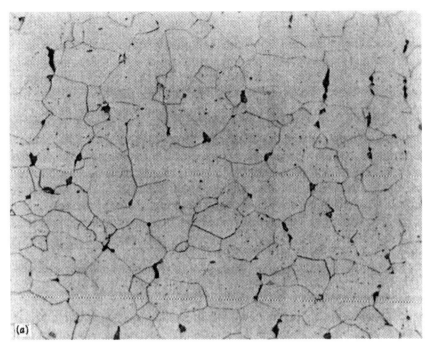

FIGURE 3.29a. Creep damage first appears in the steel as voids that form at triple points, where three grains come together (Magnification 500×, etched).

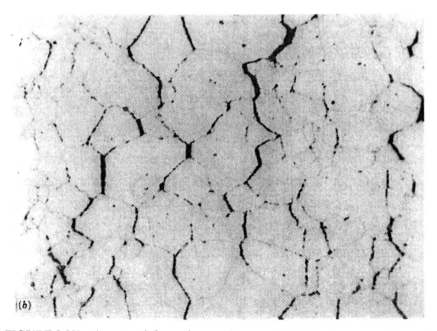

FIGURE 3.29b. As creep deformation continues, voids link to form extensive grain-boundary cracks. From a 304H secondary-superheater tube (Magnification 500×, etched).

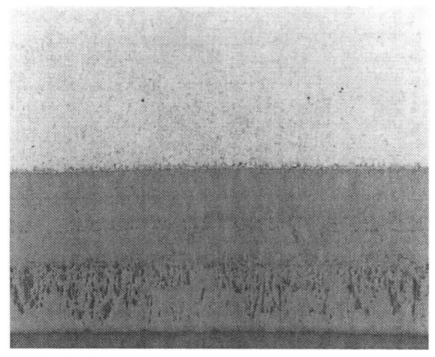

FIGURE 3.30. ID or steam-side scale with no creep damage; T-22 superheater tube (Magnification 100×, etched).

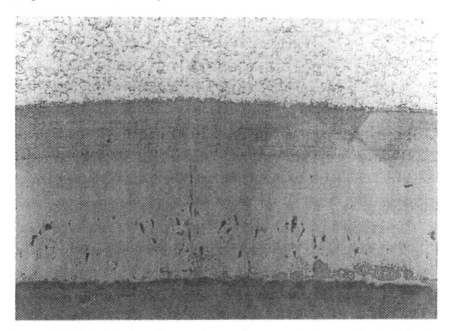

FIGURE 3.31. First sign of creep damage in a superheater or reheater tube is a longitudinal crack in the steam-side scale. The initial crack may not fully penetrate the oxide (Magnification 100×, etched).

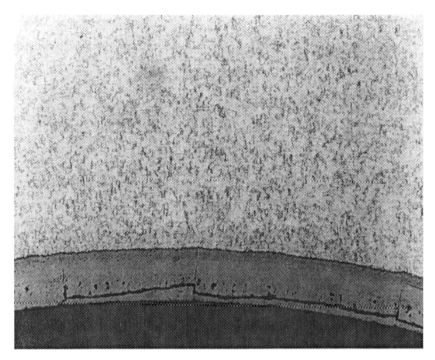

FIGURE 3.32. Several closely spaced cracks may form; the circumferential cracks in the oxide may contribute to exfoliation (Magnification 25×, etched).

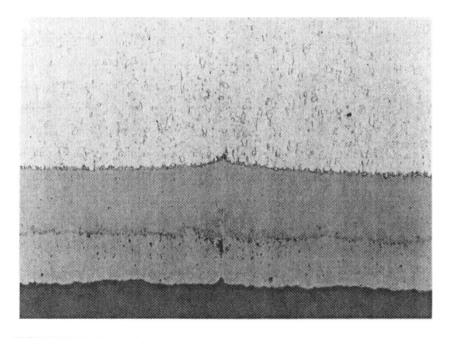

FIGURE 3.33. A cusp forms at the base of the longitudinal crack. At this point there is still no creep damage visible within the steel (Magnification 100×, etched).

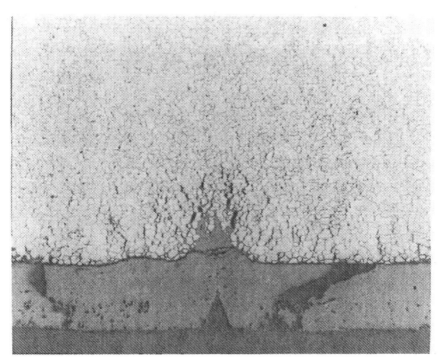

FIGURE 3.34. Creep cracks grow from the cusp (Magnification 200×, etched).

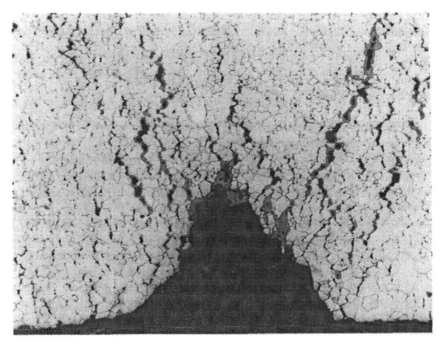

FIGURE 3.35. Cracks expand from the cusp (Magnification 200×, etched).

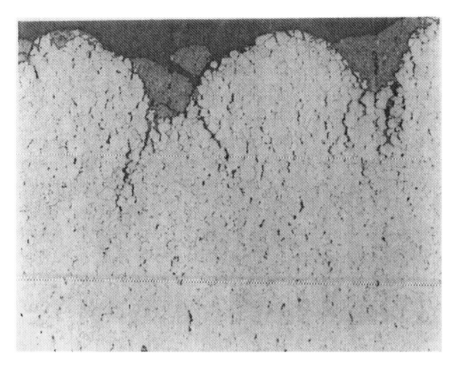

FIGURE 3.36. OD cracks form in a similar manner (Magnification 200×, etched).

stress intensification at the cusp tip is a greater factor than the higher temperature along the OD surface.

IMPACT PROPERTIES

Ferritic steels undergo a change in the mechanism of failure from ductile to brittle as the temperature is lowered. Below the transition temperature the failure mode is brittle or cleavage failure, while above the transition temperature the failure mode is ductile or shear failure. The transition temperature depends on the composition, heat treatment, and cold working. For plain-carbon steels, the transition temperature may be at or above room temperature. To prevent brittle failure from occurring, care must be exercised in design and fabrication to eliminate all possible stress-concentrating notches.

The Charpy V-notch impact test has been frequently used to evaluate the basic susceptibility to brittle fracture for a particular material. Figure 3.37 shows the impact strength versus temperature for SA-515 grade 70, a coarse-grained steel with a relatively high transition temperature. Since boilers operate at temperatures well above the transition temperature, brittle failures of this type during operation do not occur. However, during hydrostatic-test procedures following construction, brittle failures can occur if the

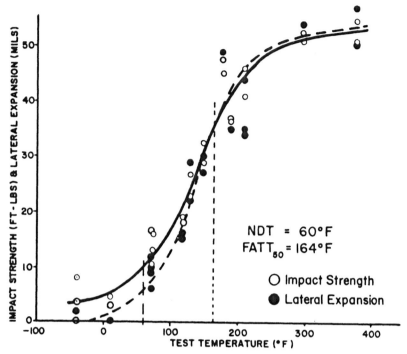

FIGURE 3.37. A Charpy impact strength versus temperature curve for SA-515 grade 70 material.

testing temperature is too low or fabrication is careless and leaves stress raisers within the structure.

Impact tests are carried out over a range of temperatures from below freezing to a few hundred degrees. Figure 3.37 shows data from −75 to 400°F (−60 to 200°C). Additional information may also be measured from the broken Charpy-test specimen: lateral expansion of the width of the piece at the fracture and the percentage shear on the fracture surface. All three measurements may be used for defining the ductile–brittle transition temperature. The fracture-appearance transition temperature is defined as the temperature at which the fracture surface exhibits 50% shear ($FATT_{50}$). The nil-ductility transition temperature (NDT) may be defined as the temperature at which the impact strength is 10 ft·lb.

FATIGUE

When a material is stressed in such a fashion that a tensile load is alternately applied and removed, failure can occur at a level much less than it can withstand under static conditions. Dynamic cyclic loading will, at very small localized areas, cause sufficient cold work and slip to develop microscopic

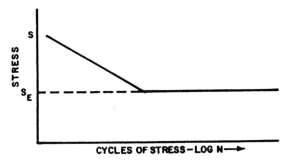

FIGURE 3.38. A schematic presentation of a fatigue curve for mild steel (Ref. 1).

cracks that grow until complete fracture occurs. The endurance limit, or fatigue strength, is the magnitude of a cyclic stress that a material can withstand without failure. Figure 3.38 shows a schematic fatigue curve for carbon steel. Note that for stress levels below a certain limit, the number of cycles to failure is nearly infinite; this is the so-called endurance limit, S_E.

The endurance limit can be affected by surface conditions—smoothness, notches—and by corrosion, which either causes the notches or magnifies their effect. Fatigue may also be caused by temperature gradients; in general, service failures usually result from a combination of these effects.

THERMAL FATIGUE AND CORROSION FATIGUE

Thermal fatigue involves the variation of stress caused by fluctuating temperatures. Room-temperature fatigue results from the variation of stress caused by changing loads, but environmental effects are considered constant. At elevated temperatures, crack propagation is enhanced by the formation of oxides or other corrosion products. In the simplest case of cyclic stress at elevated temperatures, the protective oxide cracks, exposing fresh metal to further oxidation. A surface crack is "wedged" open by the formation of these scales since the oxide occupies a greater volume than the metal from which it forms. The oxide "wedge" imposes higher stresses at the crack tip, and the crack-propagation rate increases.

Temperature gradients can also add strain related to the thermal coefficient of expansion. Perhaps the most common example is an austenitic stainless-steel support member welded to a ferritic steel with a stainless-steel welding alloy. Not only is the support at a higher operating temperature but the coefficient of expansion is greater; thermal-fatigue failures will likely occur unless special care is taken. Even without these added problems of different materials, thermal fatigue can occur as a result of low-amplitude vibrations of entire superheaters.

The same kinds of variable stress that lead to room-temperature fatigue failures can occur at elevated temperatures and under corrosive environ-

FIGURE 3.39. Thermal-fatigue cracks are usually oxide filled, transgranular, and dagger-shaped (Magnification 50×, etched).

FIGURE 3.40. Corrosion-fatigue cracks are shaped by the corrosion media and form. For example, in a waterwall tube where the corrosion may be oxygen contamination in the boiler water, the shape is more rounded and irregular (Magnification 18¾×, etched).

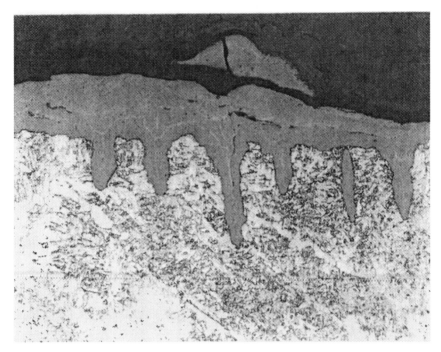

FIGURE 3.41. For an ash-corrosion-fatigue crack the shape is similar to a thermal-fatigue crack; but instead of being oxide filled, the corrosion debris is a sulfide (Magnification 200×, etched).

mental conditions. At elevated temperatures where the principal corrosion product is an oxide, these failures are usually referred to as *thermal fatigue*. Under corrosive environmental conditions, the failures are usually referred to as *corrosion fatigue*. The morphology of the cracks is somewhat different. Thermal-fatigue cracks are usually dagger-shaped, oxide-filled, and transgranular. See, for example, Fig. 3.39. Depending on the corrosion mechanism involved, corrosion-fatigue cracks are somewhat variable in cross section. For corrosion fatigue that initiates on the water-side of the boiler tube, the cracks are more rounded in shape but otherwise have the same general appearance as thermal-fatigue cracks; see Fig. 3.40. For ash corrosion, the cracks are similar in appearance to thermal-fatigue cracks; but the corrosion product, instead of being an oxide, is usually a sulfide; see Fig. 3.41.

SPHEROIDIZATION

The successful operation of a boiler requires that pearlitic steels function for long periods of time at temperatures that range from 700 to 1100°F (370 to 590°C). Changes occur in the microstructural constituents that tend to lower the overall internal energy of the metallic sytems. Pearlite is a structure of thin blades or platelets of iron carbide and ferrite that is thermodynamically unstable relative both to the shape of the iron carbide and to the decomposition of iron carbide into ferrite and graphite. A platelet of iron carbide has a large surface area-to-volume ratio and, thus, a high surface energy. The surface energy can be reduced by changing its shape from a blade or plate to a sphere. Further reduction in surface energy can be achieved by decreasing the number of iron-carbide particles from a large number of small particles to a small number of large particles. In practice, the change in morphology shows a progression from the normal pearlite structure to an *in situ* spheroidization of the iron carbide. This first stage is the change from platelets of iron carbide to spheres of iron carbide, but the pearlite colonies are still clearly defined. Further changes in the microstructure show that the fine dispersion of iron-carbide particles coalesces to form larger particles of iron carbide. Figures 3.42 through 3.45 show this progression for SA-210 A-1 plain-carbon steel.

The final stage in the change from the normalized microstructure to the one of least internal energy, or greatest thermodynamic stability, is to ferrite and graphite. Graphite particles show progressive coalescence and growth similar to the iron carbide particles: a very fine dispersion of graphite forms and, ultimately, the disappearance of the smaller particles of graphite and the appearance of much larger and fewer graphite particles. Figure 3.46 shows the ferrite and graphite structure of SA-210 A-1 material.

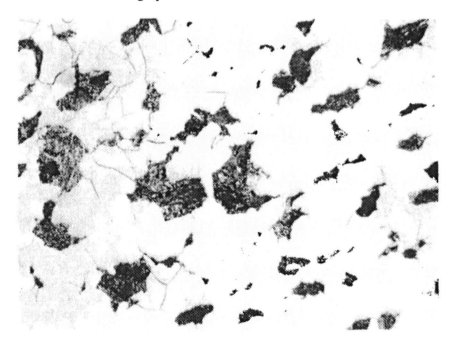

FIGURE 3.42. The normal structure for SA-210 A-1. The microstructure is composed of ferrite and pearlite (Magnification 500×, nital etch).

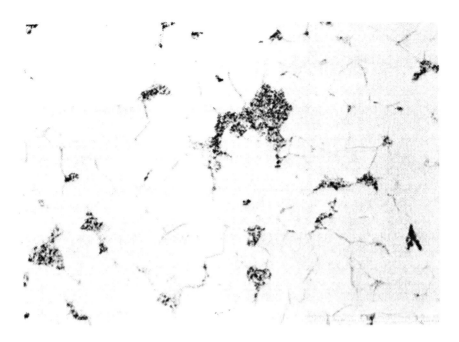

FIGURE 3.43. An *in situ* breakdown of the pearlite in SA-210 A-1. In this structure the platelets of pearlite have spheroidized, but the original pearlite colonies are still clearly defined (Magnification 500×, nital etch).

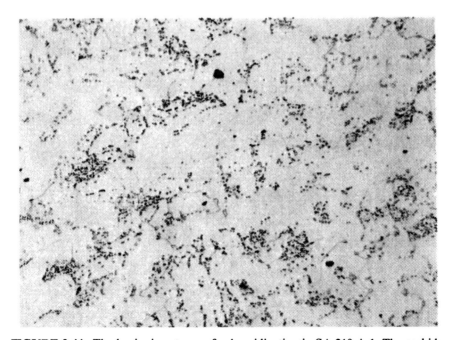

FIGURE 3.44. The beginning stages of spheroidization in SA-210 A-1. The carbide particles are now clearly spheroidized and have begun to disperse throughout the microstructure (Magnification 500×, nital etch).

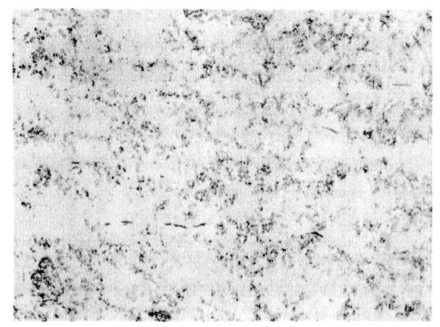

FIGURE 3.45. An advanced stage of spheroidization in SA-210 A-1. The carbide particles have begun to coalesce and grow in size (Magnification 500×, nital etch).

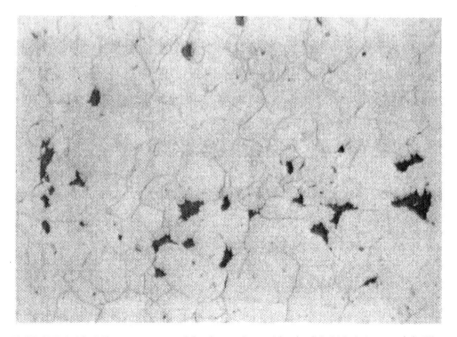

FIGURE 3.46. Microstructure of ferrite and graphite in SA-210 A-1 material. The most thermodynamically stable structure of ferrite and graphite: the final stages of progression from normalized structure to what we see in this figure (Magnification 500×, nital etch).

All of these transformations occur over a temperature range below 1341°F (727°C). However, what occurs at high temperatures in a short time also occurs at lower temperatures, but requires a much longer time.

The Larson–Miller parameter P has been used to relate this time–temperature interchangeability for diffusion-controlled microstructural degradation.[8,10]

$$P = T(20 + \log t)$$

where T = temperature, °R (°F + 460)
 t = time, hr

and 20 is an empirical constant. Small changes in temperature have a profound effect on the time for complete spheroidization to occur.

As steel is heated above 1341°F (727°C), the structure of pearlite and ferrite begins to transform to austenite and ferrite. Similar to the transformations on cooling previously discussed, the relative amounts of ferrite and austenite can be determined from the iron–carbon phase diagram. If alloys are heated above 1341°F (727°C), referred to as the lower critical transformation temperature, and are rapidly cooled, the transformation of austenite

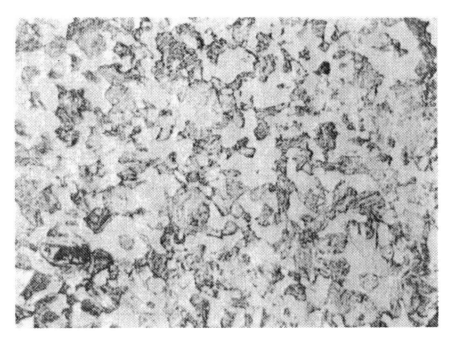

FIGURE 3.47. SA-210 A-1 quenched from a temperature of about 1400°F (760°C). The microstructure consists of ferrite, the white-etching constituent, and bainite, the dark-etching constituent (Magnification 500×, nital etch).

into ferrite and iron carbide is suppressed, and bainite or martensite forms instead. Since ferrite does not undergo any transformation during this cooling, by measuring the amount of martensite or bainite present in the microstructure, one may determine the temperature from which it was quenched. For temperatures within the two-phase field, the relative amounts of ferrite and austenite change as the temperature increases from 1341°F (727°F) to about 1500°F (816°C), in keeping with the Lever Rule. Figure 3.47 shows the microstructure of an SA-210 A-1 carbon steel quenched from about 1400°F (760°C).

GRAPHITIZATION

In plain-carbon steels and carbon–molybdenum alloys, the iron carbide is unstable relative to the decomposition into ferrite and graphite. Iron carbides can undergo two transformations at elevated temperatures: (1) change in shape from a blade to a sphere, and (2) change from iron carbide to ferrite and graphite.

Spheroidization and graphitization are competing processes; see Fig. 3.48.[11] At low temperatures, graphitization occurs before the pearlite colonies have completely spheroidized and disappeared. At elevated temperatures, spheroidization occurs before graphitization. In both cases, temperatures of at least about 850°F (455°C) for carbon steel and 875°F (470°C) for carbon–½Mo steel are necessary to transform the iron carbide to graphite. Small amounts of aluminum and higher stresses also promote graphitization.

The heat-affected zone in a weld is a rapidly cooled or quenched structure. Such structures graphitize more quickly than normal ferrite and pearlite. Thus, it is not unusual for heat-affected zones to contain graphite while the rest of the component is still ferrite and iron carbide.

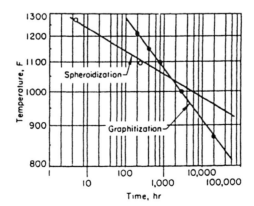

FIGURE 3.48. Temperature–time plot of pearlite decomposition by the competing mechanisms of spheroidization and graphitization in carbon and low-alloy steels.

Temperature variations also promote graphitized microstructures. Temperature cycling is another way of transforming iron carbide to ferrite and graphite.

Figures 3.49–3.51 show these variations in the graphitized microstructures found in these plain-carbon and carbon–molybdenum steels.

Plain-carbon steels are often used at the inlet end of a primary or low-temperature reheater. Steam temperatures are low, metal temperatures are usually less than 850°F, and the hoop stress may be less than 3500 psi. Figure 3.52 is the microstructure of an SA-210 A-1 tube removed from such a reheater after 90,000 hr. The structure has graphitized, but the pearlite colonies are still sharply defined. The estimated temperature is in the 850–900°F range. After 2½ more years, 110,000 hr total, the pearlite colonies are gone, replaced by a random dispersion of iron carbide particles; see Fig. 3.53. These two microstructures indicate that graphitization can occur before spheroidization, but complete spheroidization develops when there is sufficient time.

On occasion, graphite forms an aligned structure called "chain" graphite. Figures 3.54 and 3.55 present such a structure from an SA-209 T-1a reheater tube. The row of graphite is a plane of weakness, since there is very little ferrite between the individual particles; see Fig. 3.53.

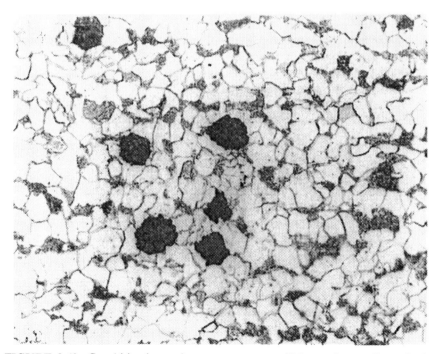

FIGURE 3.49. Graphitization at low temperatures will leave the pearlite colonies intact. From an SA-209 T-1 superheater tube (Magnification 500×, etched).

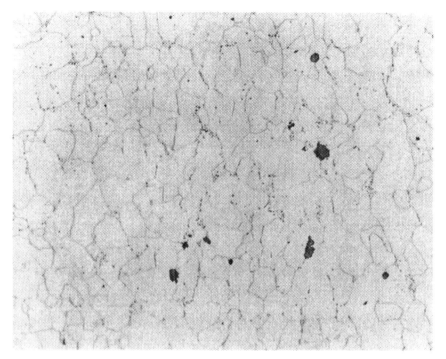

FIGURE 3.50. At higher temperatures, spheroidization precedes graphitization. From an SA-192 tube mistakenly used for T-22 (Magnification 500×, etched).

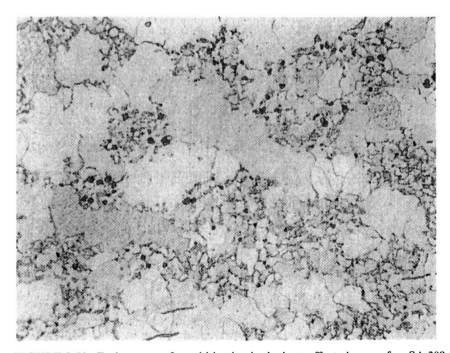

FIGURE 3.51. Early stages of graphitization in the heat-affected zone of an SA-209 T-1 superheater-tube support weld (Magnification 500×, etched).

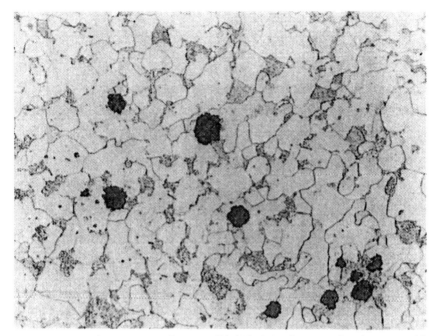

FIGURE 3.52. Microstructure from an SA-210 A-1 reheater tube shows graphite particles amid pearlite and ferrite (Magnification 500×, etched).

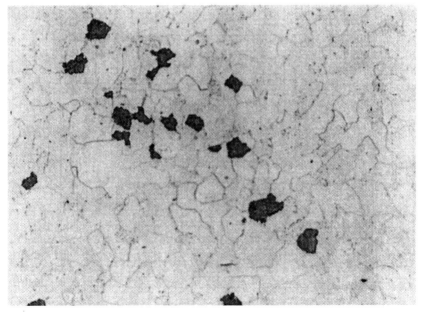

FIGURE 3.53. Same reheater 2½ years later indicates that pearlite will completely spheroidize after a long time. Figure 3.52 represents 90,000 hr, and Fig. 3.53 represents 110,000 hr in the 850–900°F temperature range (Magnification 500×, etched).

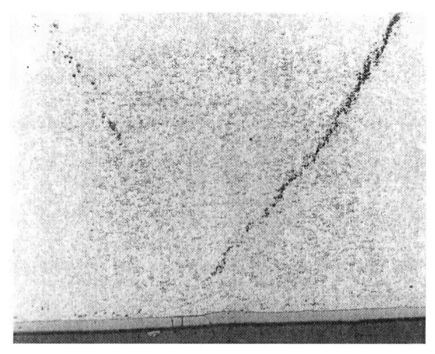

FIGURE 3.54. A longitudinal section through an SA-209 T-1a reheater tube. The graphite particles form in rows, and the structure is called "chain" graphite (Magnification 25×, etched).

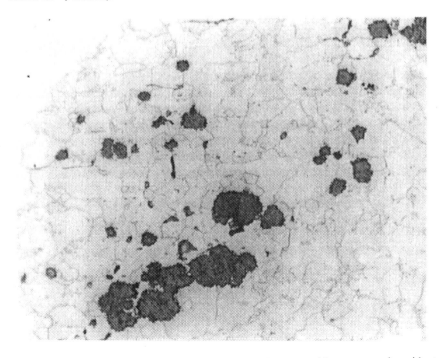

FIGURE 3.55. At higher magnification, the weakness to this structure is evident, since there is very little ferrite between the individual graphite particles (Magnification 500×, etched).

120   CHAPTER 3 METALLURGICAL PRINCIPLES: FERRITIC STEELS

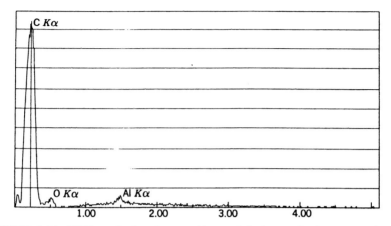

FIGURE 3.56. EDX analysis of a graphite nodule in an SA-210 A-1 tube. The carbon peak is clear; the oxygen and aluminum peaks are from alumina used in the sample preparation.

Proof that the nodules are indeed graphite is easy to obtain with the use of a scanning-electron microscope and an energy-dispersive x-ray (EDX) analysis. Figure 3.56 displays only a carbon peak. The aluminum and oxygen peaks are remnants of the alumina media used in the prepration of the metallographic specimen.

EQUILIBRIUM MICROSTRUCTURES

The solubility of carbon in ferrite at 1340°F (727°C) is 0.0218%.[5] For all temperatures below 700°F (370°C) the solubility of carbon in ferrite is essentially nil. For carbon-steel components that operate at temperatures below about 800°F (425°C) for long periods, there is sufficient time for the carbon to come out of solution in the ferrite. Thus, the equilibrium microstructures relative to the solubility of carbon in ferrite will show faint traces of iron carbide along particular crystallographic planes within the ferrite. Figures 3.57 and 3.58 are two such examples. Figure 3.57, from an SA-178 A tube shield, had been in service at temperatures in the neighborhood of 800°F (425°C) for a couple of years. Figure 3.58, from an SA-53 pipe, had been in service at 400°F (200°C) for nearly 30 years. Both of these structures show the precipitation of iron carbides along particular crystallographic planes within the ferrite.

HEAT TREATMENT

The heat treatment of metals is done by a combination of heating and cooling to achieve specific properties. The following is a partial list of heat

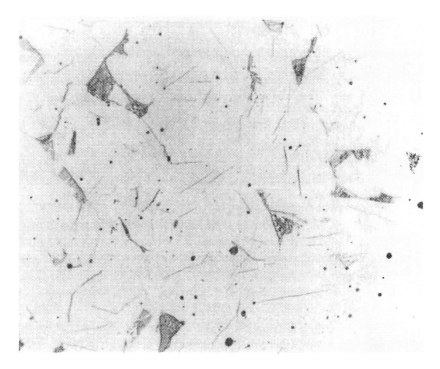

FIGURE 3.57. Microstructure from SA-178 A tube shield in service at a temperature of about 800°F for two plus years (Magnification 500×, etched).

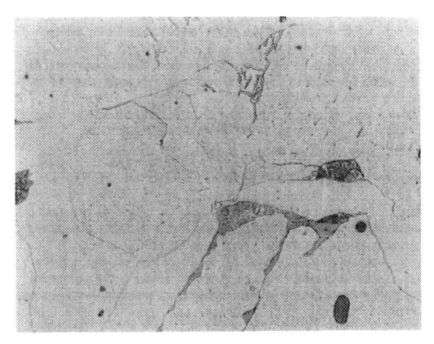

FIGURE 3.58. Microstructure from SA-53 pipe in service for nearly 30 years at about 400°F (Magnification 500×, etched).

treatments commonly used and some useful definitions; see, for example, Ref. 12.

Annealing. A heat treatment usually applied to induce softening.

Grain Growth. An increase in the grain size of a metal or alloy usually produced by extended heating.

Normalizing. Steel heated into the all-austenite region and cooled in still air to room temperature to produce the "normal" structure of ferrite and pearlite.

Quenching. Rapid cooling.

Recrystallizing. Annealing process to produce strain-free grain structure from the distorted structure resulting from cold work. Recrystallization temperature is the lowest temperature at which a strain-free structure is produced.

Solution Heat Treatment. Applied to stainless steels, whereby the alloy is heated to high enough temperatures long enough to allow all alloying elements to form a solid solution.

Spheroidizing. Process to produce rounded or globular carbides in steel; may be done by:

1. Prolonged heating below eutectoid temperature
2. Heating and cooling alternately above and below the eutectoid temperature
3. Very slow cooling from either the all-austenite or austenite-plus-ferrite temperature

Stress Relieving. Process of reducing residual stresses in metals; may be applied for relief of stresses in castings, weldments, cold-worked microstructure, and quenching.

Tempering. Process of reheating a quenched-hardened or normalized steel to a temperature below the eutectoid temperature.

SPECIFIC EFFECT OF ALLOYING ELEMENTS

Silicon is added to all steels as a deoxidizing element. It is a mild ferrite strengthener; and, when added in amounts up to about $2\frac{1}{2}\%$, the strength of the steel is increased without any adverse effects on ductility.

Manganese is also a deoxidizer of steel. It combines with sulfur to form magnanese sulfide, a more innocuous inclusion in steel than iron sulfide. It is also a ferrite strengthener and an inexpensive means of improving the tensile strength of steel.

Molybdenum is a ferrite strengthener and increases the strength of steel without any loss of ductility. It is a mild carbide stabilizer and retards the formation of graphite upon prolonged heating. Carbon steel graphitizes above about 850°F (455°C), and carbon–molybdenum steels graphitize above 875°F (470°C).

Chromium added to steel greatly enhances the oxidation resistance and improves the high-temperature strength and creep properties. The oxidation limit for T-11 (a 1¼ chromium alloy) is 1025°F (550°C), and for T-22 (a 2¼ chromium alloy) it is 1075°F (580°C). The addition of chromium also improves the hardenability of steel so that preheating is usually necessary when welding the high-chromium ferritic alloys. It is also a carbide stabilizer to such a degree that the chromium–molybdenum steels with ½% chromium do not form graphite under any condition.

SAMPLE PREPARATION AND METALLOGRAPHY

Failure analysis may be useful in the understanding of more of the problems with a steam generator than just the failure. In order to perform a satisfactory metallographic examination and thorough failure analysis, a few simple rules on sample preparation are necessary. Care must be exercised in removing the sample from the unit. It is desirable to have the entire failure plus 1 ft on either end of the rupture for study. If the sample is cut from the unit with an oxyacetylene torch another 12 in should be added to each end. Debris from the torch cut may disturb any internal or external deposits. The following pertinent information should accompany the tube sample:

1. Steam or fluid-flow direction
2. Fire side (if not obvious)
3. Orientation [especially useful in horizontal superheaters (SH) and reheaters (RH)], top, bottom, up, down, etc.
4. Location
5. Specified tube material
6. Specified tube diameter and wall thickness
7. Boiler fuel
8. Boiler operating conditions at the time of failure
9. Age or length of service of the failed tube and the hours of use of the unit

The sample should be carefully packed for shipment to protect and preserve any scale or deposits on the outside (OD) or inside diameter (ID) of the tube. Scale and deposit analyses can provide useful information as to the cause of the failures, especially corrosion-induced failures.

Visual examination of the tube failure is necessary to establish the type of failure and the necessary examination that will be performed. A photograph is usually taken of the overall tube sample and any other feature that may be useful—wall thinning, distortion of any kind, attachments, etc. The outside diameter of the tube and wall thickness are measured with calipers and a micrometer. Sometimes it is useful to measure the circumference with a tape to obtain the average diameter increases. Wall thickness is measured at several points around the tube. General conditions of the tube, the amount of deposits, color, scale, swelling, distortion of the tube, attachments, welds, etc., are noted.

Metallographic samples are taken through the failure, at the end of the tube sample on the fire side and in the same plane as the failure, and of the rear side or 180° from the rupture, and anywhere else it seems appropriate. If the failure is related to an overheating condition, an attempt should be made to determine the extent of overheating. For tube surveys on older units where the general condition of the superheater or reheater is being studied, samples should be taken from the hottest zones—not only in the finishing legs but also the hottest zones for a given alloy. For example, where carbon steel joins T-11, samples should be taken from the carbon steel even though, at this point, the steam temperature is much less than the final steam temperature.

Sample preparation should be done with the greatest of care to preserve intact any scale or deposit. If there is no scale or deposit in the vicinity, as is typical when a violent high-temperature failure occurs, a sample for metallographic examination may be saw-cut. For scale measurements, a water-cooled cutoff wheel to prevent damage to the scale is needed. Samples are mounted in thermosetting plastic, and standard metallographic practice for steels is followed.

For etching of ferritic steels, 3% nitric acid in alcohol (nital) is satisfactory. The specimen is swabbed with a cotton ball soaked with nital for a few seconds; this is repeated if necessary to clarify the microstructural details. For austenitic steels, an electrolytic etch is preferred. Ten percent oxalic acid in water at 6 V DC with a stainless-steel cathode for a few seconds is generally sufficient. For a macro etch, the same techniques may be used, but the etching time is increased to several minutes.

DEFINITIONS

Fire Side: The portion of the tube most intimately in contact with the flue gas. For waterwalls it is usually quite obvious, but may not be for superheaters, reheaters, and economizers, especially in gas-fired units. For SRE elements, the fire side is defined as the semicircle that faces into the gas stream. In oil- and coal-fired units, the fire-side wall has the most ash deposit.

Back Side (Rear Side): The half of the tube, semicircle, opposite or 180° from the fire side.

Clock Positions: The circular cross section of a tube is referenced to a clock face with the center line of the fire side given the 12 o'clock position. Typically, ash deposits will extend from about the 10 o'clock to the 2 o'clock positions.

Scale: The normal oxide of steel in contact with steam, air, flue gas, or water. It is mostly iron oxide, called magnetite (Fe_3O_4), and forms a protective layer that usually prevents further oxidation.

Deposit: Products not formed mainly from oxidation of steel. On the OD surface these can be ash or other products of combustion. On the ID surface these can be corrosion products, products brought into the boiler, or precipitates from the chemical-treatment control system.

MICROSTRUCTURES

The photomicrographs that follow are representative of the microstructures found in the commonly used ferritic steels as boiler-tube alloys age in service. They are grouped to demonstrate the expected changes in morphology from a normalized structure of ferrite and pearlite exhibited by an unused or new tube. The first stage in the transformation is an *in situ* breakdown of the pearlite: the shapes of the pearlite colonies remain intact, but the platelets of iron carbide become spheroids. Further changes to a dispersion of nearly spherical carbide particles and, finally, to graphite follow. The addition of chromium stabilizes the carbide so that even for very long times or high temperatures no graphite has been observed in the chromium–molybdenum alloy steels with at least ½% chromium.

Plain-carbon steels are the first set, typified by SA-178 A, and are representative of SA-178 C and SA-210 A-1 as well. Additions of molybdenum strengthen the alloy and raise the oxidation limit but do not prevent the formation of graphite. The microstructural changes for SA-209 T-1 are similar to the plain-carbon steels. Compare plain-carbon steel, Micros 3.1–3.4, with the carbon + ½ molybdenum SA-209 T-1, Micros 3.5–3.8. Additions of chromium alter the morphology of iron carbide such that even for an unused tube the pearlite is not as regular as in the plain-carbon steels; compare Micros 3.1, 3.9, and 3.13.

A new chromium–molybdenum–vanadium steel, SA-213 T-91, is available. It has better high-temperature strength than T-22 (see Table VII) and, with 9% Cr, much better high-temperature oxidation resistance. The microstructure is quenched and tempered martensite; see Micro 3.17. With time at elevated temperatures, the structure ultimately spheroidizes; refer to Micro 3.18.

All of the chromium–molybdenum steels end up with spheroidized carbides and ferrite. Micros 3.19 and 3.20 from SA-213 T-5 (5% Cr–½% Mo) and SA-213 T-9 (9% Cr–1 Mo), respectively, were removed from reheaters and after more than 20 years duty. These both have similar morphologies to T-11 (Micro 3.12) and T-22 (Micro 3.16).

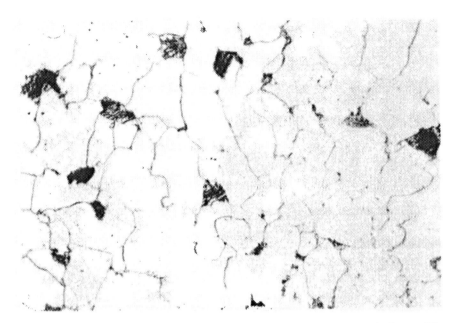

MICRO 3.1. SA-178 A, carbon steel. Normalized microstructure similar to Fig. 3.16, but with a lower carbon content and hence a smaller amount of pearlite. It is about 85% ferrite, as would be expected from low-carbon steel (Magnification 500×, nital etch).

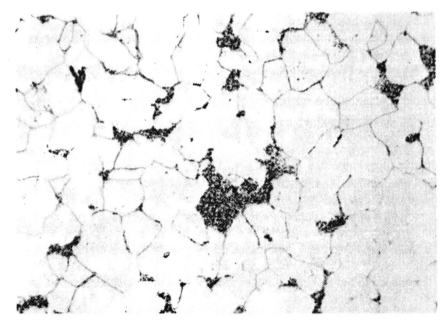

MICRO 3.2. SA-178 A, steel. An *in situ* spheroidization of the pearlite is clearly evident in this photomicrograph. The pearlite colonies are still well defined, but the morphology of the iron carbide is now small spheres rather than platelets (Magnification 500×, nital etch).

MICRO 3.3. SA-178 A, carbon steel. The structure has begun to graphitize. There are small black dots indicating the initial stages of graphitization. The iron carbide has completely coalesced into a few large carbide particles uniformly dispersed throughout the structure (Magnification 500×, nital etch).

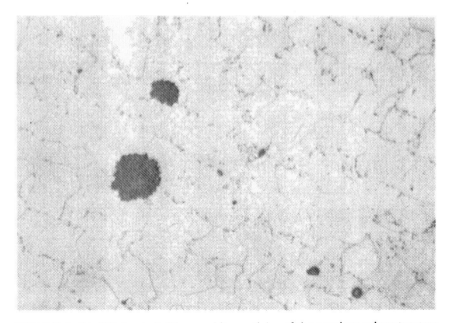

MICRO 3.4. SA-178 A, steel. The graphite particles of the previous microstructure have grown, and the amount of iron carbide is decreased; this is an advanced stage of graphitization (Magnification 500×, nital etch).

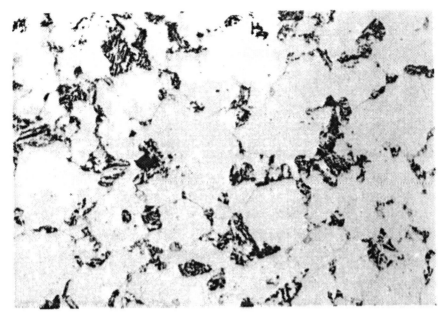

MICRO 3.5. SA-209 T-1, carbon–molybdenum steel. The normal microstructure of lamellar pearlite and ferrite. (Magnification 500×, nital etch).

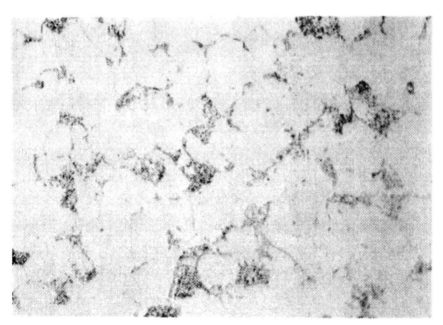

MICRO 3.6. SA-209 T-1, carbon–molybdenum steel. An *in situ* breakdown of the pearlite is evident in this microstructure. The pearlite colonies are still well defined, but the iron carbide platelets have begun to spheroidize (Magnification 500×, nital etch).

MICRO 3.7. SA-209 T-1, carbon–molybdenum steel. The final stages of spheroidization, or the initial graphitization stage. Iron carbide is uniformly dispersed throughout the microstructure, and we see the start of small graphite nodules forming (Magnification 500×, nital etch).

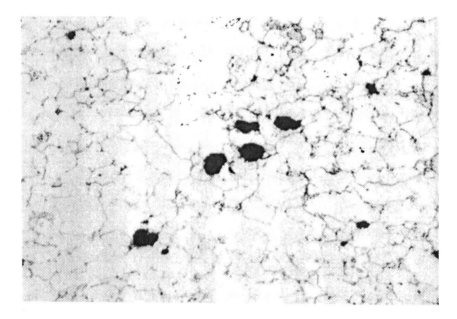

MICRO 3.8. SA-209 T-1, carbon–molybdenum steel. The final stages of graphitization. Note that there is very little iron carbide present in the microstructure, and the graphite particles have grown compared with Micro 3.7 (Magnification 500×, nital etch).

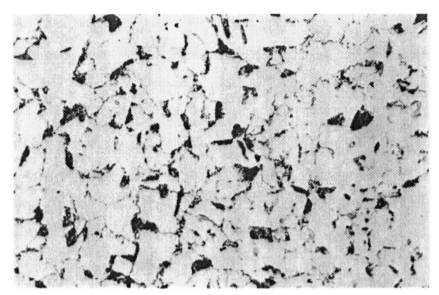

MICRO 3.9. SA-213 T-11, chromium–molybdenum steel. The normalized microstructure expected of unused boiler tubing. The structure is pearlite and ferrite. However, the addition of chromium to the alloy alters the morphology of the pearlite, so the platelets of iron carbide and ferrite are not as clearly defined as in the plain-carbon or carbon–molybdenum steels (Magnification 500×, nital etch).

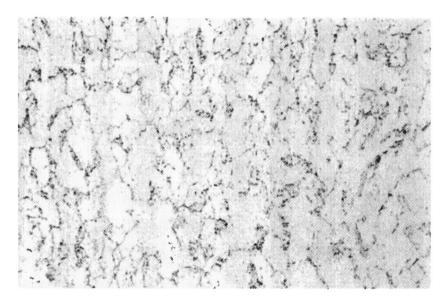

MICRO 3.10. SA-213 T-11, chromium–molybdenum steel. Similar to the plain-carbon steels, the pearlite of T-11 will show the same transformation sequence except that, chromium being a carbide stabilizer, these alloys will not graphitize. This microstructure shows the *in situ* breakdown of the iron carbide into the spheroidal form. The prior pearlite colonies are reasonably well defined (Magnification 500×, nital etch).

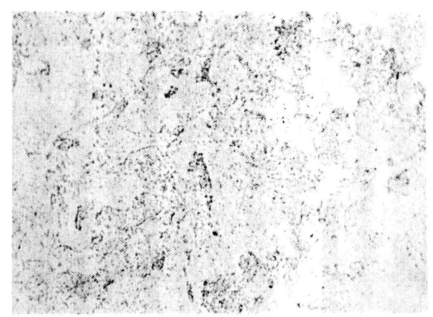

MICRO 3.11. S-213 T-11, chromium–molybdenum steel. The carbide is nearly spheroidized and has begun to disperse throughout the microstructure (Magnification 500×, nital etch).

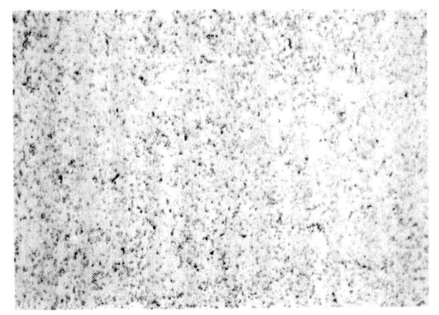

MICRO 3.12. SA-213 T-11, chromium–molybdenum steel. The carbide has uniformly dispersed throughout the microstructure. Since these steels do not form graphite, the most stable microstructure to be achieved is a uniform dispersal of spherical carbide particles (Magnification 500×, nital etch).

MICRO 3.13. SA-213 T-22, chromium–molybdenum steel. The morphology of pearlite in these higher-chromium–molybdenum steels is quite variable. This microstructure, from an unused T-22 boiler tube, shows the carbide as reasonably well dispersed platelets, but no distinct pearlite colonies (Magnification 500×, nital etch).

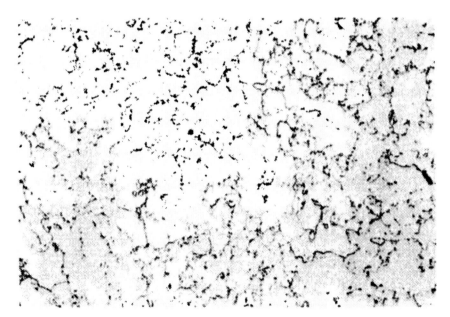

MICRO 3.14. SA-213 T-22, chromium–molybdenum steel. The dispersed carbide platelets of the previous microstructure have begun to spheroidize. The carbides tend to congregate along ferrite grain boundaries, and we see the initial stages of spheroidization (Magnification 500×, nital etch).

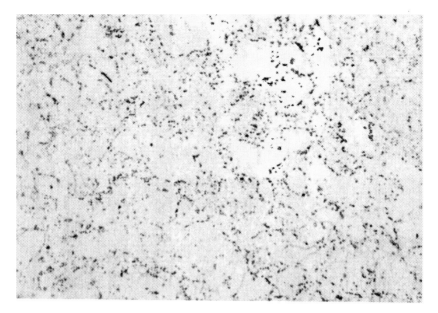

MICRO 3.15. SA-213 T-22, chromium–molybdenum steel. The carbides are now almost spherical. There is still a tendency for them to form along ferrite grain boundaries, so some of the ferrite grains appear to be carbide-free (Magnification 500×, nital etch).

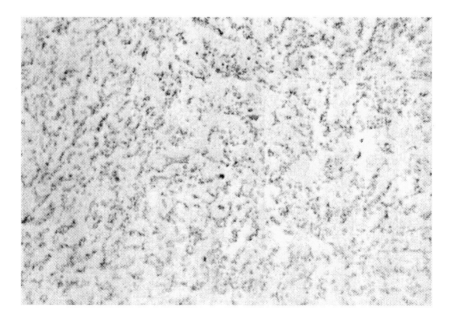

MICRO 3.16. SA-213 T-22, chromium–molybdenum steel. The carbide particles have coalesced. Note that there is now a uniform dispersion of carbide particles throughout the microstructure (Magnification 500×, nital etch).

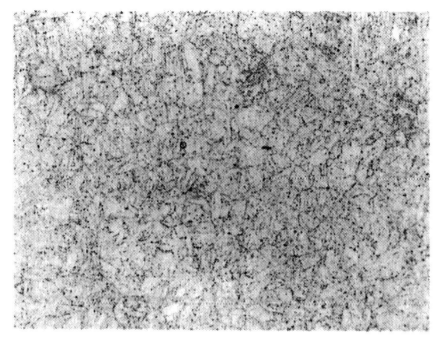

MICRO 3.17. SA-213 T-91 chromium–molybdenum–vanadium steel. The structure is tempered martensite (Magnification 500×, nital etch).

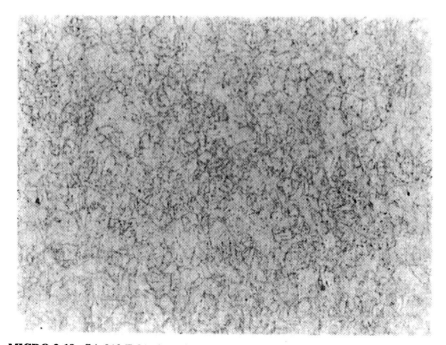

MICRO 3.18. SA-213 T-91 chromium–molybdenum–vanadium steel. The structure is a uniform dispersion of carbide in ferrite. (Magnification 500×, nital etch).

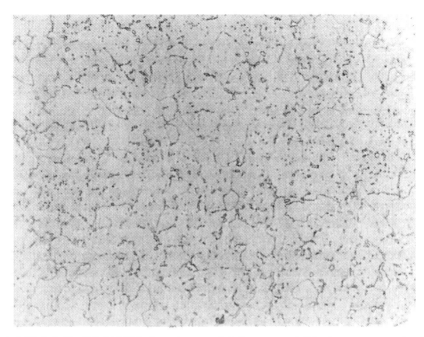

MICRO 3.19. SA-213 T-5 chromium–molybdenum steel. The long-term service in a reheater has altered the carbide appearance to spherical particles in a matrix of ferrite (Magnification 500×, nital etch).

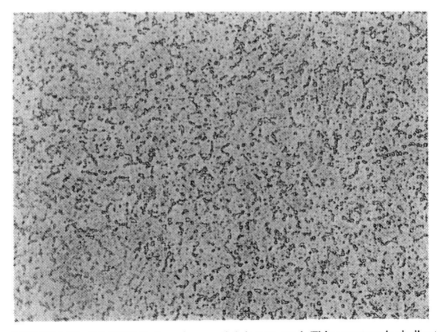

MICRO 3.20. SA-213 T-9 chromium–molybdenum steel. This structure is similar to all Cr–Mo steels after long-term, elevated-temperature service: spheroidized carbides and ferrite (Magnification 500×, nital etch).

136 CHAPTER 3 METALLURGICAL PRINCIPLES: FERRITIC STEELS

These alloys of T-5 and T-9 are often supplied in the fully annealed condition. Such a heat treatment improves the ductility, which makes bending easier for the fabricator. The microstructure is then ferrite and a very fine dispersion of carbide particles. Micro 3.21 is a T-9 tube as received from the mill in the annealed condition. The Rockwell B hardness is 79, and the elongation is over 35%. During long-term service, the carbide particles coarsen and hardness decreases.

A short-time, high-temperature failure will develop a microstructure based on the peak temperature and the rate of cooling, usually a very rapid quench. These failures are most frequent in waterwalls due to localized DNB (departure from nucleate boiling) conditions. Waterwall tubes are mainly plain-carbon steel, SA-178 A, SA-178 C, or SA-210 A-1 materials. At temperatures above 1340°F, the ferrite and pearlite transform to ferrite and austenite, and at temperatures above about 1550°F (depending on the carbon content) ferrite and austenite transform to austenite. With increasing temperature the strength decreases. When the hoop stress equals the tensile strength, sudden failures occur. These ruptures are characterized by considerable ductility at the failure lip, a wide-open burst, and microstructures that reflect the metal temperature at the instant of rupture. The next six micros present the general features observed in these short-term failures of plain-carbon steel, SA-210 A-1.

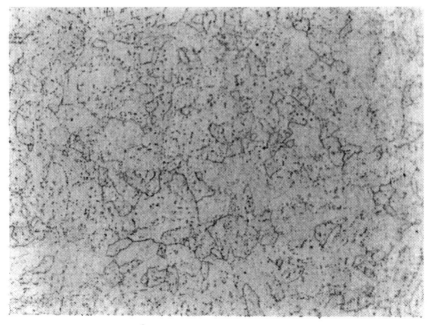

MICRO 3.21. Microstructure of T-9 as received from the mill, in the annealed condition, is ferrite and a fine dispersion of carbides. Rockwell B hardness is 79 (Magnification 500×, etched).

MICRO 3.22. Ferrite and bainite, with very little ferrite. Temperature at failure probably above upper critical transformation temperature. The black lines are cracks that form along austenite grain boundaries, similar to creep cracks. Deformation of the austenite elongates the cracks parallel to the principal strain during the rupture process (Magnification 500×, etched).

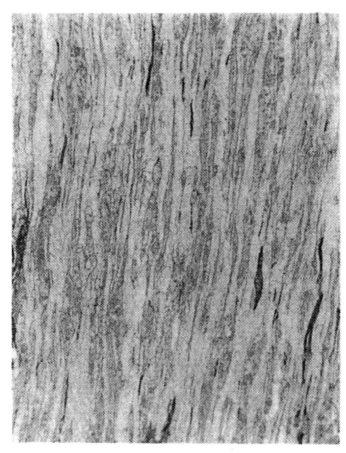

MICRO 3.23. Ferrite and bainite. Rupture temperature about 1450°F since there are nearly equal amounts of ferrite and bainite. Deformation of the ferrite and austenite is retained to room temperature as ferrite and bainite (Magnification 500×, etched).

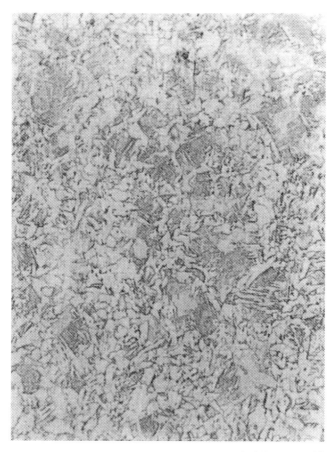

MICRO 3.24. Widmanstätten structure of ferrite and bainite (Magnification 500×, etched).

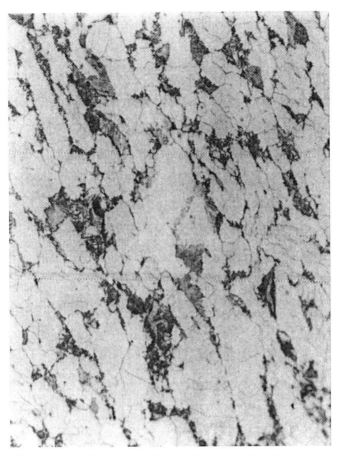

MICRO 3.25. Ferrite and bainite, and note the subgrains in the ferrite that form during plastic deformation (Magnification 500×, etched).

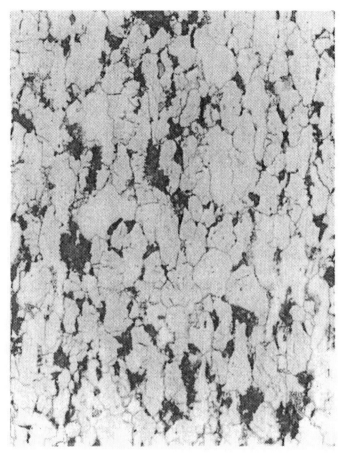

MICRO 3.26. Ferrite and bainite. Temperature and time above 1340°F long enough to completely transform the pearlite to austenite (Magnification 500×, etched).

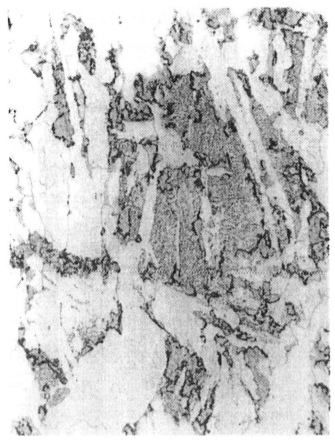

MICRO 3.27. Ferrite, pearlite, and bainite. Temperature was just above the lower critical transformation temperature but not long enough to completely transform the pearlite (Magnification 500×, etched).

Superheaters and reheaters also fail by the short-term high-temperature mode, usually during start-up when condensate blocks steam flow. Micros 3.28 and 3.29 are SA-209 T-1a superheater tubes that failed in such a fashion.

The final set shows the microstructures of pipe and plate material. These are characterized by higher carbon contents but are still plain-carbon steels. For SA-106 gr B, the carbon content is 0.30% maximum, and from the Lever Rule it contains about 37% pearlite. For SA-106 C, SA-515-70, and SA-516-70 the carbon content is 0.35% maximum and contains about 45% pearlite; compare Micros 3.30–3.33. With a fine-grained steel, the impact properties are improved and the nil-ductility transition temperature is lowered. Both SA-515-70 and SA-516-70 have the same tensile strength, 70,000 psi, but quite different grain size; compare Micro 3.32, a coarse-grained steel, with Micro 3.33, a fine-grained steel.

Microstructure 3.34 shows a banded structure typical of all of these plain-carbon steels. Segregation of alloying elements during the initial stages of steel making changes the carbon diffusion rate to give regions of varying carbon content in the austenite. Subsequent ingot rolling and metal processing elongate these regions. The final structure has bands of pearlite and ferrite.

References 13 and 14 are excellent sources of microstructures.

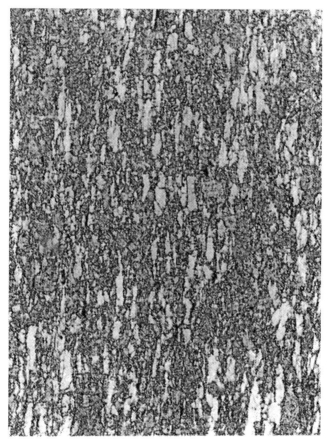

MICRO 3.28. Ferrite and bainite, peak temperature close to the upper critical transformation temperature (Magnification 500×, etched).

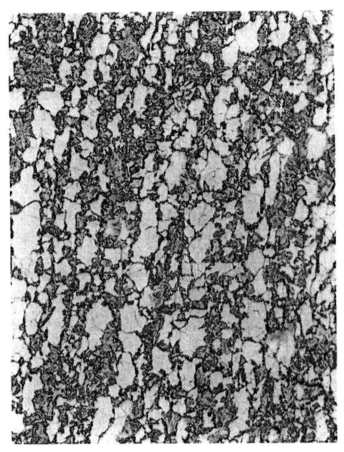

MICRO 3.29. Ferrite and bainite. The banded structure carries into the ferrite and austenite region; when quenched, it leaves a "banded" structure of ferrite and bainite (Magnification 500×, etched).

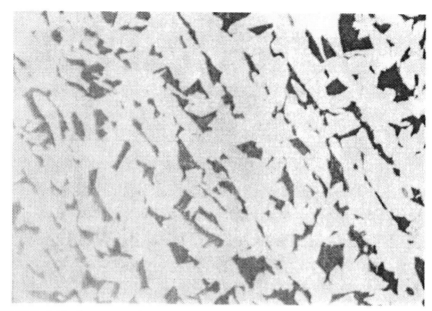

MICRO 3.30. SA-106 grade B, pipe material. The normal structure for a plain-carbon steel. The white areas are ferrite, the black areas are pearlite; but at this magnification, the pearlite platelets of carbide and ferrite are not resolved (Magnification 500×, nital etch).

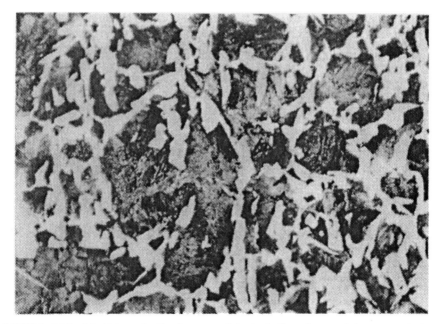

MICRO 3.31. SA-106 grade C, pipe material. A similar alloy to SA-106, grade B, shown in the previous microstructure, but with a higher carbon content. The higher carbon content results in a larger percentage of pearlite in the microstructure, as is clearly evident in comparison with Micro 3.30 (Magnification 250×, nital etch).

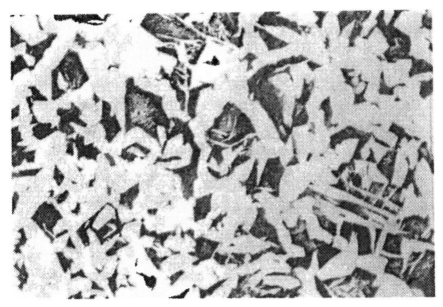

MICRO 3.32. SA-515 grade 70, plate material. The normal microstructure for a carbon–manganese steel commonly used for low-pressure drums (Magnification 250×, nital etch).

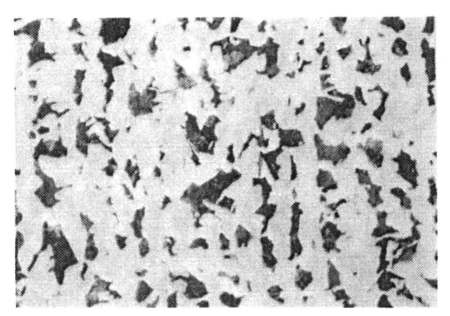

MICRO 3.33. SA-516 grade 70, plate material. A similar material to the previous microstructure except that it is manufactured by a fine-grained steel-making practice. The finer grain size is evident in comparing these two microstructures. The finer grain size of 516 grade 70 gives an improvement in the low-temperature impact properties (Magnification 250×, nital etch).

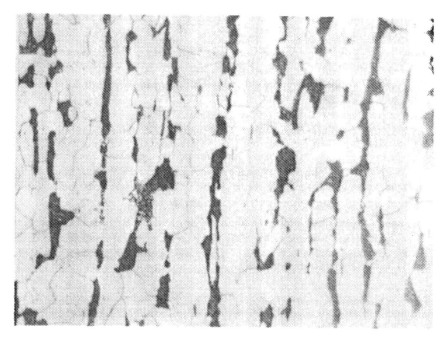

MICRO 3.34. SA-210 A-1, banded structure. The pearlite colonies align in bands to give the appearance of layers of ferrite and pearlite. Segregation of alloying elements during the casting of the ingot alters the carbon diffusion rate to give regions of higher carbon content in the austenite. Subsequent ingot rolling and tube forming processes elongate these regions in the primary direction of hot work, so that the final structure appears "banded" (Magnification 250×, nital etch).

CASE HISTORIES

The operation of a boiler is a dynamic balance between the formation of steam, or its superheating, and the flow of thermal energy from the combustion of a suitable fuel. In effect, an individual boiler tube is simultaneously heated and cooled. When this balance is maintained, the metal temperature of the boiler tube is appropriate for the material. When the balance is upset, tube-metal temperatures can rise, and failures will follow sooner than expected.

For all materials used in boiler construction, the strength decreases as the temperature increases, both for the short-time, high-temperature, tensile strength and the creep or stress-rupture strength; see, for example, Tables VII and VIII. The simplest explanation for all high-temperature, boiler-tube failures is that the stress (usually, but not always, the hoop stress) was "too high" for the operating conditions. The stress can be too high because:

1. Tube-metal temperature is higher than expected. The design stress has not changed, but operating conditions have raised the metal tempera-

ture so that the operating stress relative to metal temperature is excessive.
2. Hoop stress has increased. The tube-metal temperature is correct, but corrosion (either or both from the ID outward and from the OD inward) or erosion has reduced the wall thickness so that the actual hoop stress is too high.

Obviously both effects can occur simultaneously.

These high-temperature failures can be divided into two principal types:

1. *Short-Term, High-Temperature Failures.* The metal temperature at the instant of rupture may be several hundred degrees hotter than design, and failure occurs in a few minutes.
2. *Long-Term Creep or Stress-Rupture Failures.* The metal temperature at the instant of rupture is 50–100°F above design conditions or at design conditions in the creep range of the metal involved. Failure usually occurs in several months to several years.

In waterwall tubes, steam bubbles form at discrete points along the tube surface, known as *nucleate boiling*. When the bubble reaches some size, it is removed by the fluid flow along the inside of the tube. The process repeats. If the heat flux is "too great" or the fluid flow "too small," several bubbles join to form a steam blanket, a process known as *departure from nucleate boiling*. Under these conditions, heat transfer through the steam blanket is very small, and tube-metal temperatures can rise rapidly, and failure occurs in a short time. The times vary with the metal temperature, obviously, but these times can be exceedingly short, perhaps minutes or even seconds.

The kinds of operational upsets that can cause these failures can occur on either side of the heat-flow balance.

1. Misdirected or worn burners can lead to flame impingement, too high a heat flux.
2. On the fluid-flow side, partial blockage from foreign objects or debris within the tubes or dents from slag falls can reduce the flow below the critical amount.
3. A leak can reduce flow, which thus leads to a different type of failure downstream of the first rupture.

In a superheater, no DNB can occur because only steam superheating occurs. However, high-temperature failures can occur in both superheaters and reheaters due to steam blockage. During start-up, condensate can prevent steam flow from a particular tube, and failure will occur. Foreign objects, for example, broken attemperator spray nozzles, can block a tube and lead to a high-temperature failure.

Creep failures occur when tube-metal temperatures rise a relatively small amount, perhaps 50–100°F above design. Operational problems that lead to creep failures are:

1. Partial blockage from debris, usually iron oxide from exfoliation.
2. Steam-side scale leads to the formation of an effective insulating barrier, and tube-metal temperatures gradually rise until creep failures occur.
3. Wrong material, carbon steel substituted for T-22, for example.
4. Unbalanced steam flow. Superheater or reheater tubes with less flow will have poorer cooling and thus higher metal temperatures (see Chapter 2).
5. Change in tube material at too high a metal temperature.
6. Change in tube-wall thickness at too high a metal temperature.
7. Partial pluggage of the bundles by fly ash causes "laning" and too high a heat flux.
8. Misaligned tubes (see item 7).
9. Changes in fuel from design (oil to gas or high-slagging to low-slagging coal) can change overall heat-transfer coefficients.
10. Higher stresses at attachments. The most common are slip spacers that plug with ash and cannot accommodate differential movement between tubes.
11. Backing rings or excessive penetration at the root pass of a gas tungsten arc weld can form a venturi nozzle. Steam flow through the weld forms eddies downstream and leads to a region of poorer cooling. Creep failures occur in this region of reduced steam-side heat transfer.

Creep failures can occur in all components of the boiler, but particularly in the superheaters and reheaters.

What follows are several examples of these kinds of failures. While the metallurgical failure analyses cannot identify the precise root cause of the failure, they can point in the direction of likely operational problems that need to be corrected.

Case History 3.1

Furnace Side-Wall Ruptures 1200–1300°F (649–704°C)

BOILER STATISTICS

Size	600,000 lb steam/hr (270,000 kg/hr)
Steam temperature	1005°F/1005°F (540°C/540°C)
Steam pressure	2150 psig (150 kg/cm^2)
Fuel	Oil

The first case history involves a side-wall tube failure.

At the time of failure, the unit had been in service two years. The section received was removed from the left side wall, and was SA-178 C

150 CHAPTER 3 METALLURGICAL PRINCIPLES: FERRITIC STEELS

steel, having $2\frac{1}{2}$ in OD × 0.203-in-thick wall. Visual examination showed the following:

A. A wide-open burst along the center line of the tube at the highest heat input area.
B. The edges of the rupture were drawn to a near-knife-edge, and stretcher marks are visible on the ID surface, as shown in Fig. 3.1.1.
C. The wall thickness of the tube measured 0.197 in and 0.176 in away from the failure but in the same plane. At 180° away from the rupture, the walls measured 0.216 in. These measurements were made at the end of the tube above and below the weld.
D. No significant scale and/or deposit were noted on either the OD or the ID surface.
E. The tube above the weld, to the left in Fig. 3.1.1, had expanded from $2\frac{1}{2}$ in OD to $2\frac{21}{32}$ in OD. The tube below the rupture, to the right in Fig. 3.1.1, had expanded to $2\frac{19}{32}$ in OD.

Specimens were taken through the failure lip and the ends of the tube sample above and below the weld, and were examined in the etched conditions at 500 magnifications. The microstructure at the failure, Fig. 3.1.2, shows a spheroidized condition.

The outline of some pearlite colonies is still evident; however, the carbide morphology is one of spheroids. Also noted were elongated grains and subgrains within the grains, indicating rapid high-temperature deformation.

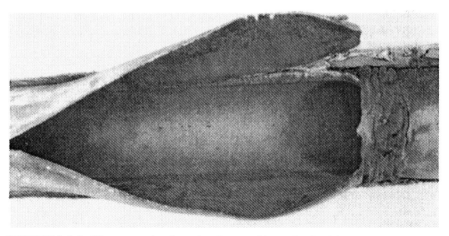

FIGURE 3.1.1. The as-received section from a side-wall tube failure. Note the weld to the right of the failure. The steam flow is from right to left.

FIGURE 3.1.2. The microstructure of the tube taken through the lip of the failure. The structure is almost completely spheroidized, indicating prolonged exposure to relatively high metal temperatures. Note also the elongated grains and some subgrains within the grains, indicating rapid high-temperature deformation (Magnification 500×, nital etch).

Specimens taken from above the weld revealed an *in situ* breakdown of the pearlite lamellae, indicating the start of elevated-temperature exposure; see Fig. 3.1.3. The pearlite colonies are still clearly defined, the microstructure is equiaxed, and only the carbide platelets have begun to spheroidize. This condition could be normal for the length of service, or due to a very short exposure to elevated temperature. Figure 3.1.4 shows the structure of the tube taken below the weld. Here we see an equiaxed ferrite grain structure and the normal pearlite. At 500× the pearlite is too fine to be resolved, but there is no change in the pearlite morphology noted. The heat-affected zone is limited to the immediate vicinity of the rupture.

The sequence of events that occurred at the time of failure, as deduced from the above metallurgical evidence, was as follows: under the low-steam-flow conditions that existed, the weld upset the smooth steam flow, formed an area of turbulence (the steam flow was from right to left in Fig. 3.1.1), and a steam blanket formed. The steam blanket acted as an insulating medium on the inside of the tube. The result was that the tube-metal temperature immediately downstream of the weld rose rapidly until the

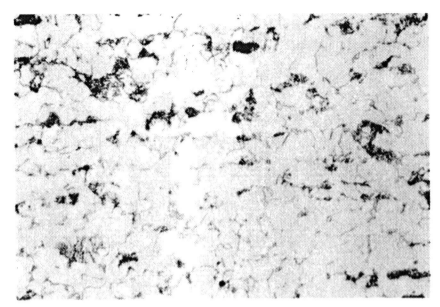

FIGURE 3.1.3. The microstructure from the fire side of the tube above (to the left in Fig. 3.1.1) the failure. This sample is taken from about 6 in away from the failure in the same plane. The microstructure shows an *in situ* breakdown of the pearlite lamellae, indicating the start of elevated-temperature exposure (Magnification 500×, nital etch).

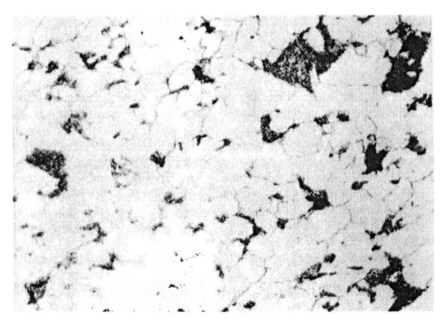

FIGURE 3.1.4. The microstructure below the failure, to the right in Fig. 3.1.1. This sample is from an area about 4 in away from the failure, and shows the structure to be ferrite and the normal pearlite (Magnification 500×, nital etch).

strength of the tube was inadequate to retain the steam pressure, and the tube ruptured. The maximum temperature was around 1300°F (700°C), since there was no evidence of any transformation to austenite. The yield strength at 1300°F (700°C) for this low-carbon steel was about 5600 psi, which was below the design stress at this location and expected metal temperature of about 750°F (400°C). As the tube-metal temperature rose, the tube began to bulge when the yield point was reached, the wall thickness decreased, and the stress increased until rupture occurred as a wide-open ductile burst.

Case History 3.2

Waterwall-Tube Failure, 1450–1500°F (788–816°C)

BOILER STATISTICS

Size	2,800,000 lb steam/hr (1,270,000 kg/hr)
Steam temperature	1005°F/1005°F (540°C/540°C)
Steam pressure	2250 psig (160 kg/cm²)
Fuel	Oil

These failures are from the waterwall tubes in the area of the burners. The unit had been in service just over two years; all four failures occurred within a single eight-day period.

Tube material is SA-209 T-1 steel, having 2½ in OD × 0.260-in-thick wall. Figure 3.2.1 shows the as-received samples. Visual examination revealed the following:

A. All tubes exhibited ductile-type bursts, but two showed a greater wall deformation than others.

B. All of the tubes had expanded over the nominal 2½ in OD.

C. The bursts or splits were on the centerline of the tube at the highest heat input zone.

D. No abnormal OD or ID deposits or scale were noted.

E. The wall thickness in line with the failures varied from sample to sample in range from 0.205 to 0.222 in. Some tube swelling was noted away from the failure, and the diameter increased to 2$\frac{19}{32}$ in.

The microscopic examination of the specimens revealed the following: The microstructure at the failure, Fig. 3.2.2, showed a mixture of ferrite and bainite (ferrite is the white- and bainite the dark-etching constituent in this figure). This microstructure and the Lever Rule indicated that the tube-metal temperature at the time of failure was between 1450°F (790°C) and 1500°F (820°C), a temperature in the two-phase austenite and ferrite region, and that the tube was cooled rapidly. In this portion of the phase

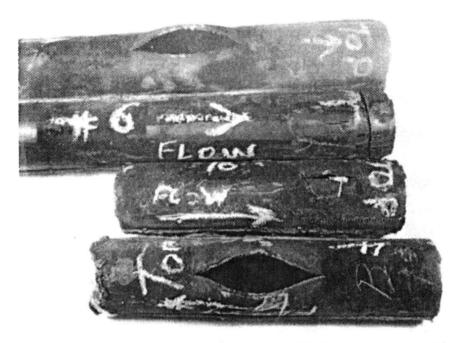

FIGURE 3.2.1. Four waterwall tubes from the area of the burners that contained the failures. Two samples show a wide-open burst and small splits. All failures were on the fire side. The tube material is SA-209 T-1 steel, having 2.5 in OD × 0.260-in-thick wall.

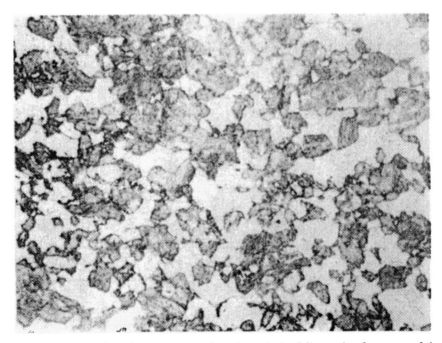

FIGURE 3.2.2. The microstructure taken through the failure edge from one of the samples showing a wide-open burst. The microstructure shows a mixture of ferrite and bainite: the ferrite is the white- and the bainite is the dark-etching constituent. From the relative amounts of ferrite and bainite, the tube-metal temperature at the time of failure was between 1450°F (790°C) and 1500°F (820°C) (Magnification 500×, nital etch).

diagram, austenite and ferrite are in equilibrium, and on rapid cooling the austenite transforms to bainite. The end of the tube section, about 3 in away from the failure, showed a similar microstructure, Fig. 3.2.3, that contained more ferrite and less bainite, indicating a lower peak temperature of about 1400°F (760°C). Our examination showed that the failures were attributable to high metal temperature. The ends of the tube sections also showed elevated-temperature exposure, but to a lesser degree than the failure area.

The microstructure 180° away from the failure, that is, on the back side of the waterwall tubes, showed the normal lamellar pearlite and ferrite; see Fig. 3.2.4. Based on the microstructures, these tubes were exposed to temperatures between 1400°F (760°C) and 1500°F (81°C) for a short time.

Since all of these tubes failed over a relatively short period, about a week, and all tubes came from the same general location of the boiler near the burners, the conclusion was that the overheating was caused by poor circulation in this high-heat-release zone. In the design of burner openings, some tubes are bent out of plane and are hidden from the flame behind other tubes over a portion of their length. These tubes are not uniformly heated, so any problems in circulation are aggravated in this high-heat-input part of the furnace. In order to solve the problem, releaser

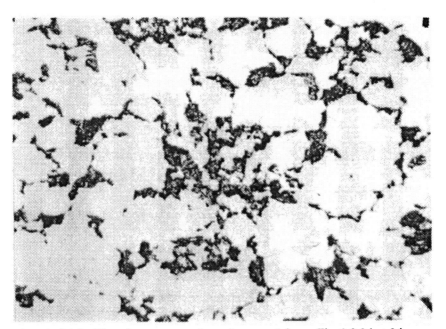

FIGURE 3.2.3. The microstructure from the same tube as Fig. 3.2.2 but 3 in away from the failure. The microstructure is similar, but contains more ferrite and less bainite, indicating a lower peak temperature of about 1400°F (760°C) at the time of failure (Magnification 500×, nital etch).

156 CHAPTER 3 METALLURGICAL PRINCIPLES: FERRITIC STEELS

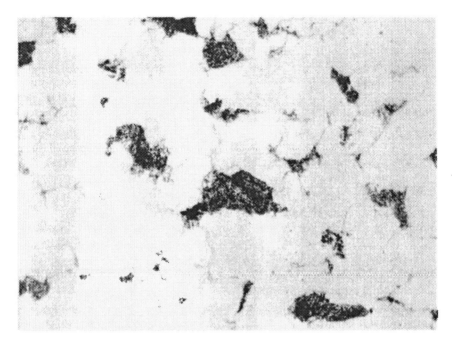

FIGURE 3.2.4. The microstructure on the back side of the waterwall tube shows the normal lamellar, pearlite, and ferrite. There is no evidence of overheating, indicating the high-temperature exposure was limited to the fire side (Magnification 500×, nital etch).

tubes at the top of the furnace were rerouted to improve overall circulation.

Case History 3.3

Waterwall-Platen Tube Failure, 1600°F (870°C)

BOILER STATISTICS

Size	3,250,000 lb steam/hr (1,480,000 kg/hr)
Steam temperature	1005°F/1005°F (540°C/540°C)
Steam pressure	2250 psig (160 kg/cm²)
Fuel	Oil

This is a study of a waterwall-platen tube failure from a utility boiler. The tube that failed is SA-210 A-1 steel, with $2\frac{3}{8}$ in OD × 0.240-in-wall thickness. The visual examination revealed the following:

A. It was a wide-open burst with the edges of the failure drawn to a near-knife-edge condition.

B. No unusual OD scale or deposit was noted.
C. No unusual ID scale or deposit was noted, but there were some ID pits less than 0.005 in deep.
D. There was the usual tightly adhering magnetite scale on both the OD and the ID.

Figure 3.3.1 shows the as-received section of the tubing. Metallographic specimens were taken from the tube at the failure area and 12 in away in the same plane of the tube. Figure 3.3.2 shows the microstructure at the failure area. Note that it is composed almost entirely of bainite, indicating that the temperature at the time of failure was close to 1600°F (870°C). Figure 3.3.3 shows that the microstructure 12 in away from the failure is the normal ferrite and pearlite.

A comparison of the microstructures presented in Figs. 3.3.2 and 3.3.3 indicates that the overheating was limited to a short section of the tube. At the time of failure, the tube-metal temperature was close to 1600°F (870°C) based on the microstructure at the rupture. The short-term high-temperature tensile strength for this carbon steel is estimated to be about 4000 psi (280 kg/cm²) at 1600°F (870°C). The internal pressure at the time of rupture may be calculated from the equation

$$P = \frac{2WS}{D - W}$$

FIGURE 3.3.1. The as-received section of failed waterwall platen tube. The failure is a wide-open burst with the edges of the failure drawn to a near-knife-edge condition.

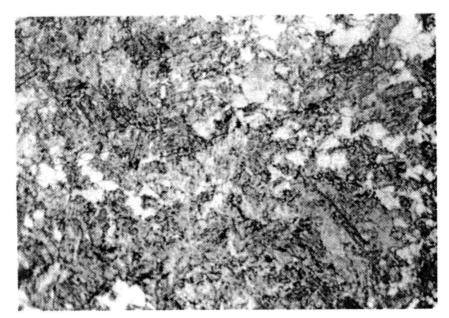

FIGURE 3.3.2. The microstructure taken through the failure area. It is composed almost entirely of bainite, indicating that the temperature at the time of failure was close to 1600°F (870°C) (Magnification 500×, nital etch).

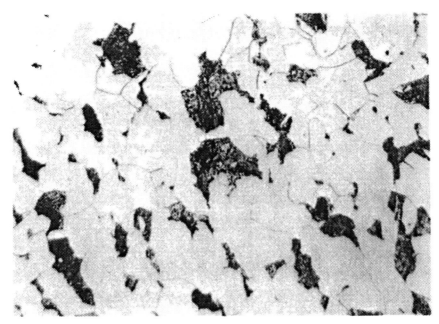

FIGURE 3.3.3. The microstructure 12 in away from and in the same plane as the failure shows the normal ferrite and pearlite structure (Magnification 500×, nital etch).

where P = pressure, psig
       S = tensile strength, psi
       W = tube-wall thickness, in
       D = tube diameter, in

For this case, $S = 4000$ psi, $W = 0.240$ in, and $D = 2\frac{5}{8}$ in, thus, $P \approx 800$ psig.

Since the normal operating pressure was 2250 psig, it is clear that for the tube to reach 1600°F (870°C), the unit had not been at the operating pressure. The failure was caused by localized overheating (metal temperature about 1600°F (870°C)), most probably the result of a too rapid start-up. At the time of rupture, the unit had reached a pressure of about 800 psig, about one-third of the normal operating pressure.

Case History 3.4

Reheater-Tube Creep Failure

BOILER STATISTICS

Size	3,000,000 lb steam/hr (1,360,000 kg/hr)
Steam temperature	1005°F/1005°F (540°C/540°C)
Steam pressure	2900 psig (205 kg/cm²)
Fuel	Oil

This failure occurred in the reheater of a utility boiler. The tube material was SA-178 grade A steel, having $2\frac{3}{4}$ in OD × 0.180-in-thick wall. The failed tube was from the primary reheater close to the division wall. Visual observation revealed the following:

A. A narrow split showing stress lines on the ID.
B. Considerable swelling at the failure area only.
C. No wall thinning was noted except, of course, at the failure.
D. A loose, flakey OD deposit, but no evidence of OD corrosion. The OD scale measured about 0.020 in thick.
E. No significant ID deposit or scale, other than the expected magnetite film.
F. Minor ID pits less than 0.004 in deep.

Figure 3.4.1 shows the appearance of the failed tube section. Specimens for metallography were cut for microstructural analysis, from the failure area, and 180° from the failure, at approximately the same plane as the failure. Figure 3.4.2 shows the microstructure at the failure. Note that the microstructure shows complete coalescence and agglomeration of the carbides. This is indicative of long-time operation at temperatures around

FIGURE 3.4.1. The as-received tube section from the reheater of an oil-fired utility boiler.

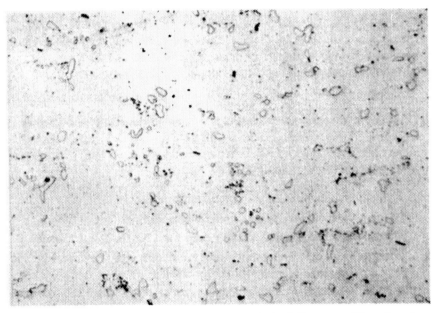

FIGURE 3.4.2. The microstructure taken through the failure area. The microstructure shows complete coalescence and agglomeration of the carbides. Tube material is SA-178 grade A steel (Magnification 500×, nital etch).

900°F (480°C). The normal peak-operating temperature for this material is 800°F (430°C). The microstructure 180° from the failure shows an *in situ* breakdown of pearlite, as can be seen from Fig. 3.4.3. Note that the pearlite colonies are still clearly visible, but the iron carbide has spheroidized. The boiler has been in operation a little less than two years; and

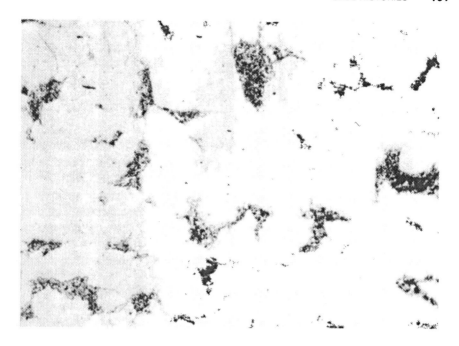

FIGURE 3.4.3. The microstructure 180°, that is, on the back side of the tubing, from the failure shown in Fig. 3.4.2. The structure shows an *in situ* breakdown of the pearlite, but the pearlite colonies are still clearly visible (Magnification 500×, nital etch).

for such a time period, the microstructures show a greater breakdown of pearlite than would normally be expected had the tube been operating at designed temperature. The conclusion based on the metallurgical evidence suggests that for this tube, either there was less steam flow than anticipated or the flue-gas temperature was higher than anticipated. In either case, the actual metal temperature was much higher than the normal design temperature for SA-178 A steel.

Case History 3.5

Superheater Failure, Graphitized Structure

BOILER STATISTICS

Size	700,000 lb steam/hr (320,000 kg/hr)
Steam temperature	1005°F/1005°F (540°C/540°C)
Steam pressure	1475 psig (105 kg/cm^2)
Fuel	Pulverized coal

This was a low-temperature superheater-tube failure from a coal-fired unit. The tube material was SA-210 A-1 steel, having 2½ in OD × 0.165-in-thick wall. The visual examination revealed the following:

A. A ductile wide-open burst with the edges drawn to a knife-edge.
B. No wall thinning was noted except at the failure.
C. No unusual OD or ID deposits other than the expected scale of magnetite.
D. A few ID shallow pits of less than 0.005 in deep.

Figure 3.5.1 shows the appearance of the as-received failure section. Specimens taken at the point of failure and 6 in away from the failure were prepared for a microscopical examination. The microstructure of the specimen at the failure showed spheroidized carbides and graphite nodules; see Fig. 3.5.2. The structure indicated exposure to a temperature slightly elevated from design, and the scattered areas of graphite nodules indicated the beginning of graphitization. Figure 3.5.3 shows the structure of the tube at 180° from the failure. This area showed an *in situ* breakdown of the pearlite and could be considered normal structure for this boiler.

The boiler has been in operation for about 10 years. The failure showed operational tube-metal temperatures in the 850°F (450°C) range. The back side of the tube, 180° from the failure, showed what should be expected as the typical microstructure for a boiler tube that has been in service this long.

The tube had been operating slightly above its safe design limit for an extended period of time, as indicated by the graphitized structure at the failure. Since safe operation is predicted on creep strength at the design temperature, any higher metal temperature will lead to premature failure.

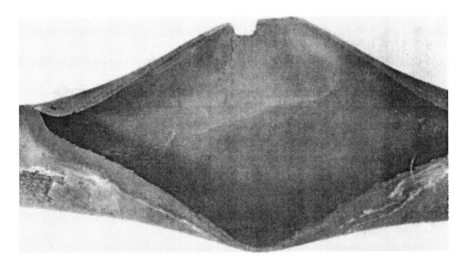

FIGURE 3.5.1. The as-received appearance of the failed superheater tube. The rupture is a wide-open burst with the edges of the failure drawn to a knife-edge condition.

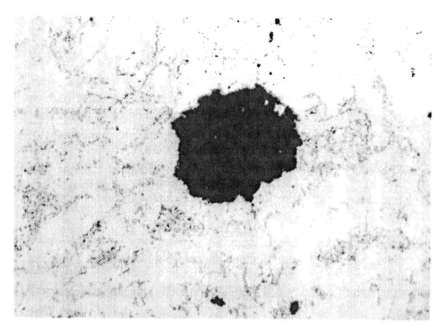

FIGURE 3.5.2. The microstructure taken through the point of failure. The pearlite has completely spheroidized, and the formation of graphite has begun. The large dark spot in center is graphite, as are two small spots at the lower edge (Magnification 500×, nital etch).

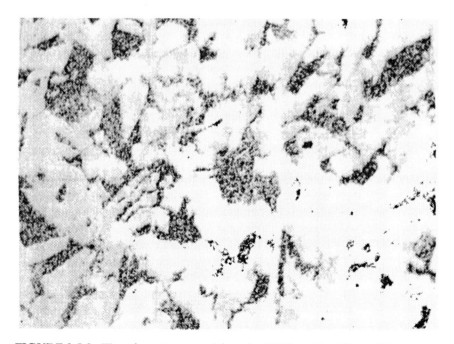

FIGURE 3.5.3. The microstructure of the tube 180° from the failure. This shows an *in situ* breakdown of the pearlite, and is considered normal structure for a boiler in service about 10 years (Magnification 500×, nital etch).

From these observations the sequence of events leading to failure was:

1. The tube expanded by creep, the wall thinned, and the diameter increased. As the tube expanded, internal cooling decreased owing to lower velocity and a flow recirculation pattern in the bulge.
2. However, the steam pressure remained constant, so the stress in the tube increased. As the tube stretched, stress increased in proportion to radius of curvature.
3. The final failure was by short-term high-temperature tensile failure; the rupture lip was drawn to a near-knife-edge.

Case History 3.6

Superheater Failure, Graphitized Structure

BOILER STATISTICS

Size	1,500,000 lb steam/hr (680,000 kg/hr)
Steam temperature	1005°F/1005°F (540°C/540°C)
Steam pressure	2075 psig (145 kg/cm^2)
Fuel	Gas

Metallurgical investigation of a low-temperature superheater tube of SA-210 steel, 2½ in OD × 0.165-in-thick wall tube taken from a gas-fired 200-MW boiler. Visual examination of the two sections of the same tube—the failure and a portion 6 ft away from the failure upstream toward the cooler condition—revealed the following:

A. The failure was a wide-open burst with a piece blown out and somewhat thick-edged failure, indicating some lack of ductility.
B. No tube swelling noted except at the failure.
C. No OD or ID deposits or scale.

Figure 3.6.1 shows the failure section. Samples were taken from the failure area and from the tube portion 6 ft away from the failure in the same plane as the failure for microstructural analysis. The microstructure from the failure area shows graphite nodules and a small amount of bainite. The microstructure 6 ft away from the failure shows a completely graphitized structure; refer to Fig. 3.6.2 for the microstructure at the failure area and to Fig. 3.6.3 for the microstructure 6 ft away.

The microstructure at the failure indicates that the tube was graphitized, but the failure was caused by localized overheating to a temperature in the 1400–1450°F (760–790°C) range. The time at this temperature was not long enough to completely dissolve the graphite particles, but did allow some austenite to form. When the rupture occurred, this austenite was quenched to form bainite.

A simple test to determine whether the particles observed in the microstructure are graphite is to heat a sample in a laboratory furnace to 1700°F (930°C) for 2 or 3 hr and cool the specimen in any convenient way. Graphite dissolves at this temperature, whereas nonmetallic inclusions and creep voids do not, as shown in Fig. 3.6.4.

FIGURE 3.6.1. The as-received tube section taken from the low-temperature superheater. The failure is a wide-open burst with a piece blown out.

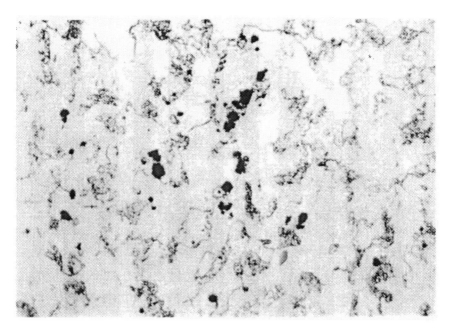

FIGURE 3.6.2. The microstructure at the failure. The structure contains graphite nodules, ferrite, and bainite. The structure indicates that the tube section had been graphitized, but had been locally heated to a temperature in the 1350–1400°F (730–760°C) range. The time was not long enough to dissolve all of the graphite, but it did allow some austenite to form, which, when rupture occurred, was quenched to form bainite (Magnification 500×, nital etch).

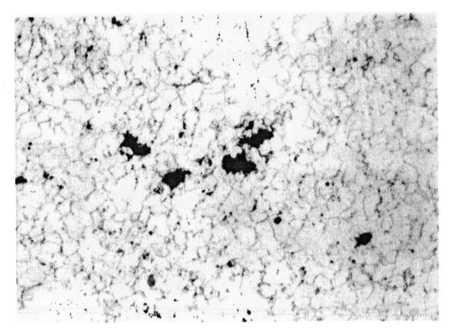

FIGURE 3.6.3. The microstructure in the same plane as the failure but 6 ft away. The graphitization of the structure is nearly complete. The structure contains graphite nodules and ferrite, with only a few carbide particles still visible (Magnification 500×, nital etch).

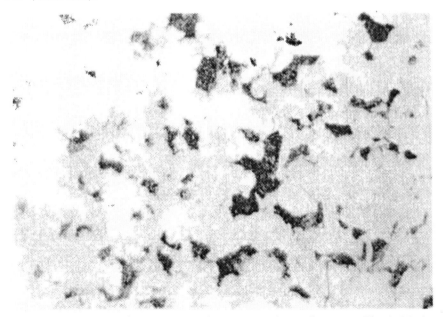

FIGURE 3.6.4. The microstructure of a sample taken adjacent to Fig. 3.6.3, but heated in a laboratory furnace to 1700°F (927°C) for 3 hr and slowly cooled. At 1700°F (930°C) the graphite completely dissolves, and the structure returns to the normal ferrite and pearlite upon slow cooling (Magnification 500×, nital etch).

This boiler had been in operation more than 12 years. The microstructure from the tube indicated that the entire tube section had graphitized, but the failure was caused by localized overheating. The cause of the failure was probably too rapid a start-up.

Case History 3.7

Superheater-Tube Failure, Improper Material

BOILER STATISTICS

Size	600,000 lb steam/hr (270,000 kg/hr)
Steam temperature	1005°F/1005°F (540°C/540°C)
Steam pressure	1800 psig (125 kg/cm^2)
Fuel	Pulverized coal

The material was reported as SA-213 T-22 steel, having 1½ in OD × 0.250-in-thick wall, and the boiler had been in operation 26 months at the time of the tube failure.

A visual examination of the failed superheater tube showed the following:

A. The failure was a wide-open burst exhibiting some gross distortion to the failure; this had been a straight tube and had bent through an angle of about 30° while failing. The edge of the failure showed that the tube wall had thinned from 0.250 in to about 0.175 in.

B. The tube itself had expanded to 1$\frac{23}{32}$ in, and there were many internal and external longitudinal stress lines running parallel to the failure.

C. There was thick OD magnetite scale, which measured 0.034 in.

D. A thick, tightly adhering ID scale measured 0.027 in.

Figure 3.7.1 shows the failure section received for examination. Microstructures were examined at the failure and 8 in away from the failure in the same plane as the failure. Figure 3.7.2 shows the microstructure at the failure, and Fig. 3.7.3 shows the structure several inches away; note that the constituents are ferrite and graphite in both figures. Since SA-213 T-22 material is not known to form graphite on prolonged heating, a chemical analysis was performed on the failed tube. Table 3.7.1 shows the results of the analysis and the specification requirements for T-22 for comparison.

The inescapable conclusion is that the wrong material was installed, confirming our suspicions that the tubing material was not T-22 as specified.

FIGURE 3.7.1. The as-received tube failure exhibiting a wide-open burst, and thick OD and ID oxide scale.

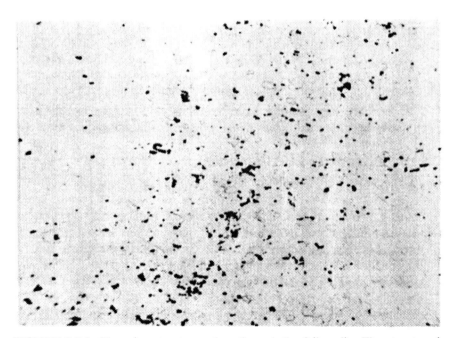

FIGURE 3.7.2. The microstructure taken through the failure lip. The structure is one of graphite and ferrite, indicating prolonged exposure to elevated temperatures (Magnification 100×, nital etch).

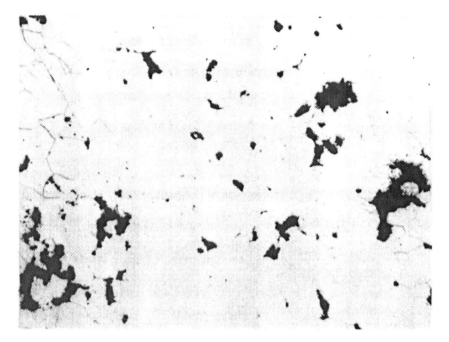

FIGURE 3.7.3. The microstructure 8 in away from the failure in the same plane. The structure at this location is also ferrite and graphite (Magnification 500×, nital etch).

TABLE 3.7.1.

Element	Composition	SA-213 T-22
Carbon	0.14%	0.15% max.
Manganese	0.48%	0.30/0.60%
Silicon	0.19%	0.50% max.
Chromium	0.03%	1.90/2.60%
Molybdenum	0.03%	0.87/1.13%

The unit had been in operation 26 months, with a design steam temperature at the superheater outlet of 1005°F (540°C). At this portion of the superheater, the expected metal temperatures are in the neighborhood of 1050°F (565°C), well above the safe operating limit of carbon steel, but well within the range for T-22. At these temperatures, carbon steel will completely graphitize in the 18,000-hr period that the boiler had been in operation.

Case History 3.8

Superheater-Tube Failure at U Bend

BOILER STATISTICS

Size	275,000 lb steam/hr (125,000 kg/hr)
Steam temperature	950°F (510°C)
Steam pressure	1330 psig (95 kg/cm^2)
Fuel	Wood and coal

Superheater-tube failure of SA-213 T-11 steel, having 2 in OD × 0.150-in-thick wall. The failure was from a small industrial boiler in service about 20 years.

Visual examination of the failure indicated the following:

A. A wide-open burst with the edges drawn to a knife-edge at the bottom of a U bend.
B. A minor amount of both OD and ID scale, each of which, when measured, was 0.005 in thick.
C. No wall thinning was observed except, of course, at the failure.
D. No tube swelling was noted away from the failure.

Figure 3.8.1 shows the failure. The microstructure at the failure, as shown in Fig. 3.8.2, indicates that the tube cooled rapidly from a peak tempera-

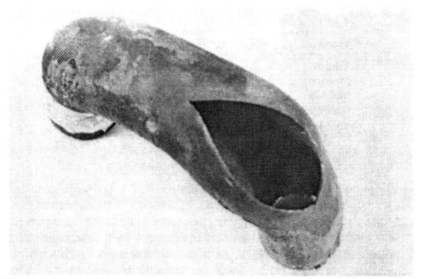

FIGURE 3.8.1. The as-received superheater failure showing the wide-open burst at the bottom of a return bend.

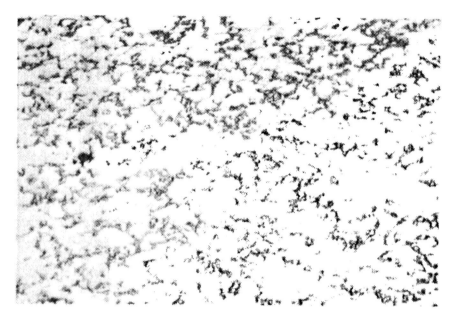

FIGURE 3.8.2. The microstructure taken from the lip of the tube failure indicates a mixture of ferrite and bainite. From the relative amount of bainite and ferrite, the peak temperature at the time of failure was estimated to be in the neighborhood of 1400°F (760°C) (Magnification 500×, nital etch).

FIGURE 3.8.3. The microstructure 10 in away from the failure and in the same plane. The structure consists of ferrite and carbides, primarily at the ferrite grain boundaries. Since the boiler has been in operation nearly 20 years, the structure is considered normal for the length of service (Magnification 500×, nital etch).

ture in the neighborhood of 1400°F (760°C). T-11 steel behaves similar to carbon steel when heated to temperatures within the two-phase austenite and ferrite region above about 1350°F (730°C), and undergoes the same phase transformations. On heating, pearlite and ferrite transform to austenite and ferrite; when rapidly cooled, the austenite then transforms into the bainitic structure observed in Fig. 3.8.2. The microstructure 10 in away from the failure, shown in Fig. 3.8.3, exhibits spheroidal carbides and ferrite, indicating long-term operation at temperatures in the neighborhood of 1000–1050°F (540–570°C).

Since the boiler had been in operation more than 15 years, the microstructure shown in Fig. 3.8.3 was considered to be a normal structure for this length of service. The tube failure, then, was caused by rapid and localized heating to temperatures around 1400°F (760°C), which led to the ductile failure.

Failures that occur in bends are not particularly common and usually indicate some sort of pluggage of the tube at this location. During a rapid start-up, either condensed steam or internal scale flakes fill the bend. Both debris and water would be blown out during the rupture. However, the minor, uniform scale thickness here noted suggests that scale buildup was not the cause of failure in this case.

Case History 3.9

Superheater High-Temperature Failure

BOILER STATISTICS

Size	3,000,000 lb steam/hr (1,360,000 kg/hr)
Steam temperature	1005°F/1005°F (540°C/540°C)
Steam pressure	2900 psig (205 kg/cm^2)
Fuel	Oil

High-temperature superheater failures of SA-213 T-22 steel, having 1¼ in OD × 0.281-in-thick wall superheater tubes that failed in service after seven months operation. Visual examination of the three superheater tubes that failed revealed the following:

A. All of the failed tube sections revealed ductile breaks, with the edges of the failures drawing to a knife-edge. Figure 3.9.1 shows these failures. All of the failures showed parallel internal stress lines and cracking of the tightly adhering scale of these areas. The swelling of the tube prior to fracture would be expected to produce these stress lines.

B. There was a thick, tightly adhering scale on both the external and internal surfaces of the tubes; it averaged 0.025 in thick on the OD and 0.020 in thick on the ID.

C. No wall thinning was observed in any of the tubes except, of course, at the failure.

The microstructures at the failure areas were slightly different from tube to tube. The microstructure at the point of failure in tube A (see Fig. 3.9.2) shows complete spheroidization of the carbides and some slightly elongated ferrite grains. This microstructure indicates that the operational

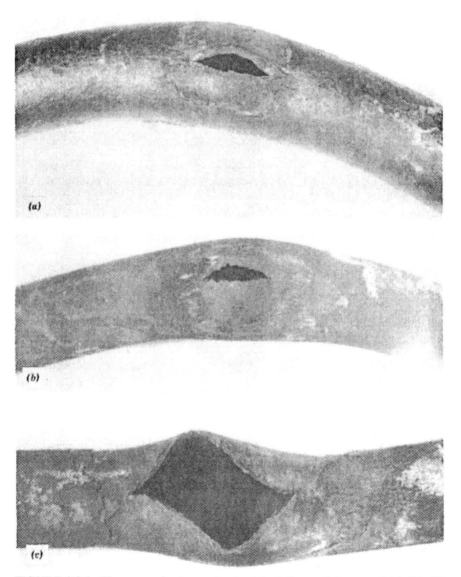

FIGURE 3.9.1. The as-received superheater tube failures. Tube material is SA-213 T-22 steel.

FIGURE 3.9.2. The microstructure through the failure from tube A. The carbides are completely spheroidized, and some slightly elongated grains are evident (Magnification 500×, nital etch).

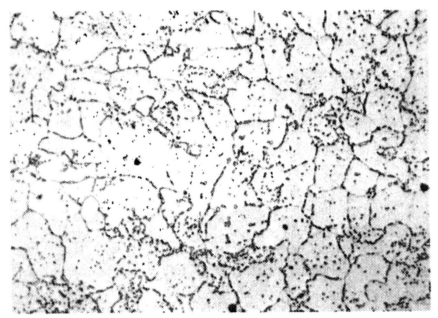

FIGURE 3.9.3. The microstructure from tube A, 180° and 10 in away from the rupture. Complete spheroidization of the carbides is evident, with most of the carbide particles concentrating along ferrite grain boundaries (Magnification 500×, nital etch).

temperature was higher than expected, and the failure occurred at a temperature of about 1150°F (620°C) and at a strain rate high enough to retain evidence of the elongated ferrite grains without allowing time for recrystallization to occur. From the same tube, 180° away and 10 in from the rupture (i.e., diametrically opposed to the failure), we see the complete spheroidization of the carbides, with a preponderance of the carbide particles along the ferrite grain boundaries. This is shown in Fig. 3.9.3.

The second tube failure, tube B, shows slightly different microstructures, both through the rupture and 10 in away, and 180° from the failure; see Figs. 3.9.4 and 3.9.5. At the failure, the ferrite grains are equiaxed, with the carbide particles principally located along the ferrite grain boundaries. The ferrite grain size is slightly larger, indicating that the tube-metal temperature for this tube was slightly higher than for the first failure. The microstructure 10 in away from the rupture, Fig. 3.9.5, shows complete spheroidization and a uniform dispersion of the carbides throughout the ferrite matrix.

The third tube failure, tube C, shows yet a third type of microstructure. At the point of failure, Fig. 3.9.6, the peak tube temperature was in the neighborhood of 1500°F (820°C), and the tube rapidly cooled at the time of

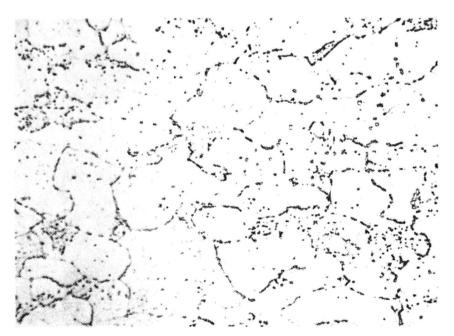

FIGURE 3.9.4. The microstructure through the failure of tube B. The carbides are principally located along ferrite grain boundaries. Note also that the ferrite grains are equiaxed. Again, complete spheroidization of the carbides has occurred, indicating exposure to elevated temperatures for prolonged periods of time (Magnification 500×, nital etch).

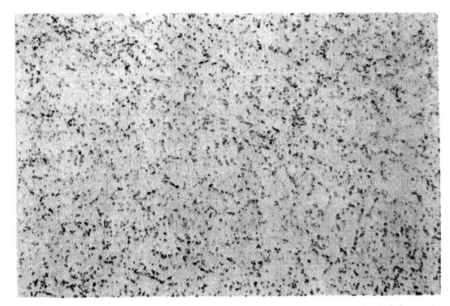

FIGURE 3.9.5. The microstructure from tube B, 10 in away and 180° from the failure. The structure shows complete spheroidization and a uniform dispersion of the carbides throughout the ferrite matrix (Magnification 500×, nital etch).

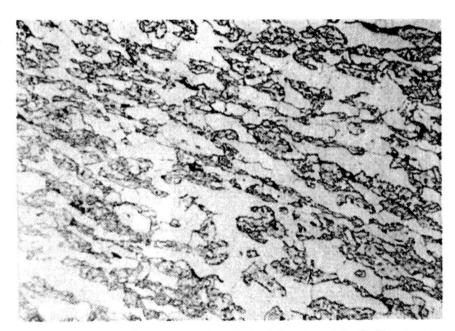

FIGURE 3.9.6. The microstructure through the failure from tube C. The structure consists of ferrite and bainite, indicating that the peak temperature was in the neighborhood of 1500°F (820°C) at the time of failure (Magnification 500×, nital etch).

rupture. The microstructure consists of ferrite and bainite, and the elongated ferrite grains indicate that the strain rate at the failure was high enough to retain the elongated ferrite grains. Figure 3.9.7 shows the microstructure 10 in from the rupture. Here we see an equiaxed ferrite grain distribution, with bainite at the ferrite grain boundaries. The peak temperature for this structure was just above the lower critical temperature, about 1400°F (760°C). Compare this ferrite grain size with that of the first tube failure, Fig. 3.9.3.

For short-term high-temperature failures, similar to this example, the temperature at the time of rupture may be estimated from the ID scale thickness. Rehn et al.[14] have correlated the thickness of scale found in high-temperature steam service with the Larson–Miller parameter.[8] Data from Rehn et al. (see Fig. 8.1) may be approximated by

$$\log X = 0.00022P - 7.25$$

where X = scale thickness, mils
$P = T(20 + \log t)$
T = temperature, °R (°F + 460)
t = time, hr

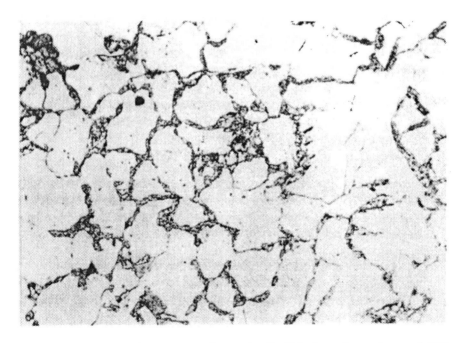

FIGURE 3.9.7. The microstructure from tube C, 10 in from the rupture and 180° away. The structure consists of equiaxed ferrite grains, with bainite at the ferrite grain boundaries. Peak temperature for this structure is just above the lower critical temperature of about 1400°F (760°C) (Magnification 500×, nital etch).

The equation may be solved for T. Use $t = 5000$ hr (seven months) and $X = 20$ mils, the ID scale thickness. From these data, the temperature is calculated to be about 1185°F (641°C). Within the accuracy of the estimate, the tube failed at a temperature between 1150 and 1200°F (620 and 650°C), consistent with the microstructural interpretation.

Since all of these failures occurred within a few weeks of each other after only seven months, or about 5000 hr of service, and all failures were from the 14th or 15th platen, the conclusion is that the steam circulation through these two platens was impaired. After a short time we are seeing microstructures that indicate long-term temperatures in the neighborhood of 1150°F (620°C) or 1200°F (650°C), temperatures that were well above the safe-operating limit for T-22 material. The heavy scale on the ID, 0.020 in thick, indicates higher than the normal, expected operating temperatures.

The failure sections from tubes A and B are typical of creep failure: some thinning of the tube wall at final failure, but no great expansion of the tube. Tube C shows a more typical wide-open burst of a short-time high-temperature tensile failure. However, the heavy ID and OD scale indicates operational temperatures were above 1150°F (620°C) or 1200°F (650°C) for some time prior to failure.

Case History 3.10

High-Temperature, 1500°F (805°C), Superheater Failure

BOILER STATISTICS

Size	1,700,000 lb steam/hr (770,000 kg/hr)
Steam temperature	1005°F/1005°F (540°C/540°C)
Steam pressure	2200 psig (155 kg/cm^2)
Fuel	Pulverized coal

The material of the failed radiant-superheater tube is SA-213 T-22 steel, having 1½ in OD × 0.248-in-thick wall. Figure 3.10.1 shows the violence of the rupture. The tube is bent back nearly 180° at the failure. A visual examination of the tube showed the following:

A. A minor amount of OD scale, which measured 0.005 in thick.
B. The ID scale measured 0.020 in thick.
C. No wall thinning or tube swelling was observed away from the failure.

Specimens taken from the point of failure for a metallographic examination revealed a structure that had been heated to a temperature in the neighborhood of 1500°F (815°C) and cooled rapidly; see note in the caption of Fig. 3.10.2. Also observed at this location were numerous oxide-

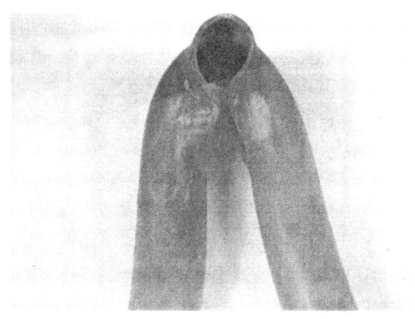

FIGURE 3.10.1. The as-received section of the failed radiant-superheater tube. Note the violence of the rupture: the tube is bent nearly 180° at the failure.

FIGURE 3.10.2. The microstructure taken through the failure edge. The structure is almost completely bainite, indicating a temperature in the neighborhood of 1500°F (871°C) at the time of rupture (Magnification 500×, nital etch).

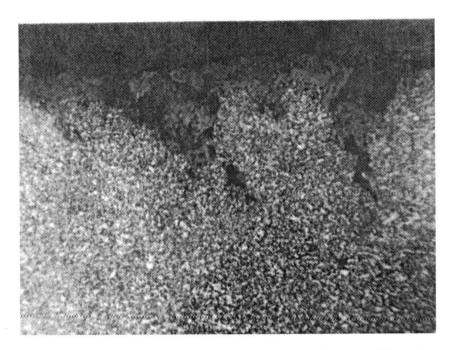

FIGURE 3.10.3. The microstructure taken in the vicinity of the tube failure shows ID cracks approximately 0.010 in deep. The cracks are the likely cause of the brittle appearance of the failure (Magnification 500×, nital etch).

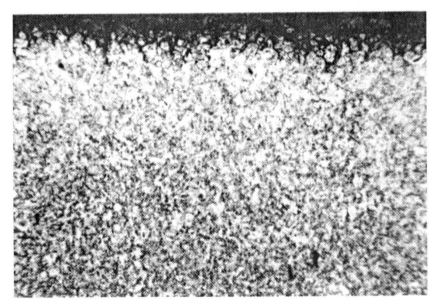

FIGURE 3.10.4. The OD surface in the vicinity of the failure. Some evidence of decarburization may be noted by the ferrite at the OD surface. There was insufficient carbon in the austenite at this location for the austenite to transform to bainite on quenching (Magnification 500×, nital etch).

and corrosion-filled OD and ID cracks penetrating into the tube-metal surfaces. The ID cracks were approximately 0.01 in deep; see Fig. 3.10.3. The OD surface at the failure location showed some evidence of decarburization, as can be seen from Fig. 3.10.4. The light-colored area at the surface was ferrite, that formed from low-carbon austenite on cooling.

There was insufficient carbon in the austenite at this location for the austenite to transform to the bainitic structure observed elsewhere. The failure of the tube was attributable to very high metal temperatures in a localized area. The lack of ductility of the tube was the result of the ID pitting, which, in effect, acted as a notch and caused the type of failure observed. However, the minor amount of ID pitting was not the cause of the failure; the cause was the localized overheating to very high temperatures, 1500°F (815°C) or so.

Case History 3.11

High-Temperature, 1600°F (870°C), Waterwall Failure

BOILER STATISTICS

Size	3,250,000 lb steam/hr (1,480,000 kg/hr)
Steam temperature	1005°F/1005°F (540°C/540°C)
Steam pressure	2250 psig (160 kg/cm^2)
Fuel	Oil

The following example is a waterwall-tube failure from an oil-fired utility boiler. The failure is in the rear waterwall between the burners, and the tube material is SA-178 C steel having 2½ in OD × 0.220-in-thick wall. The unit had been in service nearly six years at the time of failure.

A visual examination of the as-received section revealed the following:

A. It was a wide-open ductile burst, with the edges of the failure drawn to a knife-edge; see Fig. 3.11.1.
B. The external surface had a minor amount of scale.
C. No deposit or any significant scale was noted on the ID surface of the tube.
D. The tube had expanded from 2½ to 2$\frac{19}{32}$ in on the fire side surface, 10 in away from the failure.
E. The tube-wall thickness on the fire-side surface, 10 in away from the rupture, measured 0.210 in; at the same location, but on the back side of the tube, the wall thickness measured 0.230 in.
F. The rupture was just downstream of a shop weld in the waterwall panel. Fluid flow is from left to right in Fig. 3.11.1.

182 CHAPTER 3 METALLURGICAL PRINCIPLES: FERRITIC STEELS

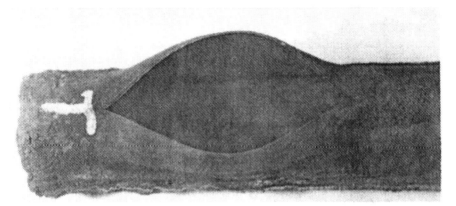

FIGURE 3.11.1. A waterwall-tube failure from an oil-fired utility boiler. The failure is a wide-open ductile burst with the edges drawn nearly to a knife-edge condition. The rupture is just upstream of a shop weld; steam flow is from left to right.

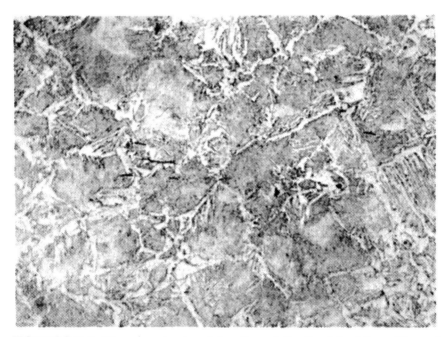

FIGURE 3.11.2. The microstructure taken through the lip of the failure. The peak temperature at the time of failure was in the neighborhood of 1600°F (870°C), high enough to be in the all-austenite region of the iron–carbon phase diagram. The cooling rate through the two-phase austenite and ferrite region was so rapid that ferrite formed principally at the prior austenite grain boundaries. Some ferrite also precipitated on particular crystallographic planes in the austenite. These arms are called Widmanstätten side plates (Magnification 500×, nital etch).

Specimens were taken from the lip of the failure and from the end of the tube sample on the fire side, 10 in away from the failure, for a microstructural examination. The microstructure for the entire 10-in tube sample that was examined showed a structure that was exposed to temperatures above 1600°F (870°C) and cooled rapidly; see Fig. 3.11.2.

It is difficult to say with certainty what the peak temperature was at the time of rupture, other than it must have been above 1600°F (870°C). The structure was all austenite. When the failure occurred, the cooling rate through the two-phase austenite and ferrite region of the iron–carbon phase diagram was rapid enough to freeze the location of the proeutectoid ferrite formed at the start of the transformation. Ferrite first forms at austenite grain boundaries and then along particular crystallographic planes within the austenite. The austenite grain boundaries are well defined by the lightly etching ferrite phase; see Fig. 3.11.2. At this cooling rate, some ferrite also precipitated on particular crystallographic planes in the austenite to form ferrite arms, called Widmanstätten side plates. The formation of the grain-boundary ferrite required a diffusion of carbon into the austenite grains; here the time necessary for this diffusion was short, since the cooling rate was rapid. As the cooling of the steel sample continued, the austenite transformed to a bainitic structure at a temperature in the neighborhood of 700°F (370°C).

The cause of the failure at this location may be related to the turbulence created by the weld. The failure may be the result of too rapid a start-up rate, that is, too high a firing rate for the amount of steam flowing through the burner area of the furnace.

Case History 3.12

Cold-Worked Tube Swage

BOILER STATISTICS

Size	150,000 lb steam/hr (68,000 kg/hr)
Steam temperature	750°F (400°C)
Steam pressure	450 psig (32 kg/cm^2)
Fuel	No. 6 oil

During the retubing of an 18-year-old industrial boiler, a swaged-tube section cracked during the initial rolling operation. The boiler bank tubes are SA-178 A steel, 0.180-in wall thickness, swaged to 2 in from the original diameter of 3¼ in.

Visual examination revealed the following:

A. A 6-in-long brittle crack in the swaged section of the tube.
B. The swaged end was shiny and free of any oxide or mill scale.

184 CHAPTER 3 METALLURGICAL PRINCIPLES: FERRITIC STEELS

C. There was no evidence of any wall thinning along the crack.
D. Hardness measurements were made in the swaged section and unswaged portions of the sample. The hardness of the swaged section averaged Rockwell B 89; through the original tube, Rockwell B 60.
E. No scale or rust was noted on the cracked surface.

Figure 3.12.1 shows the as-received tube section. The microstructure of the swaged end of the tube revealed a heavily cold-worked condition.

FIGURE 3.12.1. The as-received boiler bank tube showing the 6-in-long brittle crack in the swaged section.

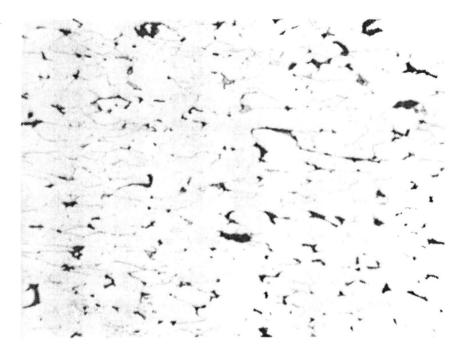

FIGURE 3.12.2. The microstructure through the swaged section. Note the elongated ferrite grains and the distorted pearlite colonies indicative of cold work (Magnification 500×, nital etch).

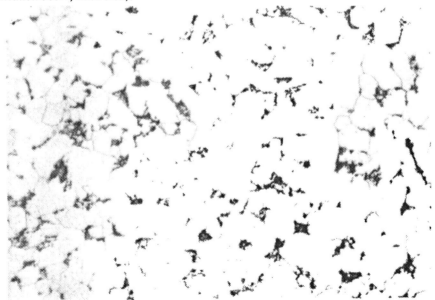

FIGURE 3.12.3. The microstructure from a laboratory heat-treated sample cut from the swaged end of the tube. The heat treatment in a laboratory furnace was at a temperature of 1150°F (620°C) for 15 min followed by a furnace cool. The microstructure shows complete removal of the effects of cold work (Magnification 500×, nital etch).

Figure 3.12.2 shows the structure; note the elongated ferrite grains and the distorted pearlite colonies.

After noting the high degree of cold work, a second section was removed from the swaged end, placed in a laboratory furnace at a temperature of 1150°F (620°C) for 15 min, and allowed to cool in the furnace. The microstructure, Fig. 3.12.3, of this specimen showed the complete removal of the effects of cold work. The hardness at this location after heat treatment was Rockwell B 55. The ferrite grains are equiaxed, indicating a removal of the effects of cold work.

Based on the metallurgical examination of the cracked swaged-tube sample, there is no doubt that it was not properly stress-relieved after swaging. Since this was the only failure in several hundred tubes, the conclusion is that this was the only tube not heat-treated.

Case History 3.13

Thermal Fatigue Failure of Reheater Header Drain Line

BOILER STATISTICS

Size	2,850,000 lb steam/hr (1,295,000 kg/hr)
Steam temperature	1005°F/1005°F (540°C/540°C)
Steam pressure	2250 psig (160 kg/cm^2)
Fuel	Oil or gas

The failure from a reheater inlet-header drain line showed the following:

A. Failure occurred in the pipe at the socket-weld connection by circumferential cracking. The circumferential crack measured about 1 in long.

B. The pipe also failed by longitudinal cracking, which opened up 90° to the weld.

C. The pipe had expanded near the weld from 1⅞-in OD to nearly 2 in. Figure 3.13.1 shows the as-received sample.

D. Visual examination of the drain line revealed a distinct line, or score mark, running longitudinally on the outside and inside surfaces. This line is similar in appearance to the longitudinal weld line of electric-resistance-welded pipe.

Specimens were taken from the weld area encompassing the circumferential crack and at the tip of the longitudinal crack. The microscopic examination of the specimen from the weld revealed some undercutting and gas porosity in the weld. The ID surface of the specimen revealed a multitude of transgranular spear-shaped corrosion-filled cracks penetrating deep into the pipe surface, propagating toward the OD surface; see Fig. 3.13.2.

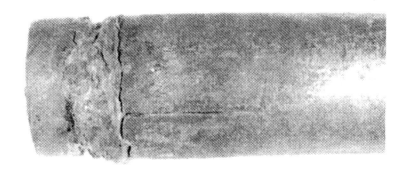

FIGURE 3.13.1. The as-received section from a reheater header drain line. Two failures are visible: a circumferential crack adjacent to the socket weld and a longitudinal crack 90° to the weld line.

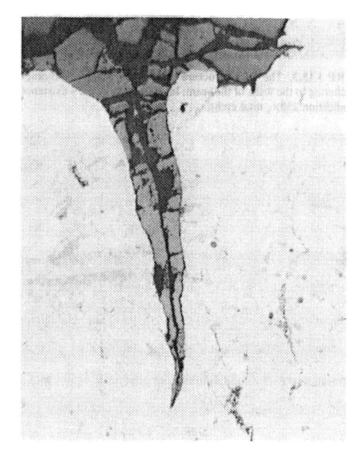

FIGURE 3.13.2. An example of the spear-shaped corrosion-filled cracks noted on the ID surface. The crack is transgranular in character (Magnification 500×, nital etch).

FIGURE 3.13.3. The microstructure through the longitudinal crack. Note the oxides adhering to the walls of the seam, indicating the crack's existence for a long time (Magnification 250×, nital etch).

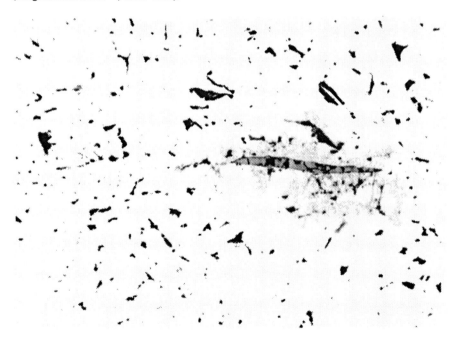

FIGURE 3.13.4. The microstructure in line with the longitudinal crack but about 1 in from the crack tip. Note the oxide entrapment and minute cracking in the same extension and direction as the inclusion. The inference is that the tube was electric-resistance-welded, and the weld line is defective by entrapped oxide (Magnification 500×, nital etch).

The specimen taken through the longitudinal crack revealed a similar condition on the OD surface, as seen in Fig. 3.13.3. Also noted was oxide entrapment and minute cracking in the same plane and direction as the longitudinal crack; see Fig. 3.13.4.

The presence of oxides in this plane suggests that the material was electric-resistance-welded, and that the longitudinal line is actually the weld seam. Figure 3.13.3 shows the large void in the electric-resistance weld. Note the oxide scale adhering to the walls of the seam.

A thermal-fatigue crack can occur either by mechanical stressing, that is, by vibration, bending loads, or stresses caused by restriction of normal expansion when there are variations in temperature in the system, or by thermal stressing, that is, repeated sudden cooling of an area of the drain line filled with water at the time the unit starts up. The condensate-filled drain line prevents the normal expansion of the drain line if the entire assembly increases uniformly in temperature.

There are two failures present: thermal-fatigue cracks of a circumferential direction and a longitudinal crack related to a poor electric-resistance weld (ERW). The material is SA-178 A ERW tubing.

Case History 3.14

Corrosion Fatigue Failure of a Roof Tube

BOILER STATISTICS

Not available

This failure is from a roof tube of a 125,000-lb/hr industrial boiler. The failure occurred in a carbon steel, 3¼ in OD × 0.177-in-thick wall, roof tube just before the offset at a three-tube support clamp near the drum. Visual examination of the tube revealed the following:

A. The tube failed by circumferential cracking on the top of the tube at the edge of the clamp.
B. The crack measured about 2½ in long and from ⅛ to ¼ in wide.
C. The outside surface of the tube contained loosely adhering deposits, most likely from the products of combustion. These deposits measured 0.010 in thick.
D. The inside surface of the tube revealed some scale and/or deposits, less than 0.003 in thick.
E. Also noted on the inside surface were two horizontal lines parallel to the tube axis, ¼ in wide, appearing at the 4 and 8 o'clock positions. These lines appeared to be water-level lines.

F. No swelling of the tube circumference was observed. However, the end of the tube at the clamp location revealed deformation on the top surface of the tube; see Figs. 3.14.1 and 3.14.2 for the appearance of the as-received tube sample.

FIGURE 3.14.1. The cross-sectional appearance of the as-received section. Note the distortion in the tube at the 12 o'clock position in the photograph.

FIGURE 3.14.2. The longitudinal section of the as-received tube. The view is from the 12 o'clock position in Fig. 3.14.1.

The deformation was most probably caused by the clamp pressure. The outside surface was flattened by the clamp to a considerable degree. The specimens were taken from the top of the tube through the crack and 180° away from the crack. The microstructural examination of the specimens taken from the crack revealed a number of small oxide-filled cracks penetrating into the tube's OD and ID surfaces. These cracks are transgranular in appearance and are typical of a thermal-fatigue condition. The structure did not indicate that it had been exposed to any high metal temperatures, since it showed the normal equiaxed grains of ferrite and lamellar pearlite. Also observed in the microstructure were what appeared to be subgrains in the ferrite. See Fig. 3.14.3 for the OD cracks, and Fig. 3.14.4 for the ID cracks.

Here we see the start of subgrains developing during the fatigue process. Deformation caused by the temperature cycling and by the constraint of the clamp, led to the formation of these subgrains. This deformation is similar to the cold working of metal at room temperature.

The corrosion-fatigue cracks are transgranular in character. The oxidation products formed at elevated temperatures tend to widen a crack at its mouth, giving rise to a spearlike shape, increasing the stress at the crack tip. The oxide has a lower density than the metal from which it forms and,

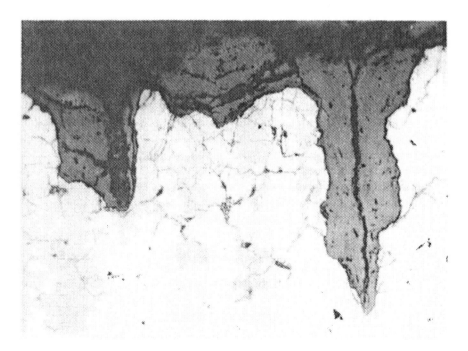

FIGURE 3.14.3. The OD cracks penetrating into the tube surface. Note the formation of subgrains within the ferrite, indicating some deformations, as would be associated with fatigue (Magnification 500×, nital etch).

192 CHAPTER 3 METALLURGICAL PRINCIPLES: FERRITIC STEELS

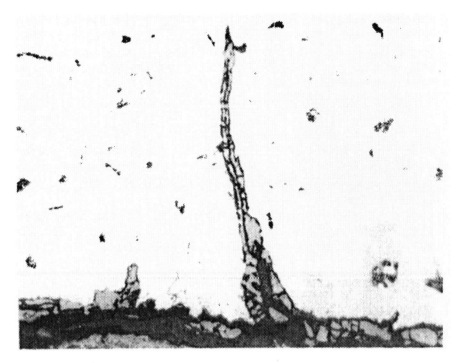

FIGURE 3.14.4. The ID cracks, transgranular in appearance and typical of corrosion-fatigue condition. No evidence of overheating, as the pearlite colonies are still well defined (Magnification 500×, nital etch).

therefore, occupies a greater volume. The conditions that cause thermal-fatigue cracks are mechanical and/or thermal stresses in a corrosive or oxidizing environment. The temperatures noted here were not high enough or long enough to alter the microstructure. Nevertheless, there easily could have been a temperature difference between the top and the water-filled bottom of the tube. The normal expansion of the top of the tube was restricted by the clamp, which most probably deformed the tube and set up the stresses necessary for failure.

Case History 3.15

Low-Water Upset

BOILER STATISTICS

Size	2,600,000 lb steam/hr (1,180,000 kg/hr)
Steam temperature	1005°F/1005°F (540°C/540°C)
Steam pressure	2250 psig (160 kg/cm^2)
Fuel	Oil

CASE HISTORIES   193

This case history is of an investigation of several tubes from a utility boiler involved in a low-water upset. Samples from several locations were submitted for examination: waterwall-platen tubes, a three-tube section from the waterwall, and a single tube from the bottom of the high-temperature reheater.

Several tubes were received from the waterwall platens; one contained a failure; and the others were rupture-free. The platen tubes are SA-120 A-1 steel, having $2\frac{1}{2}$ in OD × 0.220-in-thick wall. Visual examination of these tubes revealed the following:

A. A wide-open burst with the edges drawn nearly to a knife-edge; see Fig. 3.15.1.
B. No substantial OD or ID deposits or unusual scale.
C. The hardness of the failed tube was Rockwell C 30.

The microstructure at the point of failure was essentially all bainite; see Fig. 3.15.2. The peak temperature was in the neighborhood of 1600°F (870°C), followed by a rapid quench at the time of fracture. The microstructure 2 ft away from the failure and in the same plane is shown in Fig. 3.15.3. The peak temperature was lower than at the failure, in the neighborhood of 1500°F (820°C), followed by rapid cooling. Here the structure

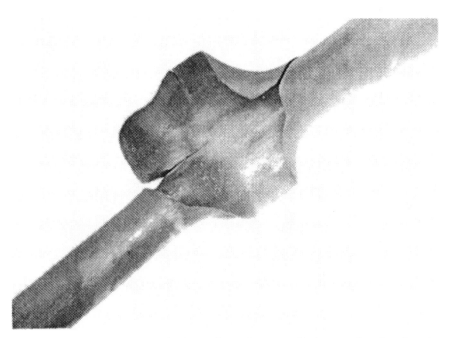

FIGURE 3.15.1. The as-received section of waterwall-platen tube showing the wide-open burst.

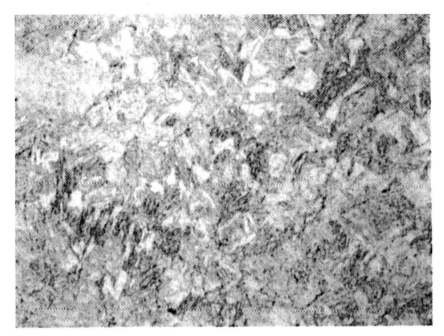

FIGURE 3.15.2. The microstructure at the point of failure is essentially all bainite. The peak temperature at the time of failure was in the neighborhood of 1600°F (870°C), followed by a rapid quench (Magnification 500×, nital etch).

FIGURE 3.15.3. The microstructure 2 ft away from the point of failure and in the same plane as the failure. The peak temperature was lower than the failure, in the neighborhood of 1500°F (820°C). The structure is a mixture of ferrite and bainite, with some tendency toward a Widmanstätten structure (Magnification 500×, nital etch).

is a mixture of ferrite and bainite, with evidence of a Widmanstätten structure.

A nonfailed waterwall-platen tube from the platen adjacent to the failure was also examined microstructurally. The structure in this platen is an *in situ* breakdown of the pearlite, but the pearlite colonies are still clearly defined; see Fig. 3.15.4. Considering the boiler had been in service more than seven years at the time of the low-water upset, this structure is considered to be normal. The material is SA-210 A-1 steel and had expanded from 2½ to 2⅝ in in diameter.

A 3-ft section of waterwall panel containing three tubes from the west wall identified as Nos. 49, 50, and 51 were also examined. The material is SA-178 C steel with a 2½ in OD × 0.220-in-thick wall. No failures were observed in any of the waterwall sections. The visual examination of this waterwall showed the following:

A. The tubes had expanded from 2½ to 2⅝ in OD.
B. Wall thickness was reduced to 0.198 in.
C. The 3-ft-long section had about a ¼-in bow in it.
D. The fire side of all three tubes had a heavy blue–black scale typical of high-temperature exposure.

FIGURE 3.15.4. The microstructure from a nonfailed waterwall-platen tube adjacent to the point of failure. The structure in this platen is an *in situ* breakdown of the pearlite, but the prior pearlite colonies are still clearly defined (Magnification 500×, nital etch).

The microstructures of tube no. 50 indicate a peak temperature of around 1600°F (870°C), with a rapid quench. From the microstructure it appears that there are slightly different cooling rates between the OD and ID, as can be seen from Figs. 3.15.5 and 3.15.6. There are differing amounts of ferrite in the Widmanstätten structure, with more ferrite in the OD than in the ID of the tube.

Tube no. 51 had expanded to $2\frac{19}{32}$ in. The microstructure showed large coarse grains of ferrite and pearlite, indicating a renormalization structure; see Fig. 3.15.7. The peak temperature was in the neighborhood of 1550°F (840°C), with a slow enough cooling rate to give a more normal microstructure. The hardness of the tube was Rockwell B 80.

The microstructure on the back side of the waterwall tubes is shown in Fig. 3.15.8. The structure is normal ferrite and pearlite with a Rockwell B hardness of 75 and is considered to be the original tube structure. It seems unlikely that the back side of the waterwall tubes would be heated hot enough and long enough to alter the pearlite structure.

The last tube was from the very bottom of the high-temperature reheater and is SA-213 T-22 steel with a $1\frac{3}{4}$ in OD × 0.148-in-thick wall. The visual examination of this tube revealed nothing unusual—no swelling, no

FIGURE 3.15.5. The microstructure from a nonfailed waterwall tube. The peak temperature at the time of quench was in the neighborhood of 1600°F (870°C). The microstructure is from the OD of the tube; note the amount of ferrite at the prior austenite grain boundaries (Magnification 500×, nital etch).

FIGURE 3.15.6. The microstructure from the same tube as the previous microstructure, but from the ID surface. Compare the amount of ferrite at the prior austenite grain boundaries in this structure with Fig. 3.15.5 (Magnification 500×, nital etch).

FIGURE 3.15.7. The microstructure from the tube adjacent to the previous two microstructures shows coarse grains of ferrite and pearlite, indicating a renormalized structure. Peak temperature was in the neighborhood of 1550°F (843°C), with a slow enough cooling rate to give a more normal microstructure (Magnification 500×, nital etch).

FIGURE 3.15.8. The microstructure on the back side of the waterwall tubes. The structure is normal ferrite and pearlite, and is considered to be the original structure (Magnification 500×, nital etch).

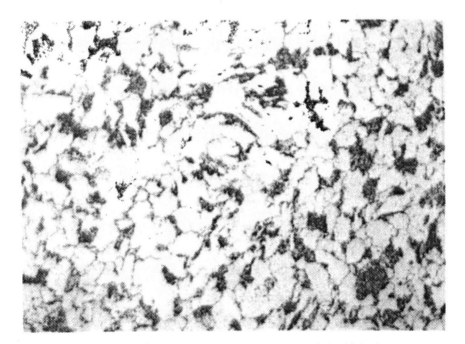

FIGURE 3.15.9. The microstructure from the bottom of the high-temperature reheater shows an *in situ* breakdown of the pearlite colonies, but is considered normal for a boiler of this age (Magnification 500×, nital etch).

unusual deposits either on the ID or OD, and no unusual color. The microstructure, presented in Fig. 3.15.9, shows an *in situ* breakdown of the pearlite colonies, and is considered normal for a boiler of this age.

This one low-water upset has provided microstructures that range from essentially no-effect of the upset to quenched structures resulting from temperatures in excess of 1600°F (870°C).

Case History 3.16

Thermal Fatigue Damage—Waterwall Tubes

BOILER STATISTICS

Size	2,900,000 lb steam/hr (1,320,000 kg/hr)
Steam temperature	1005°F/1005°F (540°C/540°C)
Steam pressure	2650 psig (185 kg/cm²)
Fuel	Pulverized coal

This example is a window section taken from the fire-side crown of a waterwall tube located at the midpoint on the side wall 10 ft above the burners. The tube material is SA-210 A-1, 2¾ in OD by 0.280-in-minimum wall. Figure 3.16.1 shows the overall view of the ID of the sample received. Figure 3.16.2 shows the OD surface of the same sample. A cross-sectional view of the crown is shown in Fig. 3.16.3. Visual observation revealed the following:

A. The internal surface showed circumferential cracks and a "water" mark about 1 in wide at the crown.

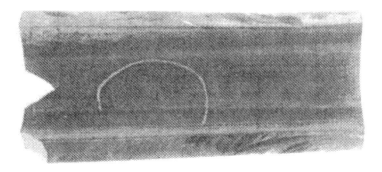

FIGURE 3.16.1. The as-received tube sample viewed from the inside. Note the circumferential crack, typical of thermal fatigue caused by thermal shock, and the "water" mark about 1 in wide.

200 CHAPTER 3 METALLURGICAL PRINCIPLES: FERRITIC STEELS

FIGURE 3.16.2. The fire side of the same tube sample. The crown has a crazed appearance caused by the expansion of the tube (Magnification 1.5×, nital etch).

FIGURE 3.16.3. Cross-sectional view of the tube section. The wall thinning at the crown is clearly visible (Magnification 2×, nital etch).

B. The fire side showed a multitude of fine surface cracks.
C. The wall thickness varied from 0.310 in at the corner corresponding to the 10 o'clock position on the tube to 0.198 in at the crown.

Microstructural analysis of the tube section revealed the following. The microstructure on the fire side at the thinnest portion of the tube is completely spheroidized and decarburized; see Fig. 3.16.4. On the ID surface at the crown is an *in situ* breakdown of the pearlite and elongated ferrite

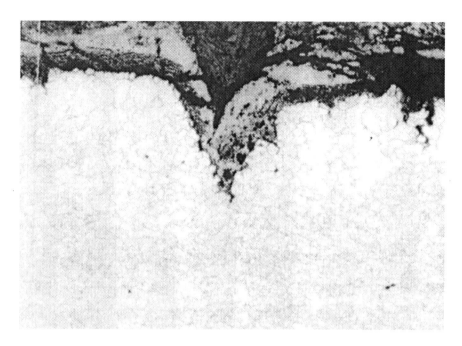

FIGURE 3.16.4. The microstructure through the crown of the tube on the fire side; note the complete spheroidization and decarburization (Magnification 250×, nital etch).

grains, indicating that the tube is expanding in diameter, Fig. 3.16.5. The elongated grains are evidence of deformation of the tube. At the 10 o'clock position we see nearly normal ferrite and pearlite, perhaps just the early stages of an *in situ* spheroidization of the iron carbide; see Fig. 3.16.6.

From the microstructure and visual observations the following conclusions may be drawn. The OD surface temperature is around 1000°F (540°C), and the crazed appearance suggests rapid expansion. The watermark on the ID indicates that a steam blanket or film boiling is occurring on the crown of the tube. The multitude of circumferential cracks on the ID implies a thermal-fatigue problem exists as well.

From all of the metallurgical information and a careful study of the plant records of operation, the cause of the wall thinning can be explained this way: The load was increased at a rapid rate. During the rapid load change, the drum water level became unstable and dropped below the control range. With the drum level too low, there was an increase in the amount of steam entrained in the downcomer that feeds the sidewall. The steam and water remained well mixed in the vertical downcomer but separated under the influence of gravity in the horizontal lower side-wall header. Thus, the fluid that fed the center tubes on the side-wall contained

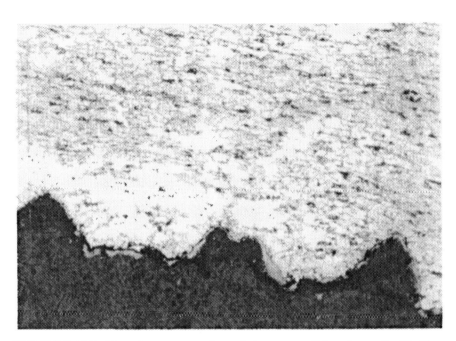

FIGURE 3.16.5. The microstructure through the crown of the tube on the ID. Here the structure displays an *in situ* breakdown of the pearlite and elongated ferrite grains (Magnification 250×, nital etch).

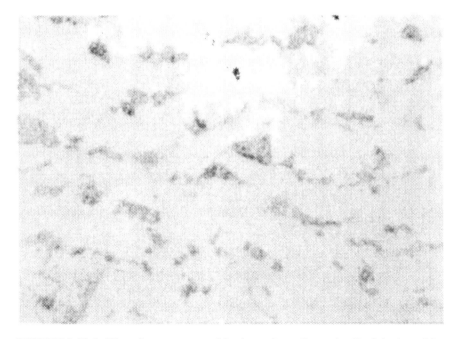

FIGURE 3.16.6. The microstructure of ferrite and pearlite at the 10 o'clock position of the tube. From Fig. 3.16.4, the high temperature of the tube is limited to the region of film boiling.

too much steam and had too low a density to keep the tube properly cooled. The high firing rate and high heat-transfer rate led to the formation of a steam blanket or film boiling on the inside of the tube for a relatively short time. From the microstructures, the tube temperature increased to about 1000°F (540°C). Once the unit was "up to speed," the firing rate was reduced, the drum water level returned to normal, the steam blanket collapsed on the inside of the tube, and the tube-metal temperature decreased from about 1000°F (540°C) to about 750°F (400°C).

The sudden quench led to a thermal-shock condition on the tube ID. Repeated cycles of thermal stress caused the circumferential thermal-fatigue cracks noted on the bore; see Fig. 3.16.1. Since the tube was also expanding due to the overheating (the wall thickness had been reduced by 36%), the microstructure shows the evidence as elongated grains; see Fig. 3.16.5. Without question, this tube was very close to failure, the calculated stress was up to 18,500 psi, well above the safe allowable limit of 13,000 psi at 750°F (400°C).

The corrective action is to maintain proper drum water level during rapid load changes. With the correct water level, the fluid entering the downcomers will not contain too much steam, and proper circulation will be maintained in all furnace-wall tubes.

Case History 3.17

Two Failures—Same Waterwall Tube

Two portions of the same waterwall tube from a gas-fired unit were submitted for metallurgical analysis. The tubes are specified as 2.5 in OD × 0.260 in MWT SA-210 A-1 material. The failure locations were given as being near the side-wall observation port, and there are two types of failures:

1. A wide-open burst from a tube bent out of the plane of the waterwall
2. A narrow bulge or small fissure-type failure in the general vicinity of several bulges in the waterwall tube

The open burst is at a higher elevation than the small, fissure-type failure.

Figure 3.17.1 shows the as-received tube sample with the wide-open burst. Note at this elevation that there are no fins welded to the tube, since this tube formed a portion of the observation port. The fracture edge, Fig. 3.17.2, shows considerable elongation and distortion of the microstructure. The microstructure at the failure, Fig. 3.17.3, is a mixture of ferrite and bainite, both severely elongated. The microstructure some 8 in away from the rupture, in line with the failure, is ferrite and bainite; see Fig. 3.17.4. The microstructure 180° around the tube perimeter, Fig. 3.17.5, shows ferrite and pearlite with the carbide phase spheroidized within the pearlite colonies.

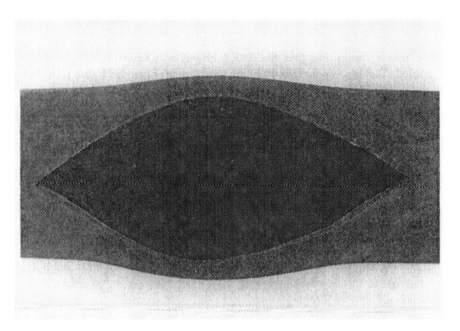

FIGURE 3.17.1. As-received tube, rupture is a wide-open burst. Note also the absence of membrane welds (Magnification 0.8×).

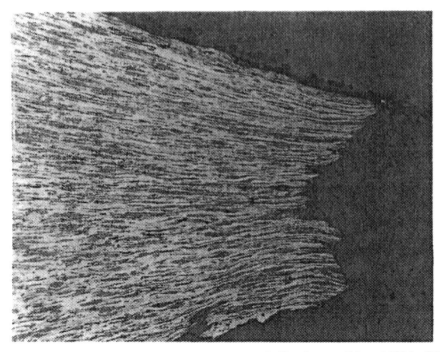

FIGURE 3.17.2. Fracture lip is drawn to a near-knife-edge condition, 0.027 in (27 mils), at the instant of rupture (Magnification 75×, etched).

FIGURE 3.17.3. Microstructure near the fracture lip is a mixture of ferrite and bainite, indicative of a metal temperature in excess of 1350°F at the instant of rupture (Magnification 500×, etched).

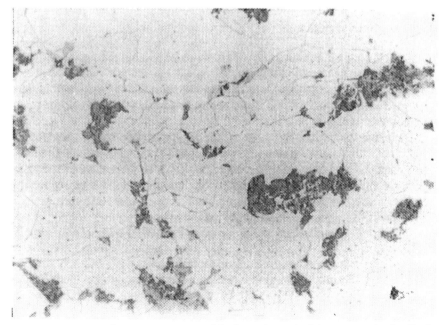

FIGURE 3.17.4. Microstructure some 8 in from the end of the rupture is also ferrite and bainite but without the elongation of the failure lip (Magnification 500×, etched).

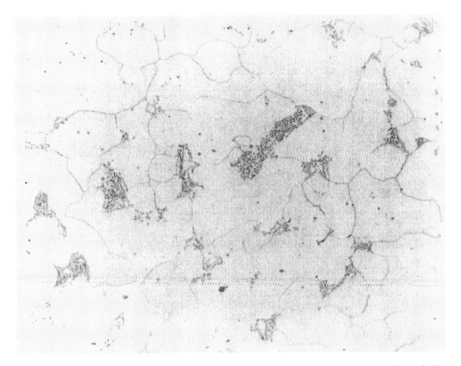

FIGURE 3.17.5. Microstructure 180° around tube perimeter from the failure is ferrite and spheroidized carbides, with the pearlite colonies still well defined (Magnification 500×, etched).

This failure is a good example of a short-term, high-temperature failure that occurred at a metal temperature, at the instant of rupture, in excess of 1350°F. The microstructure has been transformed to ferrite and austenite at the moment of failure, and the rupture rapidly cooled the austenite to bainite, as shown in Figs. 3.17.3 and 3.17.4.

The microstructure 180° around the tube perimeter from the rupture, Fig. 3.17.5, indicates some overheating, since the normal lamellar pearlite has spheroidized. For this tube location, the heat is applied over 360° of the tube perimeter, and is the most likely location for DNB to occur. The tube microstructure suggests some higher-than-normal temperatures during operation, perhaps during rapid load swings.

The other failure in this tube, which occurred lower in the furnace, is shown in Fig. 3.17.6. Here the appearance shows a narrow, fissurelike failure and a small window removed from the surface. The ID surface, Fig. 3.17.7, shows severe corrosion and extensive cracking. The cross section through a bulge, Fig. 3.17.8, shows extensive ID longitudinal cracking.

The microstructure along the ID surface, Fig. 3.17.9, shows the extensive intergranular cracks and decarburization indicative of hydrogen dam-

FIGURE 3.17.6. Fissure-type failure (Magnification 1).

FIGURE 3.17.7. ID surface contains considerable corrosion debris and multiple longitudinal cracks (Magnification 1.5×).

FIGURE 3.17.8. Cross section through bulge shows extensive cracking and an ID scale thickness of 0.013 in (13 mils) (Magnification 18.75×, etched).

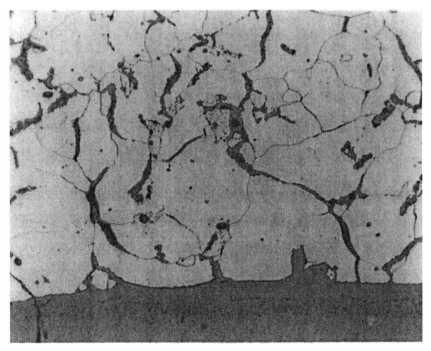

FIGURE 3.17.9. Microstructure along ID shows intergranular cracking and decarburization, characteristics of hydrogen damage (Magnification 500×, etched).

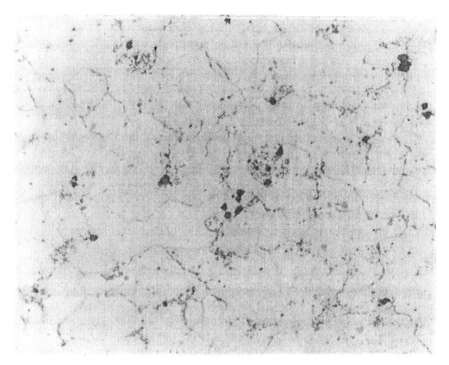

FIGURE 3.17.10. Midwall microstructure has begun to graphitize, indicating prolonged metal temperatures above 850°F (Magnification 500×, etched).

age. The microstructure midwall is graphitized, indicative of operation at elevated temperature for a long enough time to convert iron carbide to graphite, Fig. 3.17.10.

The second failure is caused by hydrogen damage; see Chapter 6, page 335. The thick ID scale has also effectively raised the tube-metal temperature for long enough periods to convert some of the iron carbide to graphite; see Fig. 3.17.10. Whether there is a creep component to this hydrogen-damage failure is subject to question. The graphitized structure does suggest a metal temperature in the creep range for carbon steel, but the decarburized structure, Fig. 3.17.9, indicates hydrogen damage.

The high-temperature failure ($T > 1350°F$) is related to the low-temperature failure ($T < 900°F$) by loss of steam flow through the hydrogen-damage failure lower in the furnace. Loss of fluid flow reduces the effective cooling of the tube, and the second failure occurred in the vicinity of an observation-port window. The tube was bent out of plane from the waterwall, so the heat flux was higher, since heat was applied to the tube over most of the 360° circumference. This led to excessive steam formation in a local region, and the high-temperature failure followed as a result of the formation of a steam blanket.

Case History 3.18

Waterwall-Tube Creep Failure

A eight-foot-long section from a furnace-wall tube was submitted for metallurgical analysis. The tube contained three blisters with small, fissure-type failures on two of them. The tubes are 2½ in OD × 0.167 in AWT carbon steel, but the grade was not specified.

The single 8-ft-long tube was visually examined, and three bulges with two small fissure-type failures were noted along the fire side. Four ring sections were taken:

1. Near the end of the tube, bottom
2. Through the lower failure blister
3. Through the bulge that contained no through-wall cracks
4. The upper bulge that contained a second failure

These four ring sections are displayed in Fig. 3.18.1. Dimensional measurements were made in four equally spaced positions around the tube perimeter. Rockwell B, R_B, hardness was measured at the 12 o'clock position. These data are given in Table 3.18.1. For reference, the 12 o'clock position is taken as in line with the center line of the blisters.

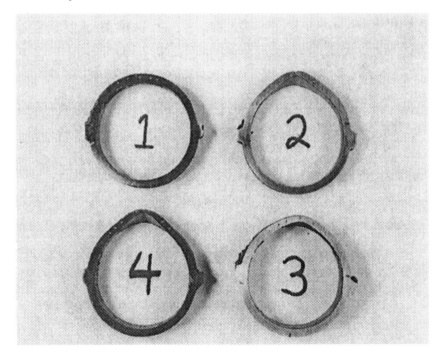

FIGURE 3.18.1. Ring sections from a waterwall tube, Nos. 1–4, from bottom up: 1. bottom end of tube sample; failure; bulge but no failure; failure.

TABLE 3.18.1. Dimensional Measurements

Ring	Position	OD, in	Wall, in	R_B Hardness
1	12:00	2.505	0.162	72.5
	3:00	2.497	0.164	
	6:00		0.167	
	9:00		0.167	
2	12:00	2.669	0.087	63
	3:00	2.480	0.163	
	6:00		0.160	
	9:00		0.172	
3	12:00	2.618	0.125	66
	3:00	2.479	0.169	
	6:00		0.167	
	9:00		0.164	
4	12:00	2.746	0.071	64
	3:00	2.490	0.171	
	6:00		0.171	
	9:00		0.163	

Figure 3.18.2 presents a cross section through the failure in ring 2, and the majority of the longitudinal cracking initiates at the OD surface of the tube. The microstructure shows spheroidized carbides and creep voids, Fig. 3.18.3. The OD surface along the bulge from ring 3 contains numerous longitudinal cracks and creep voids; see Fig. 3.18.4. The microstructure through the bulge that contains no failure is graphitized, as shown in Fig. 3.18.5. The cross section of the failure from ring 4 is presented in Fig. 3.18.6. Again, the failure initiates from the OD and progresses inward; and the microstructure contains ferrite and bainite, Fig. 3.18.7, suggesting that this was not the first failure. However, there is considerable creep damage; thus, the failure in ring 4 is likely to have been a blister, prior to the first failure, which was lower in the waterwall.

The midwall microstructure on the fire side of ring 1 is normal ferrite and pearlite, Fig. 3.18.8. This ring shows no significant swelling, and the microstructure indicates normal operating temperatures. The microstructure 180° around the perimeter is also normal ferrite and pearlite, as in Fig. 3.18.8.

The blisters and failures are caused by creep, that is, long-term overheating of the waterwall tube. While there was not much scale left within the blisters, an examination of Figs. 3.18.2 and 3.18.6 shows about 5 mils of scale intact. Figure 3.18.9 shows the ID deposits from ring 1, which contain considerable scale as well. Reference to Fig. 6.45 shows that 5 mils of scale can raise the metal temperature of a waterwall tube by 100°F, depending on the heat flux. Thus, the likely cause of failure was excessive steam-side scale, which acted as an effective insulating barrier to heat

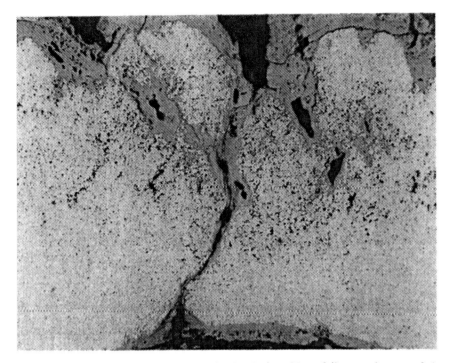

FIGURE 3.18.2. Cross section through No. 2 ring. Note failure and most of the damage initiates at OD (Magnification 37.5×, etched).

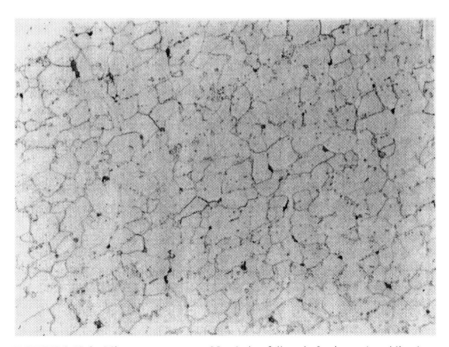

FIGURE 3.18.3. Microstructure near No. 3 ring failure is ferrite, spheroidized carbides, and creep voids (Magnification 500×, etched).

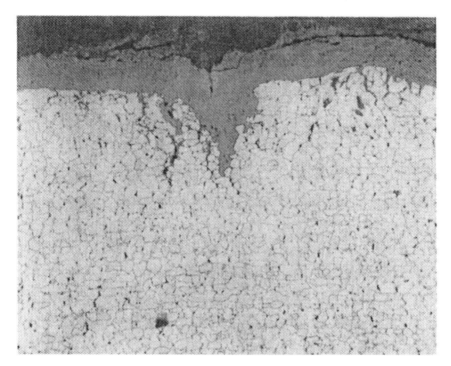

FIGURE 3.18.4. OD surface of bulge from No. 3 ring contains numerous longitudinal cracks and voids (Magnification 200×, etched).

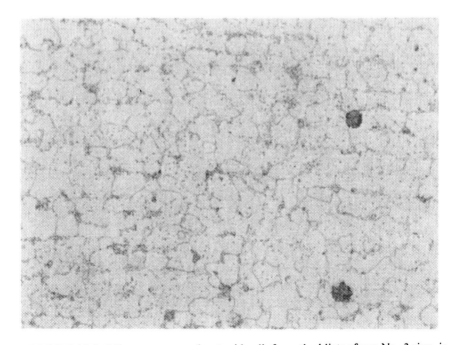

FIGURE 3.18.5. Microstructure, about midwall, from the blister from No. 3 ring, is graphitized (Magnification 500×, etched).

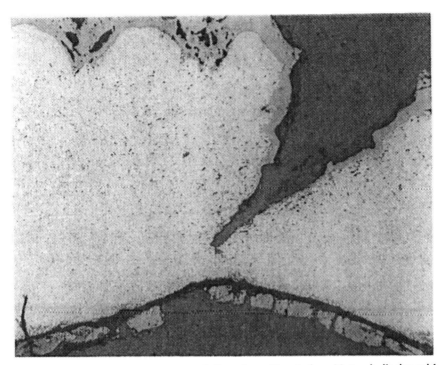

FIGURE 3.18.6. Cross section near failure from No. 4 ring. Note similarity with Fig. 3.18.2 (Magnification 37.5×, etched).

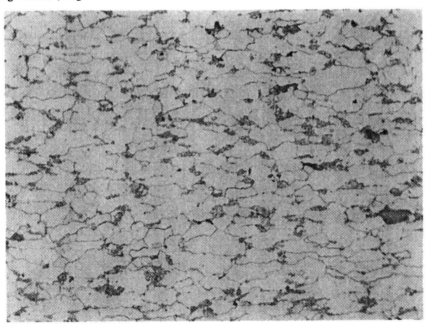

FIGURE 3.18.7. Midwall microstructure is ferrite and bainite, indicating that No. 4 ring failure was not the first; and metal temperature was >1350°F at final rupture (Magnification 500×, etched).

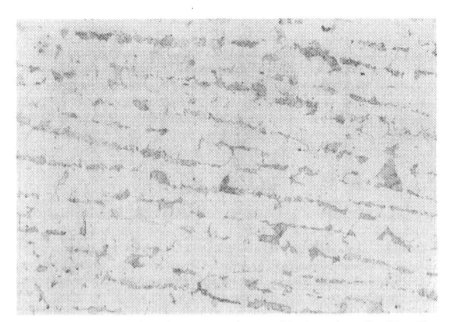

FIGURE 3.18.8. No. 4 ring microstructure is normal ferrite and pearlite (Magnification 100×, etched).

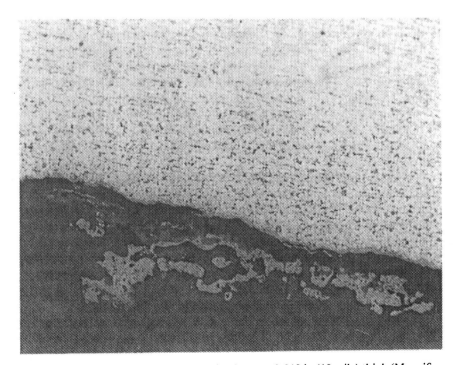

FIGURE 3.18.9. ID scale from No. 1 ring is up to 0.010 in (10 mils) thick (Magnification 100×, etched).

transfer. The net effect was to raise the tube-metal temperature into the creep range. The first failure occurred lower in the furnace, which reduced fluid flow in the tube, and other blisters failed as tube-metal temperatures rose; see, for example, Fig. 3.18.7. Chemical cleaning on a regular basis can prevent excessive scale formation and waterwall-tube creep failures.

Case History 3.19

Steam-Cooled Tube Failures: Wall Thickness and Alloy Changes

A common location for creep failures in steam-cooled circuits of a superheater or a reheater is at the change from one alloy to another, or where the wall thickness of a given alloy changes. The design conditions of tube-metal temperature changes increase from inlet to outlet; see Chapter 2. Material changes are made based on oxidation limits; see Table VI. Wall thickness changes are made based on ASME Code-allowable stresses at the design temperature; see Table VII. When the location of the transition is at too high a metal temperature, creep failures occur sooner than expected.

This example deals with two failures: one of a superheater tube at a change in wall thickness, the other with a superheater-tube failure at the change of alloy. Figure 3.19.1 shows the failure of a superheater tube (note the weld at the left). Ring samples were cut from the unfailed tube section to the left of the weld and from the end of the failed tube section toward the right of Fig. 3.19.1. These ring data are given in Table 3.19.1. Note that the failure has a wall thickness that varies from 0.215 to 0.240 in, while the nonfailed tube variation is 0.255 to 0.260 in. Chromium and molybdenum analyses show the failed tube to be T-22 as required; see also Table 3.19.1.

The microstructure of the failure lip is shown in Fig. 3.19.2. There is considerable cracking and creep damage within the microstructure. The microstructure of the fracture edge is presented in Fig. 3.19.3 and shows ferrite and spheroidized carbides and the expected creep voids. The microstructure of the nonfailed tube, in line with the failure, is presented in Fig. 3.19.4. In this microstructure, the remnants of the pearlite colonies are still visible. There is no evidence of any creep damage in the nonfailed tube sample; in fact, the ID scale in line with the failure, Fig. 3.19.5, shows no evidence of any creep expansion of the thicker tube.

The wall thicknesses given in Table 3.19.1 show some wastage on the failed tube, from 0.240 to 0.215 in; the unfailed tube has a 0.260-in wall. Nevertheless, at the time of initial operation, the differences in wall thickness, about 0.260 in versus 0.240 in, translate into a 10% higher hoop stress in the thinner tube. The higher stress was sufficient to cause a creep

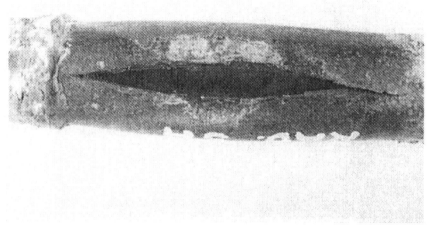

FIGURE 3.19.1. As-received tube sample from a radiant superheater; note weld at left. Steam flow is from right to left (Magnification 0.6×).

failure well before the thicker tube showed any outward signs of creep damage.

Figure 3.19.6 shows the failure of a superheater tube. Again, there is a weld in the vicinity, and the transition is from T-11 to T-22. The microstructure through the fracture edge, Fig. 3.19.7, shows extensive creep

TABLE 3.19.1. Dimensional Measurements

Tube	Position	OD, in	ID, in	Wall, in	%Cr	%Mo
Failed	12:00	1.775	1.330	0.215	2.47	1.00
	2:00	1.795	1.315	0.235		
	4:00	1.785	1.320	0.225		
	6:00			0.225		
	8:00			0.240		
	10:00			0.230		
Nonfailed	12:00	1.750	1.230	0.260	—	—
	2:00	1.750	1.225	0.260		
	4:00	1.750	1.230	0.260		
	6:00			0.255		
	8:00			0.260		
	10:00			0.260		

FIGURE 3.19.2. Cross section through fracture lip; wall thickness is about 0.175 in (Magnification 18¾×, etched).

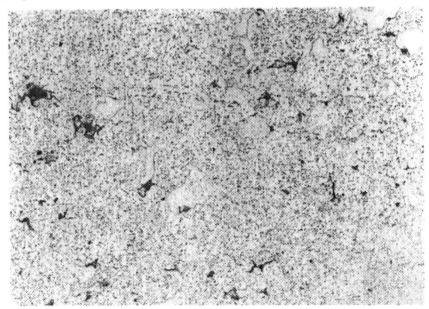

FIGURE 3.19.3. Microstructure contains spheroidized carbides and creep voids. Figures 3.19.2 and 3.19.3 confirm the failure as a creep or stress-rupture type (Magnification 500×, etched).

FIGURE 3.19.4. Midwall microstructure of the unfailed tube in line with the rupture is ferrite and spheroidized carbides, with the remnants of the pearlite colonies still vaguely discernible. Figures 3.19.3 and 3.19.4 indicate higher stresses will promote more rapid spheroidization (Magnification 500×, etched).

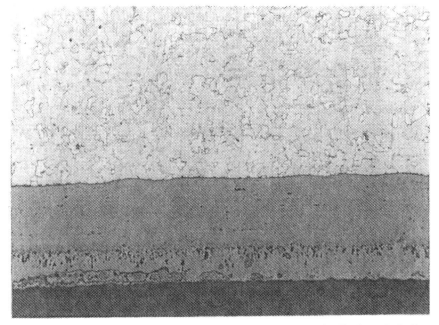

FIGURE 3.19.5. ID scale of the nonfailed tube shows no longitudinal crack, indicating no creep expansion (Magnification 100×, etched).

FIGURE 3.19.6. As-received tube sample from a secondary superheater; note weld. Steam flow is from left to right (Magnification 0.2×).

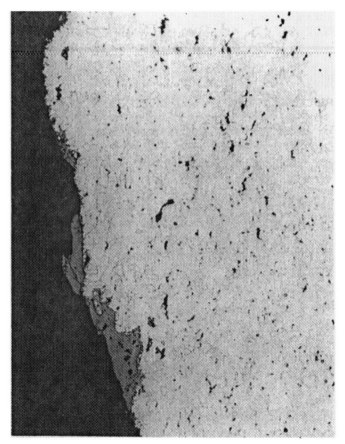

FIGURE 3.19.7. Fracture lip contains extensive creep damage (Magnification 100×, etched).

along the failure edge; and, again, the fracture was caused by a creep or stress-rupture mechanism.

These two failures occurred at the transition from thin- to thick-walled material of the same composition (T-22) or at the transition from one material (T-11) to another (T-22). The transitions were made at tube-metal temperatures that were excessive for the wall thickness or material selected. The transition should have been made earlier within the superheater to ensure proper metal temperature for the thickness and material selected.

Case History 3.20

Thermal-Fatigue–Corrosion-Fatigue Examples

The locations of potential thermal-fatigue and corrosion-fatigue cracking within the boiler are numerous. Sharp notches or stress raisers, alternating temperature cycles, or differential thermal expansion that can lead to changes in the local stress are the most likely locations. What follows are several examples.

Figure 3.20.1 shows the tube samples removed from the throat of a cyclone burner. The tubes are refractory covered and have separate studs

FIGURE 3.20.1. As-received, refractory-covered, cyclone burner tubes with the refractory removed and studs visible (Magnification 0.4×).

222   CHAPTER 3 METALLURGICAL PRINCIPLES: FERRITIC STEELS

to hold the refractory in place. Thermal-fatigue cracks initiate at the stud-to-tube interface; see Fig. 3.20.2. The cause of these kinds of failures is the loss of refractory. The ends of the studs are heated by the molten coal ash. A temperature cycle is set up as the studs are alternately covered with frozen ash and in contact with molten ash.

Waterwall tubes can be subjected to a variable stress along the membrane as the boiler expands and contracts. Figure 3.20.3 presents a photographic print of a radiograph; the white line is a crack within the waterwall tube. Figure 3.20.4 shows the metallographic mount of this defect, and there is a water-side crack on the fire side that is larger than the crack on the cold or casing side. These two cracks are presented in Figs. 3.20.5 and 3.20.6.

The next example shows both thermal-fatigue cracks along the OD surface at the toe of a fillet weld and ID corrosion-fatigue cracks opposite. Figure 3.20.7 shows the ID surface, and Fig. 3.20.8 shows the cross section through the tube, note that there are cracks on both the OD and ID. Again, these are waterwall tubes, but this time at buck-stay attachments. The relative motion of the waterwall tube and the buckstay leads

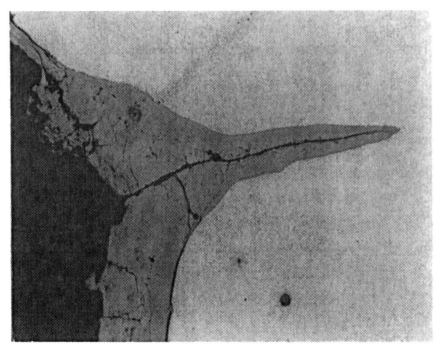

FIGURE 3.20.2. Thermal-fatigue crack has formed at the weld line between the tube and stud (Magnification 37.5×, etched).

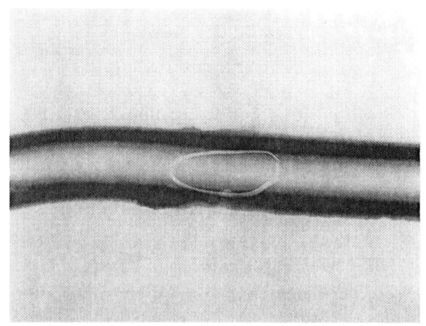

FIGURE 3.20.3. Photographic print of a radiographic film shows a crack (white line within the circle) in a waterwall tube (Magnification 1×).

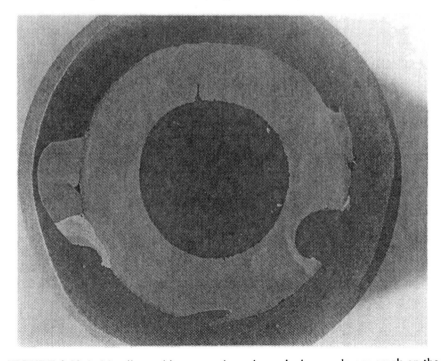

FIGURE 3.20.4. Metallographic mount through crack shows a larger crack on the ID of the fire side than on the ID of the cold side (Magnification 3.1×).

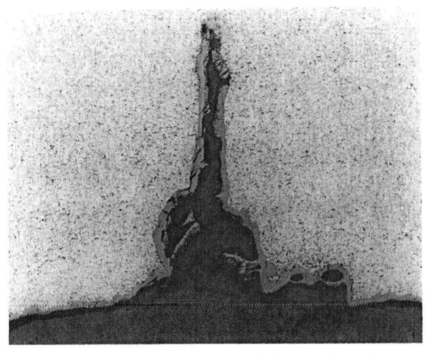

FIGURE 3.20.5. Fire-side ID crack; crack depth is about 0.060 in (60 mils) (Magnification 50×, etched).

FIGURE 3.20.6. Cold- or casing-side ID crack; crack depth is about 0.015 in (15 mils) (Magnification 50×, etched).

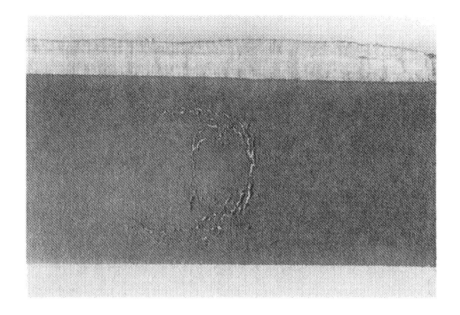

FIGURE 3.20.7. ID surface of a waterwall tube at a buckstay attachment. The weld outline is quite clear (Magnification 0.8×).

FIGURE 3.20.8. Cross section through the waterwall tube at the buckstay weld shows the thermal-fatigue cracks at the OD (toe of weld) and corrosion-fatigue cracks at ID of the tube (Magnification 6.4×, etched).

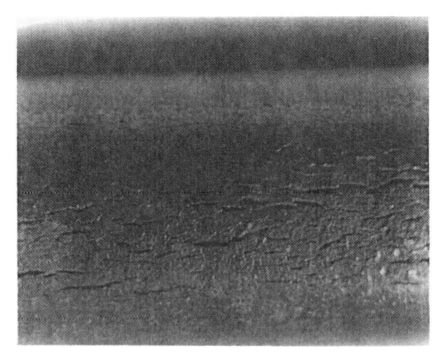

FIGURE 3.20.9. Closeup of crack pattern along the OD surface of a superheater tube in the sootblower travel path (Magnification 2×).

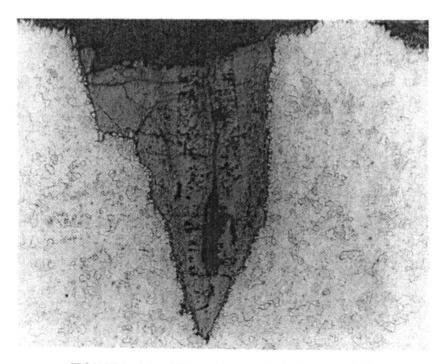

FIGURE 3.20.10. OD crack (Magnification 100×, etched).

to the formation of thermal-fatigue cracks from the OD surface and corrosion-fatigue cracks from the ID surface.

Sootblower operation can cause thermal-fatigue cracks. At the start of the sootblowing cycle, any condensate in the line strikes the tube before steam flow through the sootblower is established. The condensate rapidly cools the tube. Figure 3.20.9 shows a closeup of the crack pattern along the OD surface. Cross sections through the OD and ID show the thermal-fatigue cracks that grow from the OD surface, Fig. 3.20.10, and the ID surface, Fig. 3.20.11.

One final example deals with a waterwall flow instability. Under some operating conditions, inadequate and variable fluid flow produced a hot–cold, heating and cooling of the tube. The failure, Fig. 3.20.12, is a circumferential crack, and the ID surface contained several circumferential cracks as well, Fig. 3.20.13. The midwall microstructure is graphitized; see Fig. 3.20.14.

While this is not an exhaustive tabulation of corrosion fatigue and thermal fatigue, it presents several examples of the kinds of damage that can occur. The common denominator to all is a variation in either temperature or stress coupled with a corrosive environment on the water side or on the OD.

FIGURE 3.20.11. ID crack (Magnification 100×, etched).

FIGURE 3.20.12. Circumferential through-wall crack from a waterwall tube with a fluid-flow instability (Magnification 1.3×).

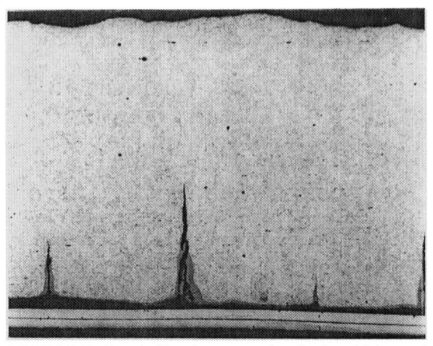

FIGURE 3.20.13. Along the ID surface are several smaller thermal-fatigue cracks (Magnification 25×, etched).

FIGURE 3.20.14. The microstructure is graphitized, indicating an operating temperature, for at least a part of the cycle, above 850°F (Magnification 500×, etched).

Case History 3.21

Erosion Damage

Erosion can occur along the ID surface, as a result of solid-particle erosion from exfoliated steam-side scale, and along the OD surface, due to high-velocity fly ash. At velocities above about 65–70 ft/sec, removal of material from tube surfaces begins.

Figure 3.21.1 shows the ID surface of a tube at a bend (note the localized wastage along the ID surface). Figure 3.21.2 shows some fly-ash erosion on a stainless-steel superheater tube.

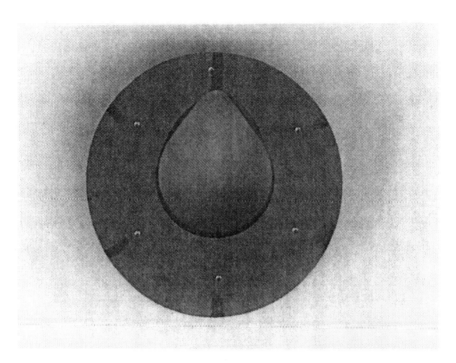

FIGURE 3.21.1. Cross section at a tube bend shows the ID wastage due to steam-side scale exfoliation (Magnification 2×).

FIGURE 3.21.2. Fly-ash erosion on a stainless-steel secondary-superheater tube (Magnification 2×).

Case History 3.22

Waterwall-Tube Failure of SA-178 A

This example discusses the failure of an SA-178 grade A waterwall tube from an oil-fired boiler. The tube is specified as 3.25 in OD × 0.180 in MWT. The unit suffered a low-water-level excursion, and the tube failed by a short-term overheating. Estimated metal temperature is above 1500°F. The failure was through the electric-resistance-weld fusion line.

Figure 3.22.1 shows the as-received tube sample. The tube has swollen from 3.25 in OD to 3.75 in OD 6 in away from the end of the failure. Figure 3.22.2 presents ring sections removed from the failure and the end of the tube sample. Note that the failure is not centered between the membranes, which is expected to be the hottest portion of a waterwall tube.

The fracture lip, Fig. 3.22.3, is drawn to a virtual knife-edge. Wall-thickness reduction is more than 95%. The microstructure ½ in away from the failure edge is ferrite and bainite. The estimated tube-metal temperature, based on the relative amounts of ferrite and bainite, is in excess of 1500°F; see Fig. 3.22.4. The microstructure 180° around the tube perimeter from the failure on the cold or casing side, Fig. 3.22.5, is ferrite and pearlite, the expected microstructure of SA-178 A.

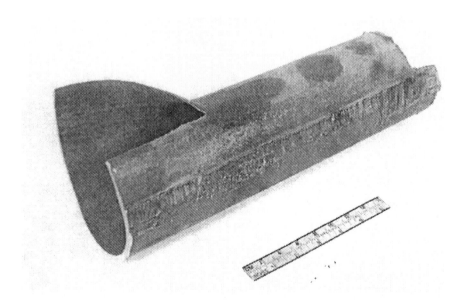

FIGURE 3.22.1. As-received waterwall tube (Magnification 0.25×).

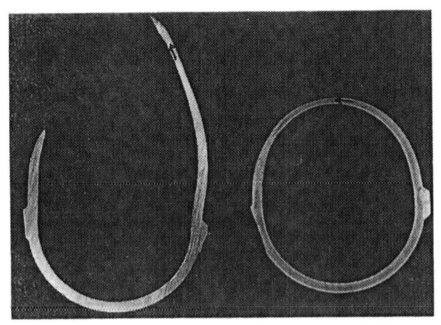

FIGURE 3.22.2. Ring sections through rupture and 6 in away show localized reduction at the fracture lip and swollen tube away from failure. Failure is not centered on waterwall membranes (Magnification 0.5×).

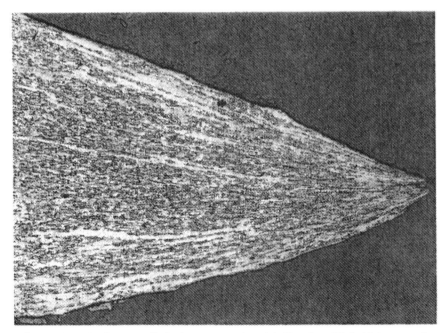

FIGURE 3.22.3. Fracture lip is drawn to a virtual knife-edge. Wall thickness reduction is more than 95% (Magnification 50×, etched).

FIGURE 3.22.4. Microstructure ½ in from failure edge is ferrite and bainite (Magnification 500×, etched).

FIGURE 3.22.5. Microstructure 180° around perimeter from rupture is normal ferrite and pearlite, the expected structure for SA-178 A (Magnification 500×, etched).

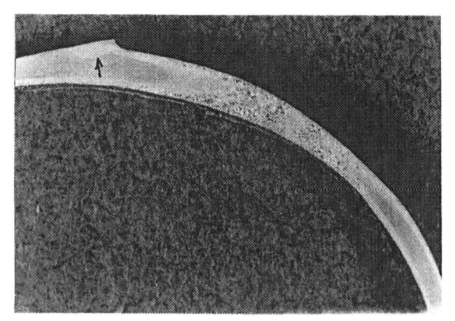

FIGURE 3.22.6. ERW line at the end of the tube sample in line with the rupture shows decarburized region and grain growth. Arrow points to membrane weld (Magnification 2×, etched).

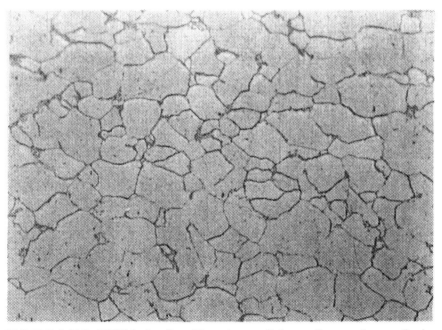

FIGURE 3.22.7. ERW fusion line. There is very little pearlite, nearly pure ferrite. The weld line is the weakest and failed even though the hottest portion of a waterwall tube is usually centered between the membranes (Magnification 500×, etched).

Figure 3.22.6 shows the ERW fusion line at the end of the tube sample in line with the rupture. The fusion line shows considerable decarburization and ferrite grain growth. The arrow points to the membrane weld. The microstructure of the ERW fusion line, Fig. 3.22.7, shows nearly pure ferrite; almost no pearlite is present. The fusion weld line is very low carbon and is weaker than the rest of the tube. Thus, the failure occurred through the ERW fusion line even though the hottest portion of a waterwall tube is usually centered between the membranes.

When SA-178 grade A tubes are made from strip material rolled from a single ingot, the decarburized surface of the ingot carries through the entire length, width, and thickness of the strip. Thus, the ERW is made between decarburized edges of the strip. The weld line shows the remnants of the decarburized surface layer of the ingot. With virtually no carbon present, this fusion line has a lower strength than the pearlitic structure in Fig. 3.22.5. Under normal operating conditions, this, relatively speaking, weak link on the perimeter of the tube does not usually cause any problem, because the strength of the ferrite is adequate. However, when the low-water upset occurred, the tube failed at the weakest point, the decarburized ERW fusion line.

Case History 3.23

Economizer-Tube Failure SA-178 C

An economizer tube failed with a wide-open burst but with very little obvious ductility to the fracture surface. The material is specified as 2.5 in OD × 0.220 in MWT SA-178 grade C, ERW tubing. The operating or drum pressure is 2100 psi, and the boiler had in excess of 150,000 hr of service at the time of the failure.

The failure was caused by a lack of fusion in the ERW. Only approximately 20% of the wall thickness at the failure was actually welded at the time of the tubing manufacture. The weld defect does not run the full length of the tube but is confined to the area of the failure.

Figures 3.23.1 and 3.23.2 show the fracture edge near the center of the failure opening and at the end. The fracture surface is quite smooth toward the middle of the rupture opening, Fig. 3.23.1, and is suggestive of a lack of fusion along the ERW. Wall-thickness measurements along the fracture edge vary from 0.225 to 0.235 in. All measurements are greater than the specified minimum. The Rockwell B hardness is 71 to 72, the expected hardness range for SA-178 C material.

The microstructure along the fracture edge near the ID of the tube, Fig. 3.23.3, is quite smooth and shows a thin decarburization layer. Note also the distortion of the banded structure at the fusion line. Compare this with

FIGURE 3.23.1. Fracture edge is smooth and suggests a lack of fusion of the ERW (Magnification 3.1×).

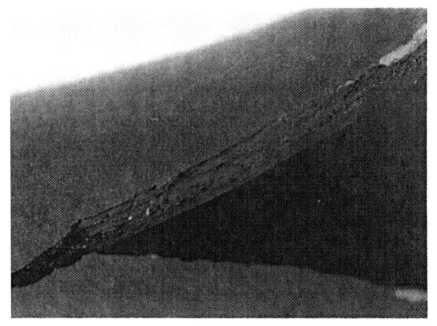

FIGURE 3.23.2. Fracture edge at end of the opening is rougher than in the preceding figure, suggesting more complete fusion. At the left, open end, the smooth region is visible and covers about ½ the wall thickness (Magnification 3.1×).

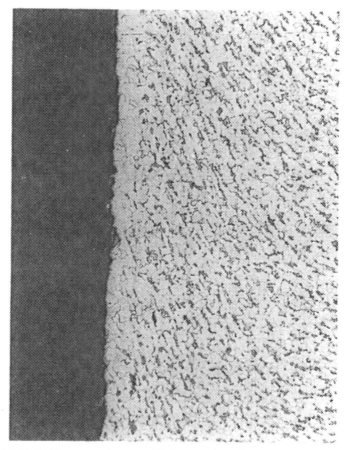

FIGURE 3.23.3. Fracture edge near ID is smooth and decarburized (Magnification 100×, etched).

Fig. 3.24 for a view of an intact ERW. The fracture surface, near the OD surface, Fig. 3.23.4, is rougher and shows no decarburization layer. The electric-resistance fusion line at the end of the sample, Fig. 3.23.5, shows complete fusion with a decarburized layer along the ID surface. The microstructure throughout the tube is ferrite and pearlite, Fig. 3.23.6, the normal structure for SA-178 grade C material.

The results of this examination may be summarized as follows:

1. The cause of the failure is an incomplete-fusion defect over a portion of the tube length. The evidence for this is the smooth fracture surface shown in Figs. 3.23.1 and 3.23.3. Also, the fracture surface in this area is decarburized, as is the ID surface, suggesting that the lack of fusion existed when the tube was given its final normalization heat treatment.

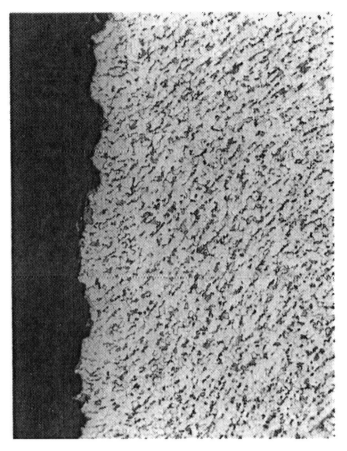

FIGURE 3.23.4. Fracture surface near OD is rough and shows no decarburized layer (Magnification 100×, etched).

2. The defect is limited to the length of the sample, because the weld line at the end of the sample shows complete fusion.

The hoop stress in a pressurized cylinder is given in Eq. (3.23.1) and may be rearranged to solve for the wall thickness, as given in Eq. (3.23.1):

$$S = \frac{P(D - W)}{2W} \qquad (3.23.1)$$

$$W = \frac{PD}{2S + P} \qquad (3.23.2)$$

where S = hoop stress, psi
 P = operating pressure, 2100 psi
 D = tube diameter, 2.5 in
 W = wall thickness, in

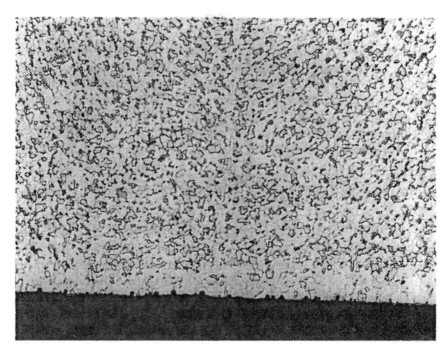

FIGURE 3.23.5. ERW fusion line at end of sample shows complete fusion (Magnification 100×, etched).

FIGURE 3.23.6. Microstructure is normal ferrite and pearlite, the expected structure for SA-178 C material (Magnification 500×, etched).

The metal temperature in an economizer is well below the creep range. The microstructure shown in Fig. 3.23.6 indicates no elevated-temperature exposure. Thus, failure occurs when the hoop stress is equal to the tensile strength. For SA-178 grade C, the strength is about 60,000 psi at 600°F; see Table VIII. Equation (3.23.2) can be solved for W with the hoop stress equal to 60,000 psi and P and D equal to 2100 psi and 2.5 in, respectively. Equation (3.23.3) shows the result:

$$W = \frac{(2100)(2.5)}{2(60,000) + 2100} = 0.043 \text{ in} \quad (3.23.3)$$

The wall thickness at failure is estimated to be 0.043 in. The wall thickness, with complete fusion, needs to be only 0.043 in or about 20% of the specified minimum wall to ensure adequate strength to contain the operating pressure. At a 1½ design-pressure hydrostatic test, the applied pressure is 3410 psi, and the room-temperature tensile strength is 64,000 psi. Using these values, in Eq. (3.23.2) gives a wall thickness of 0.065 in (30% of MWT) needed to contain the test pressure. Failure occurred when normal oxidation or corrosion wastage reduced the wall thickness to 0.043 in or so. Thus, it is not surprising that this tube had a service life of 150,000 hr, with a substantial lack-of-fusion weld defect, before failure.

This also suggests that the hydrostatic test is a poor method of finding defects in ERW tubing. Even with a hydrostatic-test pressure of 1½ times design, the amount of wall thickness required is still only a small fraction, 30%, of the specified minimum. It is far better to examine ERW tubes with ultrasonic or eddy-current or flux-leakage tests as a more reliable method for identifying lack-of-fusion defects.

REFERENCES

There are several textbooks that cover the subject of elementary physical metallurgy from which the section of metallurgical principles is taken.

1. J. Wulff, H. F. Taylor, and A. J. Shaler, *Metallurgy for Engineers*, Wiley, New York, 1952.
2. L. H. Van Vlack, *Elements of Material Science*, 2nd ed., Addison-Wesley, Reading, Massachusetts, 1964.
3. J. Wulff, ed., *Structure and Properties of Materials*, 4 Vols., Wiley, New York, 1964.
4. A. G. Guy, *Elements of Physical Metallurgy*, Addison-Wesley, Reading, Massachusetts, 1958.
5. *ASM Metals Handbook*, 8th ed., Vol. 8, "Metallography, Structures, and Phase Diagrams," ASM International, Metals Park, Ohio, 1973.
6. U. S. Steel Corp., *Atlas of Isothermal Transformation Diagrams*, 1951.

7. W. F. Simmons and H. C. Cross, *Report on the Elevated-Temperature Properties of Chromium-Molybdenum Steels,* Special Technical Publications No. 151, ASTM, Philadelphia, PA, 1953.
8. F. R. Larson and J. Miller, "A Time-Temperature Relationship for Rupture and Creep Stresses," *Trans ASME,* July 1952, pp. 765–775.
9. F. Cjarofalo, *Fundamentals of Creep and Creep-Rupture in Metals,* Macmillan, New York, 1965.
10. A. Grogli, "Creep Fractures on Tubes from Steam Generating Plants," *ASM Source Book in Failure Analysis,* ASM International, Metals Park, Ohio.
11. *ASM Metals Handbook,* 9th ed., Vol. 11, "Failure Analysis and Prevention," ASM International, Metals Park, Ohio, 1986, p. 613.
12. M. A. Grossman and E. C. Bain, *Principles of Heat Treatment,* ASM International, Metals Park, Ohio, 1968.
13. *ASM Metals Handbook,* 8th ed., Vol. 7, "Atlas of Microstructures of Industrial Alloys," ASM International, Metals Park, Ohio, 1973.
14. R. M. Brick, R. B. Gordon, and A. Phillips, *Structure and Properties of Alloys,* McGraw Hill, New York, 1965.
15. I. M. Rehn, W. R. Apblett, Jr., and J. Stringer, "Controlling Steamside Oxide Exfoliation in Utility Boiler Superheaters and Reheaters," *Materials Performance,* June 1981, pp. 27–31.

General References Useful in the Study of Failures

H. Thilsch, *Defects and Failures in Pressure Vessels and Piping,* Von Nostrand Reinhold, New York, 1965.

Case Histories in Failure Analysis, ASM International, Metals Park, Ohio, 1979.

J. L. MacCall and P. M. French, *Metallography in Failure Analysis,* Plenum, New York, 1978.

Steam/Its Generation and Use, 38th ed., Babcock and Wilcox, 1972.

L. E. Samuels, *Optical Microscopy of Carbon Steels,* ASM International, Metals Park, Ohio, 1980.

ASM Metals Handbook, 9th ed., Vol. 11: "Failure Analysis and Prevention," ASM International, Metals Park, Ohio, 1986.

F. R. Mutchings and P. M. Unterweiser, eds., *Failure Analysis: The British Engine Technical Report,* ASM International, Metals Park, Ohio, 1981.

R. D. Port and H. M. Herro, *The NALCO Guide to Boiler Failure Analysis,* Nalco Chemical Co., McGraw-Hill, New York, 1991.

Manual for Investigation and Correction of Boiler Tube Failures, EPRI, EPRI CS-3945 Project 1890-1. Final Report, April 1985.

CHAPTER FOUR

STAINLESS STEELS

Stainless steels are alloys of iron that contain at least 11% chromium. In normal environments the addition of 11% chromium to iron prevents the formation of rust and, hence, gives the popular name stainless to these alloys. Figure 4.1[1] shows the iron–chromium phase diagram. The addition of chromium restricts the austenite region of the phase diagram so that at compositions of more than 13% chromium the alloys are entirely ferritic.

As a class of materials, the ferritic stainless steels, are weldable, formable, and have good corrosion resistance. For the most part, they find applications in duct work, baffle plates, and other areas where improved corrosion resistance is desirable, but where the higher price of austenitic stainless steel is not economical. The ASME Code recognizes the use of ferritic-stainless-steel boiler tubing, but its use is limited to temperatures below 700°F (370°C).[2]

The ferritic stainless steels have a body-centered-cubic crystal structure similar to the room-temperature form of iron. They are magnetic at room temperature and up to about 1400°F (760°C). These alloys contain from 11 to 27% chromium, and the low-chromium grades are among the least expensive stainless steels.

AUSTENITIC STAINLESS STEELS[1,2]

Austenitic stainless steels contain sufficient chromium, to give them good corrosion resistance, and nickel, to retain the face-centered-cubic crystal structure of high-temperature iron down to room temperature. Austenitic stainless steels are nonmagnetic. The most common austenitic alloys are the

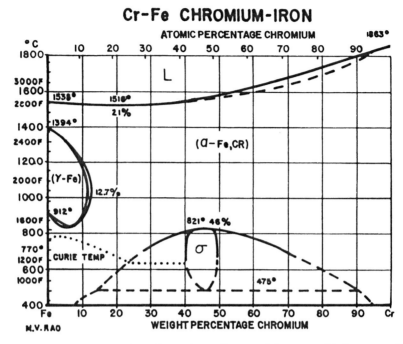

FIGURE 4.1. The iron–chromium phase diagram. The essential features of this alloy system are: First, the restriction of the austenite region by the addition of chromium, so at about 13% chromium the alloys of iron and chromium are ferritic; second, the formation of sigma-phase at temperatures below about 1550°F (845°C). From Rao, M. V., Cr-Fe diagram, Hawkins, Donald T. and Hultgren, Ralph, "Constitution of Binary Alloys," *Metals Handbook,* 8th ed., Vol. 8, Editor, Lyman, Taylor, American Society for Metals, 1973, page 291.

familiar 18 chromium–8 nickel, of which Types 304, 321, and 347 are the stainless-steel alloys frequently used in boiler tubing. Both chromium and nickel contents may be increased to improve corrosion resistance, for example, in the Type 309, 310, and 316 alloys. Additional elements, usually molybdenum, may be added to further enhance corrosion resistance. Table IX of the appendix gives the composition of some ferritic and austenitic stainless steels.

WORK HARDENING

If an alloy is deformed plastically beyond its yield point by any suitable metalworking operation, such as hammering, bending, or rolling, the material becomes harder and stronger. The microstructure of cold-worked stainless steel will reflect this deformation by revealing elongated grains, slip lines

FIGURE 4.2. Slip bands in Type 304 stainless steel are visible in the top center of the figure (Magnification 500×, electrolytic etch).

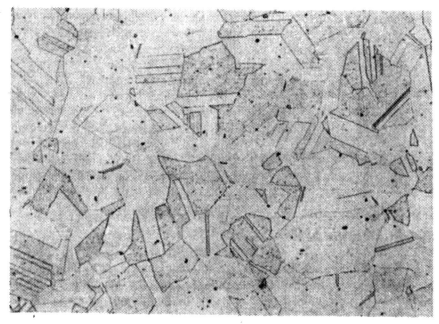

FIGURE 4.3. Twin boundaries in a 321H austenitic stainless steel (Magnification 500×, electrolytic etch).

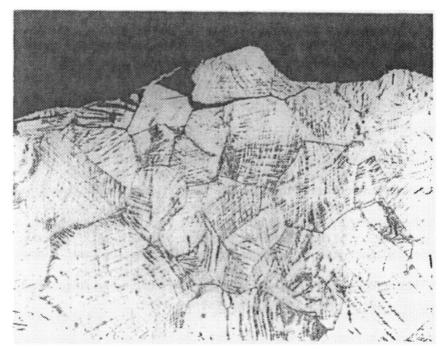

FIGURE 4.4. Cold-worked microstructure of 304H taken from a swage (Magnification 200×, etched).

within the grains, and twinning. The deformation of individual crystals occurs by the relative movement between planes of atoms. Figure 4.2 shows the slip lines in a 304 stainless steel. Deformation by twinning also involves the movement of planes of atoms. The reorientation of a plane of atoms so that it bears a mirror-image relationship to the adjacent layer of parent crystal is called twinning. Usually, twin boundaries appear as straight lines across a crystal. Figure 4.3 shows such twin boundaries in austenitic stainless steel.

The more severe the cold work is, the greater is the amount of deformation evident in the microstructure. Figure 4.4 presents the microstructure of 304H stainless steel, cold-worked about 20%, with a Rockwell C hardness in the 25–30 range. Fully annealed 304H has a Rockwell B hardness in the 75–85 range. For comparison, Rockwell B 100 is about equal to Rockwell C 20.

HEAT TREATMENT

The effects of cold work may be removed by suitable heat treatment. Stress relieving removes residual stresses remaining in the material as the result of forming or welding processes, but may not lead to appreciable softening. To stress relieve the austenitic stainless steels, a minimum temperature of about

1300°F (700°C) is required. One hour at 1300°F (700°C) is sufficient to reduce residual stresses by 50% or more.[3] However, this temperature is not recommended because of the formation of grain-boundary carbides, which will lead to sensitization—a phenomenon discussed later in the chapter.

SOLUTION ANNEAL

All effects of cold work and any precipitated carbides may be removed by a solution anneal at temperatures of 1900–2000°F (1040–1090°C). The ASME Code requires the heat treatment of all austenitic alloys. Table X of the appendix shows the ASME Code heat-treatment requirements for austenitic stainless steels covered under ASME Spec. SA-213.[2]

Creep strength is enhanced by a large grain size, since creep deformation usually occurs by the grain-boundary sliding mechanism. Fine-grain-size materials have a greater grain-boundary surface area and, thus, are a weaker material at elevated temperature. Remembering that the Code requires an ASTM grain size of 7 or coarser, a benefit of the high-temperature solution anneal is that the grain size of stainless steels is increased.

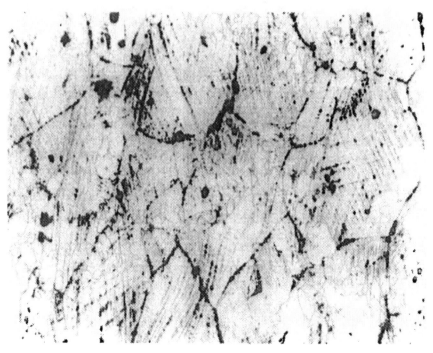

FIGURE 4.5. Initial stage of recrystallization of 304H stainless steel. The microstructure shows a few small recrystallized austenite grains in a sea of cold work (Magnification 500×, etched).

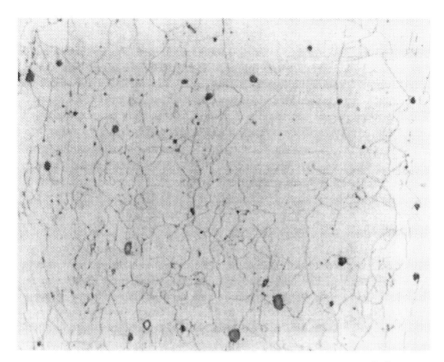

FIGURE 4.6. Later stage of recrystallization of 304H stainless steel shows some grain growth and the relaxation of the cold-worked structure (Magnification 500×, etched).

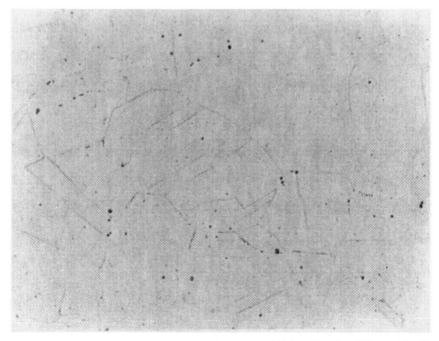

FIGURE 4.7. Fully annealed 304H stainless steel (Magnification 200×, etched).

248 CHAPTER 4 STAINLESS STEELS

Solution annealing starts at about 1600°F with the stress relief of cold-worked structures. With a long enough wait, a fully recrystallized microstructure could be expected at this temperature, but for the sake of convenience a 2000°F ± temperature gives results in a few minutes. The changes in microstructure are shown in Figs. 4.5 and 4.6. The cold-work microstructure begins to relax, and recrystallized austenite grains begin to form, Fig. 4.5. At longer times, the twinned structure is completely removed, and the small recrystallized grains begin to grow, Fig. 4.6. When the grain growth is complete, the structure is as shown in Fig. 4.7.

SIGMA-PHASE

Figure 4.1 shows the phase diagram for iron–chromium alloys. At temperatures below 1500°F (815°C) for a 50% chromium–50% iron alloy, a hard, brittle, intermetallic compound, sigma, forms. Sigma forms slowly, first developing at grain boundaries, and is readily identifiable by microscopy and x-ray diffraction. Its formation is favored by high chromium contents. The presence of sigma-phase increases hardness slightly and decreases ductility, but not by enough to be of serious consequence in most boiler applications. Sigma formation is possible between about 1050°F (565°C) and 1700°F (925°C) in austenitic stainless steels containing more than 16% chromium

FIGURE 4.8. Sigma-phase in 304 stainless steel (Magnification 500×, electrolytic etch).

and less than 32% nickel. Figure 4.8 shows sigma-phase in 304 stainless steel.

The effect of sigma-phase on mechanical properties is given in Table XIV of the appendix. If anything, the strength increases, but the most dramatic effect is on impact strength. Since boiler components usually do not fail by impact, the effect of sigma-phase on the performance of austenitic stainless steels in boiler applications is minimal.

SENSITIZATION

When austenitic stainless steels are heated between 800°F (425°C) and 1500°F (815°C), sensitization occurs; and intergranular attack occurs in corrosive environments. The most critical temperature is about 1200°F (650°C), since the rate of chromium-carbide formation is fastest in this temperature range. Unfortunately, this is the temperature range at which most high-temperature superheaters and reheaters operate.

It seems reasonable to suppose that sensitization occurs through chromium depletion in a region immediately adjacent to grain boundaries. Both chromium and carbon are distributed throughout the austenitic structure and can combine to form chromium carbide. The temperature at which this occurs most rapidly is close to 1200°F (650°C). At lower temperatures, diffusion rates of the atoms are too slow to cause the formation of chromium carbide, and at higher temperatures, decomposition of chromium carbide occurs. The precipitation of the carbide occurs most readily at grain boundaries, because the diffusion rates of carbon and chromium are greatest along grain boundaries.

The formation of chromium carbide, $Cr_{23}C_6$, occurs by the combination of 23 atoms of chromium and 6 of carbon. Carbon, a small atom, diffuses rapidly through the crystal structure, while chromium, a much bigger atom, diffuses much more slowly. The carbon atoms can migrate to the grain boundary from all parts of the crystal more easily than the chromium can; therefore, chromium is depleted from more localized regions near the grain boundary, forming an envelope of chromium-depleted material. This region, therefore, is susceptible to corrosion. As we have noted, stainless steels lose their excellent corrosion resistance below 11% chromium. The depletion of chromium by the formation of chromium carbide forms a thin layer adjacent to the grain boundaries of material containing less than 11% chromium, and, therefore, a sensitized alloy is not as corrosion resistant.

The above discussion explains the mechanism of sensitization for both the austenitic and ferritic stainless steels. Sensitization can be reduced or avoided in one of two ways: either by maintaining a low-carbon level in the alloy or by the addition of a strong carbide former. If the carbon level is maintained below 0.035% maximum, there is limited carbon present to form the undesirable chromium carbides. These grades are referred to as the "L" grades, such as Type 304L. In the second way, a strong carbide former may be added to the basic stainless-steel composition to combine with the carbon that is present. The composition of Type 321 is similar to Type 304, but contains titanium as the carbide former; in Type 347, the carbide former is columbium. The amount of titanium required is usually 5 times the carbon content, while the amount of columbium is 10 times the carbon content. In addition, note that high-temperature properties of the stabilized grades show improved resistance to creep and stress rupture over the unstabilized Type 304.

In order to take advantage of the carbide formers added to Types 321 and 347 stainless steel, a stabilization heat treatment is necessary. At temperatures of about 1650°F (900°C), titanium and columbium carbides form. This temperature is too high for the formation of the undesirable chromium carbides. The titanium and columbium carbides are more stable than chromium carbide; thus, the carbon is essentially removed from the alloy, preventing the formation of chromium carbide at grain boundaries.

Both 347 and 321 develop sensitized structures at superheater or reheater metal temperatures if not given a stabilization heat treatment. Figure 4.9 presents a sensitized 321H stainless-steel superheater tube. Intergranular oxidation can occur on sensitized structures; see Fig. 4.10. The grain bound-

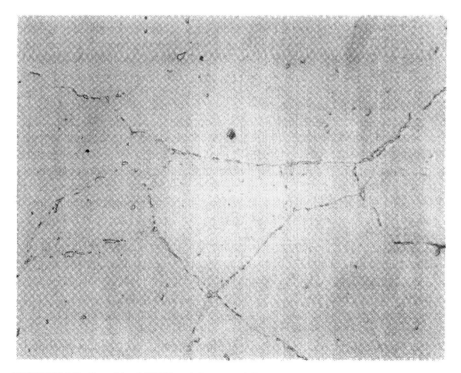

FIGURE 4.9. Sensitized 321H stainless steel. Note that carbide particles form preferentially at grain boundaries (Magnification 1000×, etched).

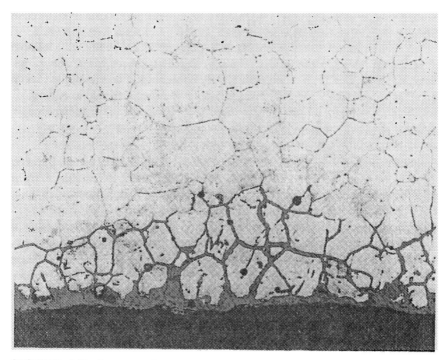

FIGURE 4.10. Intergranular corrosion in sensitized 304H (Magnification 100×, etched).

aries are not as oxidation-resistant as the austenite grains, and oxidation (or corrosion) penetrates the grain boundaries more rapidly.

WELDING

The austenitic grades of stainless steel are the most easily welded. There are two problems in welding these alloys, and, for the most part, they are considered minor. The first relates to the precipitation of grain-boundary carbides in the heat-affected zone adjacent to the welds. Carbide precipitation can be minimized by using titanium- or columbium-stabilized grades in the stabilization-annealed condition. Following welding, a high-temperature solution anneal can be given the component to redissolve any grain-boundary-precipitated carbide. The other problem relates to the high-temperature strength of the weld metal itself. An all-austenitic weld deposit has a fine-grain structure, which has poorer creep properties than a coarse-grain structure, as previously noted. To counteract this effect, it is desirable to have a small amount of delta-ferrite present in the weld. Typically, 4% or 5% ferrite is an effective means of offsetting this grain-boundary weakness. However, an excessive amount of ferrite leads to a lowering of the high-temperature strength and to poorer corrosion resistance. A second danger of excessive ferrite is the promotion of sigma-phase formation at high temperatures.

Postweld heat tratment is not usually performed on austenitic stainless alloys except to relieve residual stresses. Because sensitization occurs between 800 and 1500°F (425 and 815°C), the stress-relief temperature should be higher than 1500°F (815°C), except in the stabilized annealed condition. For Types 321 and 347 alloys that have been given a stabilization anneal, stress relief can be achieved in the 1300–1400°F (700–760°C) range without fear of detrimental chromium-carbide precipitation at grain boundaries.

The welding of austenitic stainless steels to ferritic alloys is an entirely different matter. The thermal coefficients of expansion are quite different and lead to high thermal stresses during operation. Other factors contribute to the premature failure of these dissimilar welds: carbon diffusion and externally applied stress, for example. The ferritic alloys typically have a higher carbon content than does the weld metal, whether the weld metal is an austenitic stainless steel or a high-nickel alloy. As the carbon diffuses from the ferritic alloy into the weld metal, the creep strength in a narrow zone adjacent to the weld interface is reduced by the carbon loss. The high thermal stress resulting from the differences of coefficients of expansion and the reduced creep strength of the ferrite are probably the prime cause of the dissimilar weld-material failures. Externally applied loads will also have an effect, for example, a bending stress in a horizontal superheater. Whether horizontal superheaters and reheaters are more prone to this type of failure than pendant types is not known.

HIGHER-ALLOY AUSTENITIC MATERIALS

As service conditions have become more severe, the search for more effective corrosion-resistant high-temperature alloys has extended to the high-nickel–chrome alloys. Since these alloys contain more than 50% nickel and chromium, they are not technically steels, but since they have the face-centered-cubic crystallographic structure of austenite, and have many of the same qualities as austenitic stainless steels, these alloys are included in this section.

Table XI of the appendix lists the composition of these higher alloys. Figure 4.11 shows the microstructure of Inconel 601®, and Figure 4.12 shows the microstructure of Inconel 625®. These structures are typical of all of these higher alloys. The structure is single phase, with no carbide or nitride precipitates visible. However, on occasion, precipitates of titanium nitride are seen. Figure 4.13 shows Inconel 601 microstructure after prolonged exposure to high temperature as a superheater-tube shield. The estimated exposure temperature is 1400 or 1500°F (760 or 815°C) for 10 months. The structure shows discrete particles of precipitate and nearly complete grain-boundary coverage of the same precipitate, presumably titanium nitride or a complex carbide.

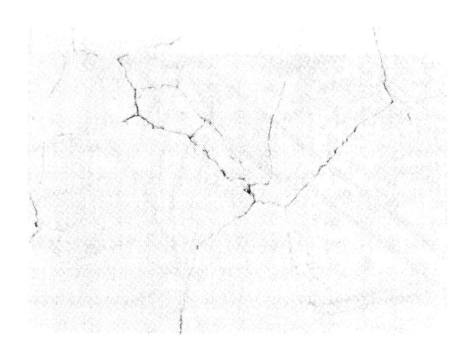

FIGURE 4.11. The microstructure of Inconel 601® (Magnification 500×, electrolytic etch).

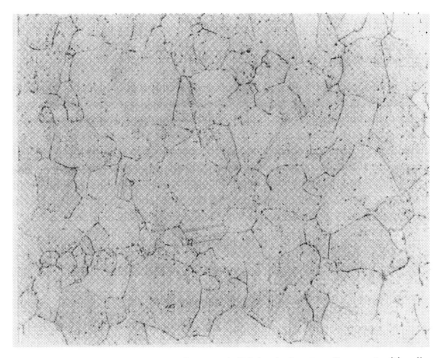

FIGURE 4.12. Microstructure of Inconel 625 is similar to other austenitic alloys (Magnification 500×, etched).

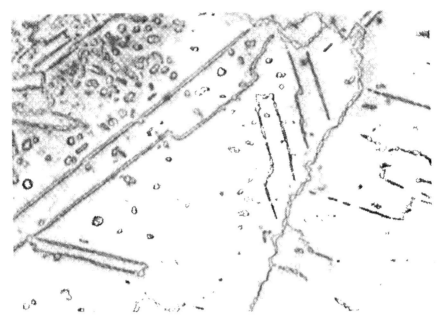

FIGURE 4.13. Inconel 601 after prolonged exposure to temperatures around 500°F (820°C). The structure shows discrete particles of a titanium-nitride precipitate and nearly complete grain-boundary coverage of the same precipitate (Magnification 500×, electrolytic etch).

CASE HISTORIES

For the next several pages, examples of failures of austenitic alloys are presented. The prediction of peak temperatures in stainless steel is more difficult than for ferritic steels, since there is no phase transformation similar to the ferrite to austenite change in low-alloy and plain-carbon steels. Grain growth occurs over a range of temperatures starting at about 1700°F (925°C). Sigmas and carbide phases also form over a range of temperatures from about 1050°F (565°C) to 1700°F (925°C), and the amount and rate of formation are not predictable. In spite of these difficulties, it is possible to determine useful information from service failures of the stainless steels.

Case History 4.1

High-Temperature Failure of 321H Stainless Steel

BOILER STATISTICS

Size	2,700,000 lb steam/hr (1,225,000 kg/hr)
Steam temperature	1005°F/1005°F (540°C/540°C)
Steam pressure	2525 psig (175 kg/cm²)
Fuel	Natural gas

The failure occurred in a pendant-type high-temperature superheater in the straight portion just above the bend near the bottom of the loop above the furnace nose-arch. The tube material is SA-213 321H stainless steel, having a 1¾ in OD × 0.188-in-thick wall. Two other tubes from the same location in adjacent tube bundles were also received for metallographic examination, but these tubes did not contain any failures. The visual observation of the failure area showed the following:

A. The failure was an open burst, ductile in appearance, with the rupture edges drawn to a near-knife-edge condition. The violence of the rupture folded the tube back on itself, as can be seen in Fig. 4.1.1.
B. There was no measurable swelling of the tube circumference away from the failure area.
C. There was no significant wall thinning away from the failure either.
D. At the point of failure, there was a small ¼-in tube-guide rod welded to the tube, but was not connected with the failure.
E. The failure was on the front, or hotter side, of the tube.

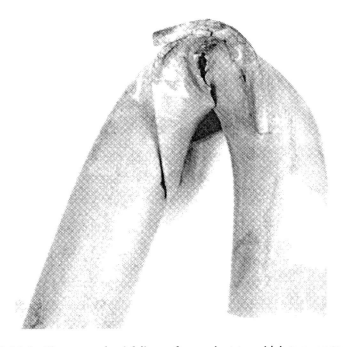

FIGURE 4.1.1. The as-received failure of a pendant-type high-temperature superheater tube. Material is SA-213 321H stainless steel with a 1¾ in OD × 0.188-in-thick wall.

256 CHAPTER 4 STAINLESS STEELS

F. Some dents were also observed, but these were probably caused by reaction to the violent nature of the rupture.
G. No unusual scale on either the OD or ID was observed.

The austenitic stainless steels, unlike ferritic steels, do not have a high-temperature transformation. For ferritic steels, transformation from ferrite and pearlite to ferrite and austenite at about 1340°F (725°C) leaves a "calling card" to aid identification of peak temperatures at the time of failure. It is not so with austenitic stainless steels. Microstructural changes of carbide precipiation and sigma-phase formation occur over a wide range of temperatures, require long periods of time, and are sensitive to composition and, to a lesser extent, structure. Grain growth occurs only at very high temperatures, 1700–2000°F (925–1090°C). A rapid high-temperature failure leaves no microstructural evidence to help estimate peak temperature.

The microstructure at the point of failure shows large equiaxed grains that are nearly completely surrounded by a grain-boundary carbide precipitate. Small carbide precipitates are also noted within the individual austenite grains, as seen in Fig. 4.1.2. The microstructure 4 ft away and in

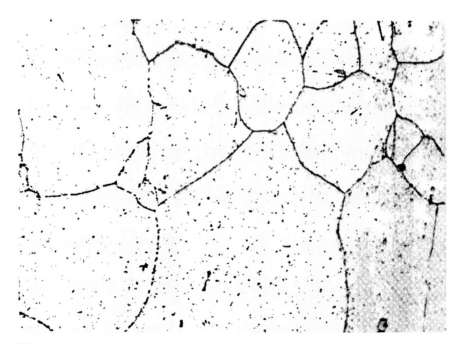

FIGURE 4.1.2. The microstructure in the vicinity of the rupture shows large equiaxed grains nearly completely surrounded by a grain-boundary carbide precipitate. Within the individual austenite grains are small carbide precipitates (Magnification 500×, electrolytic etch).

the same plane as the rupture (see Fig. 4.1.3) also shows the carbide precipitate surrounding the austenite grains, but no intergranular precipitate is evident. Microstructures of the fire side of the adjacent tubes show similar large carbide precipitates surrounding the austenite grains. A small amount of intragranular carbide precipitate was also noted; see Fig. 4.1.4.

The microstructure 180° away and in the same plane as Fig. 4.1.4 shows a carbide precipitate at the grain boundaries, but to a lesser extent; see Fig. 4.1.5.

The boiler had been in operation about six years at the time of failure. The carbide precipitates noted in all of the specimens are typical of stainless steel for normal service conditions. While 321H stainless steel can be given a stabilization anneal at 1600°F (870°C) to precipitate titanium carbide and, thus, prevent further carbide precipitation during service, these particular tubes were not given that stabilization anneal. Chromium carbides will form at grain boundaries in the temperature range of approximately 900–1500°F (480–815°C), with a maximum precipitation rate at about 1200°F (650°C). However, these microstructures are considered normal for the length of service.

FIGURE 4.1.3. The microstructure 4 ft away and in the same plane as the rupture. The grain-boundary precipitate is not as continuous as in the previous microstructure, and there are no carbide precipitates noted in the austenite grains (Magnification 500×, electrolytic etch).

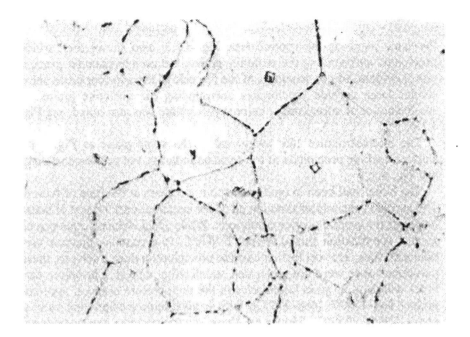

FIGURE 4.1.4. The microstructure of the adjacent tube on the fire side shows carbide precipitates surrounding the austenite grains similar to those in Fig. 4.1.3. Discrete particles can be seen at the grain boundaries, giving the appearance of a string of pearls (Magnification 500×, electrolytic etch).

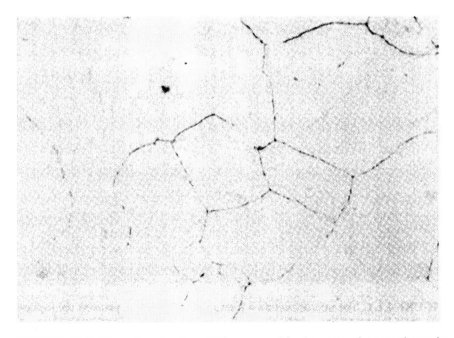

FIGURE 4.1.5. The microstructure 180° away and in the same plane as shown in Fig. 4.1.4. However, the carbide precipitates at the grain boundaries are not as large and can hardly be resolved at this magnification (Magnification 500×, electrolytic etch).

The differences noted between the fire side and 180° opposite are to be expected, since the fire side will operate slightly hotter, perhaps 50°F (280°C), than the back side of the tube.

The rupture was caused by a temporary blockage of the tube, most probably condensate, and by too rapid a start-up. The overheating was localized to the area of the rupture of that single tube.

Case History 4.2

Superheater Support Failure and Sigma-Phase Formation

BOILER STATISTICS

Size	3,250,000 lb steam/hr (1,480,000 kg/hr)
Steam temperature	955°F/955°F (515°C/515°C)
Steam pressure	2025 psig (145 kg/cm²)
Fuel	No. 4 fuel oil

Visual examination of these small 1½in pieces of reheater support-lug material showed the following:

A. Extensive metal thinning along the leading edge of the lug material.
B. These particular sections failed after only five months of operation.

Samples for microstructural analysis were taken from both the trailing and leading edges of these samples. This is typical of stainless steel heated to about 1000 to 1200°F (540 to 650°C), and exhibits intergranular carbide precipitation, the "ditched" structure (see Fig. 4.2.1). The microstructure of the leading edge of the lug, shown in Fig. 4.2.2, reveals the formation of sigma-phase. The presence of sigma-phase indicates that the lug reached a temperature in the range of 1200–1550°F (650–840°C), as would normally be expected in this application. The absence of sigma-phase in the trailing edge means the specimen was taken closer to the tube metal, which would be at a lower temperature because of the contact with the steam-cooled tube.

Since sigma-phase redissolves at a temperature of 1600°F (870°C) or higher, it is apparent the lugs were exposed to temperatures no higher than about 1550°F (840°C).

The lug deterioration indicates surface erosion from flue gas, and possible attack by the vanadium, sulfur, and sodium present in the flue gas. The deterioration of the lugs was related to the location and design of the lugs, which allows them to be exposed to the flue gas and become overheated. The lug failures are primarily associated with surface erosion and heating of the lugs to temperatures above 1250°F (675°C).

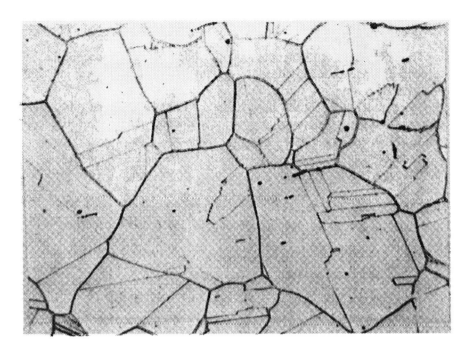

FIGURE 4.2.1. The grain-boundary carbide precipitate is nearly continuous and has grown in thickness so that the grain boundaries appear as a double line. This is the "ditched" structure (Magnification 500×, electrolytic etch).

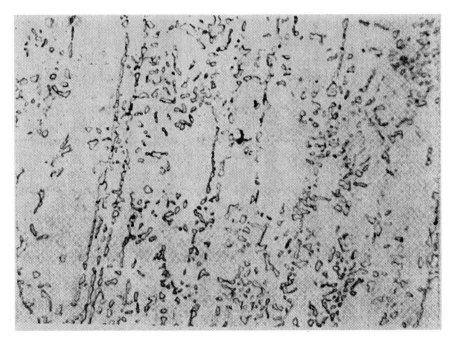

FIGURE 4.2.2. Formation of sigma-phase in Type 309 stainless steel (Magnification 500×, electrolytic etch).

Case History 4.3

Failure of a 321H Stainless-Steel Superheater Tube with Sigma-Phase

BOILER STATISTICS

Size	2,600,000 lb steam/hr (1,200,000 kg/hr)
Steam temperature	1005°F/1005°F (540°C/540°C)
Steam pressure	2400 psig (170 kg/cm^2)
Fuel	Pulverized coal

This failed high-temperature superheater tube was SA-213 TP-321H stainless steel of dimension 1½ in OD × 0.180-in-minimum wall. Visual examination of the tube sample revealed the following:

A. A small section, about ¾ in wide was blown out of the fire side of the tube leaving a blunt-edged, somewhat jagged-appearing fracture surface; see Fig. 4.3.1.

B. The remaining section revealed several longitudinal oxide-filled stress lines running parallel to the fracture.

C. The OD scale measured 0.012 in, and the ID scale measured 0.007 in thick.

D. Wall-thickness measurements were 0.192 and 0.190 in, indicating that no appreciable wall thinning had occurred.

E. The tube diameter, however, had expanded from 1½ to 1$\frac{19}{32}$ in.

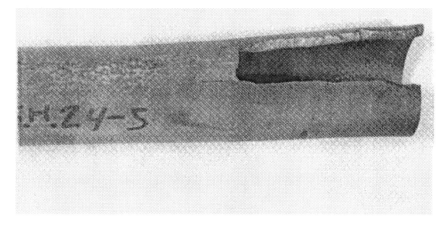

FIGURE 4.3.1. The as-received 321H stainless-steel superheater tube. Note the blunt-edged jagged appearance to the fracture surface where a small piece has been blown out.

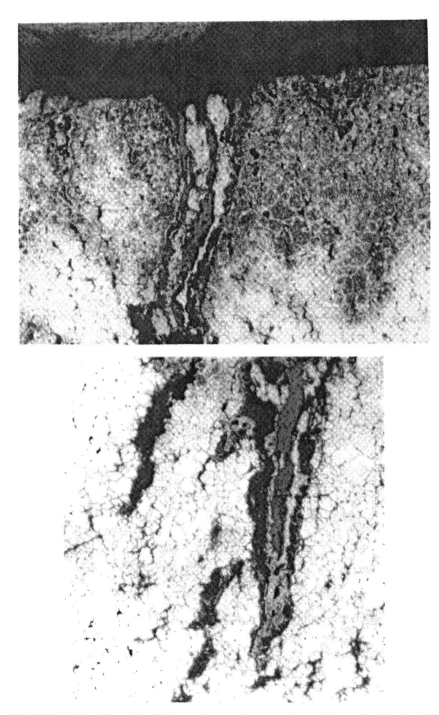

FIGURE 4.3.2. OD oxide-filled crack; one of many observed near the fracture lip. At the tip, the intergranular nature of the crack may be noted (Magnification 250×, electrolytic etch).

Microstructural analysis through and adjacent to the fracture lip showed a multitude of deep oxide-filled cracks on the OD and a structure that included the hard, brittle sigma-phase; refer to Fig. 4.3.2. Steam-side grain-boundary attack was noted throughout the tube, both on the fire side and 180° from the fire side; the penetration was perhaps four or five grains deep; see Fig. 4.3.3. The structure at the midwall shows the presence of sigma-phase (note Fig. 4.3.4).

The presence of oxide-filled OD cracks, a minor but significant ID grain-boundary penetration, and the presence of appreciable amounts of sigma-phase suggest a metal temperature of approximately 1400°F (760°C). Sigma-phase formation occurs between about 800 and 1600°F (425 and 870°C), but forms most rapidly at about 1400°F (650°C). Once sigma-phase forms, two effects may be noted: a substantial loss of ductility and a slight loss of corrosion resistance. From the tube-diameter measurements before and after expansion and rupture, the tube displayed only about 6% expansion close to the failure. Normally 321H stainless steel exhibits 50% or so ductility at 1400°F (760°C) in a simple hot tensile test. The formation of the chromium-rich (nominal composition about 50% Cr) sigma-phase may remove sufficient chromium, especially in the grain-boundary region, to reduce the overall corrosion resistance. Both OD and ID attack are manifestations of this reduced corrosion resistance.

FIGURE 4.3.3. Steam-side grain-boundary attack; this condition was noted throughout the circumference of the tube ID (Magnification 100×, electrolytic etch).

FIGURE 4.3.4. The microstructure at the midwall of the tube; note the extensive appearance of sigma-phase, the dark-etching constituent (Magnification 250×, electrolytic etch).

Under normal conditions, 321H stainless steels have very good oxidation resistance up to 1600°F (870°C) or so, but at such a high temperature, sigma-phase is not apt to be present to the extent noted in Fig. 4.3.4. The failure is typical of a high-temperature creep or stress-rupture fracture. Void formation by grain-boundary sliding may be observed in the vicinity of the crack tip in Fig. 4.3.2. The brittle, low-ductility fracture was caused by higher temperatures than expected. A combination of oxide-filled OD and ID cracks that acted as stress raisers and the inherent lower ductility of stainless steels with sigma-phase present in the microstructure caused a piece to be blown out of the tube rather than the more typical fissure observed in most creep failures.

Examination of an adjacent tube in the same superheater bundle showed no evidence of either grain-boundary oxidation or sigma-phase formation. The conclusion is that this tube alone was partially blocked, for reasons unknown, and had been operating at higher-than-design temperatures for several months or a few years.

Case History 4.4

Thermal-Fatigue Failure of a 347H Reheater Lug

BOILER STATISTICS

Size	4,650,000 lb steam/hr (2,110,000 kg/hr)
Steam temperature	995°F/995°F (515°C/515°C)
Steam pressure	1980 psig (140 kg/cm^2)
Fuel	Crude oil

The failure, as shown in Fig. 4.4.1, was located between two SA-210 A-1 carbon-steel tubes, and failed through the Type 347H stainless-steel lug without damaging either tube. The visual examination revealed the following:

A. The lug was measured and found to be 3½ in long; it appeared to be in good physical condition.
B. Only tightly adhering oxide scale was noted, with no evidence of any corrosion.
C. The welds appeared to be acceptable, with no evidence of any excessive undercut or other deleterious welding defects. Each lug-to-weld was made in four passes: one along either side of the lug,

FIGURE 4.4.1. The failure of a Type 347H stainless-steel lug.

parallel with the tube, and one at either end to enclose the lug with weld metal completely.

D. One tube was bowed more than $\tfrac{1}{16}$ in (in a 12-in-long section), implying a severe loading condition.

E. The fracture appeared to originate at the toe of the weld at one end of the lug.

The microstructure of the failure, Fig. 4.4.2, shows the tip of a crack associated with, but not part of, the final failure. No evidence of plastic deformation can be seen in the grains adjacent to the crack tip. Figure 4.4.3, some $\tfrac{1}{8}$ in or so away from the failure, shows a normal microstructure for a Type 347H material.

The cause of the failure was determined as follows. This unit is a daily peaking unit that came on and off the line every day. The unit had been in service less than two years when the failure of several reheater lugs was noted. A careful examination of the actual reheater showed considerable distortion in several of the tubes at the lower portion of the reheater. Careful analysis of the operating conditions and temperature measurements of several tubes in the reheater showed that the lower tubes would

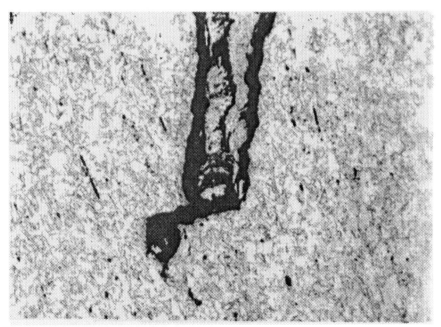

FIGURE 4.4.2. The microstructure of a tip of a crack associated with, but not part of, the final failure. No evidence of plastic deformation can be seen in the grains adjacent to the crack tip. The lack of any distortion of the grains at the crack tip is suggestive of a failure mechanism by fatigue (Magnification 250×, electrolytic etch).

FIGURE 4.4.3. The microstructure of the Type 347H material some distance away from the failure. The microstructure is normal for this material (Magnification 100×, electrolytic etch).

fill with condensate. When the boiler was restarted, the plugged tubes would stay cold, while adjacent tubes were heated to a higher-than-normal temperature. The high strain at the lug caused by one tube expanding as it was heated, while the other remained nearly stationary, induced a high stress on the tube-to-tube attachments. The failure was one of thermal fatigue caused by the cycling operation of this unit and by the pluggage with condensate of some of the tubes.

Case History 4.5

High-Temperature Failure of 304 Stainless Steel J-Strap

BOILER STATISTICS

Size	3,250,000 lb steam/hr (1,480,000 kg/hr)
Steam temperature	955°F/955°F (515°C/515°C)
Steam pressure	2025 psig (145 kg/cm²)
Fuel	No. 4 fuel oil

This study is of a J-strap of Type 304 stainless steel, 1½ in wide × ¼ in thick and nearly 4 ft long, a portion of which is shown in Fig. 4.5.1. The straps

FIGURE 4.5.1. A portion of the as-received J-strap of Type 304 stainless steel.

were removed from the bottom of the high-temperature reheater section during an inspection of the unit, for examination after they were found to be wasting away. A visual examination of this strap revealed the following:

A. Severe wastage on one end.
B. The opposite end, where it wrapped around the tube, showed no evidence of deterioration.
C. Micrometer measurements showed that the thickness decreased from 0.270 in, where the strap wrapped around the tube, to 0.165 in at the opposite end.

Specimens were taken from both ends of the tube for microstructural analysis. The specimen from the end, where it wrapped around the tube, showed the usual grain-boundary carbide network with a minor amount of carbide precipitates within the matrix; see Fig. 4.5.2. The specimen from the thinnest end exhibited some oxidation of the grain boundaries at the specimen surface, and showed abnormal grain growth; refer to Fig. 4.5.3. The cooler end of the tube had a grain size of ASTM 7–8; at the hotter end, the predominate grain size was ASTM 4, with some grains as large as ASTM 3.

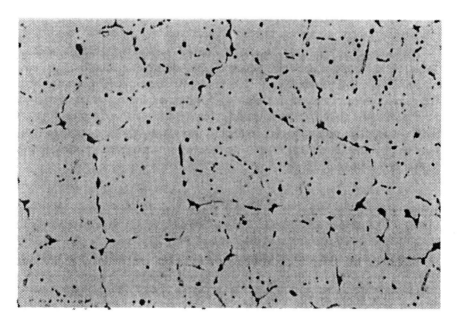

FIGURE 4.5.2. The microstructure from the "cold" end of the J-strap. The microstructure shows the usual grain-boundary carbide network with a minor amount of carbide precipitates within the matrix. The grain size is ASTM 7–8 (Magnification 500×, electrolytic etch).

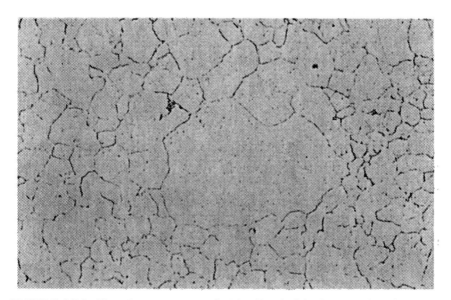

FIGURE 4.5.3. The microstructure at the "hot" end of the J-strap. The microstructure shows very little grain-boundary precipitate, and exhibits abnormal grain size. The grain size is predominately ASTM 4, with some grains as large as ASTM 3. The large grain size observed in the microstructure indicates that a temperature of nearly 2000°F (1090°C) was reached, because grain growth does not occur much below this temperature (Magnification 100×, electrolytic etch).

270 CHAPTER 4 STAINLESS STEELS

The wastage of the strap can be attributed to extremely high temperatures. Type 304 stainless has good scaling resistance for continuous duty up to 1700°F (925°C) or about 1550°F (840°C) for intermittent use. The large grain size observed in Fig. 4.5.3 indicates that a temperature of nearly 2000°F (1090°C) was reached, since grain growth does not occur much below this temperature. Based on this observation, it appears that the likely cause of wastage was the excessive temperature of the material at this location.

Case History 4.6

In order to improve performance of high-temperature serpentine clamps used for low-temperature superheater and reheater supports, the use of Inconel 601 has been investigated. It was found that unless precautions were taken in the treatment of the corners of the bar stock used, severe cold work would lead to minute edge cracks. Figure 4.6.1 shows such a serpentine clamp. The microstructure at the corners of the clamp where the cold work is the most severe is shown in Fig. 4.6.2. The austenite grains have been elongated, and a considerable number of slip and twin planes are visible in the structure. The normal grain structure is shown in Fig. 4.6.3. The structure is composed of equiaxed austenite grains, with a few titanium-nitride particles evident as a second-phase precipitate. The three straight lines through the center grain in Fig. 4.6.3 are twin boundaries.

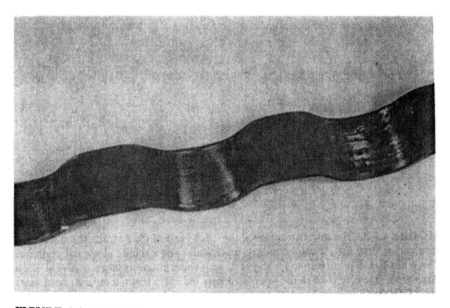

FIGURE 4.6.1. A high-temperature serpentine clamp fabricated of Inconel 601.

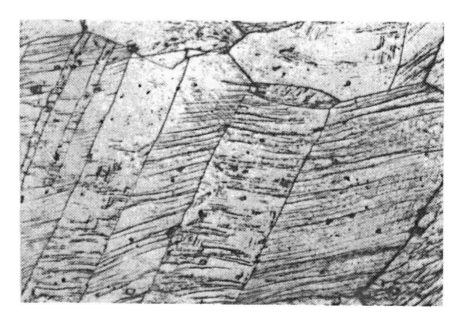

FIGURE 4.6.2. The microstructure of cold-worked Inconel 601. Note the elongation of the austenite grains, and the extensive amount of slip and twinning within the grains (Magnification 500×, electrolytic etch).

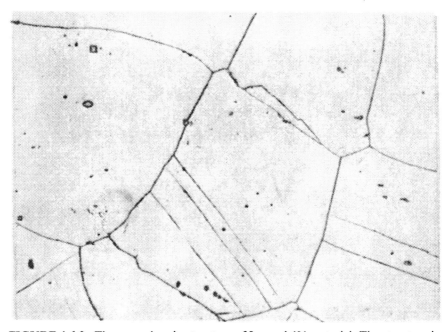

FIGURE 4.6.3. The normal grain structure of Inconel 601 material. The structure is composed of equiaxed austenite grains with a few titanium-nitride particles evident as a second-phase precipitate. The three straight lines through the center grain are twin boundaries (Magnification 500×, electrolytic etch).

Case History 4.7

Creep Damage Microstructures—321H Stainless Steel

Creep failures occur in austenitic stainless steels by the same grain-boundary sliding mechanism found in ferritic steels. At temperatures in the creep range for stainless steel, sigma-phase will also form. Thus, the microstructures often contain both creep damage and sigma-phase. Figure 4.7.1 shows an as-received tube sample that failed after nearly 30 years (>200,000 hr). Note that metallographic samples have been removed from the fracture edge, and a ring section has also been removed. Chemical analysis was performed to confirm the 321 material, and these results are given in Table 4.7.1.

Figures 4.7.2 and 4.7.3 show the OD and ID surfaces near the fracture edge. These figures are not etched to highlight the extensive creep damage

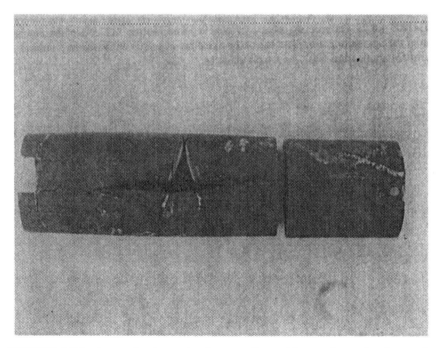

FIGURE 4.7.1. As-received sample after metallographic samples and a ring section have been removed.

TABLE 4.7.1. Chemical Analysis

% Cr	17.14
% Ni	10.61
% Ti	0.42

HIGHER-ALLOY AUSTENITIC MATERIALS 273

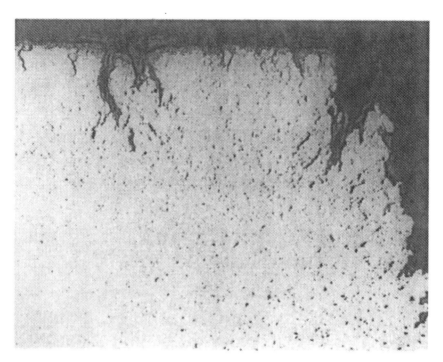

FIGURE 4.7.2. OD surface near fracture edge (Magnification 25×, not etched).

FIGURE 4.7.3. ID surface near fracture edge (Magnification 25×, not etched).

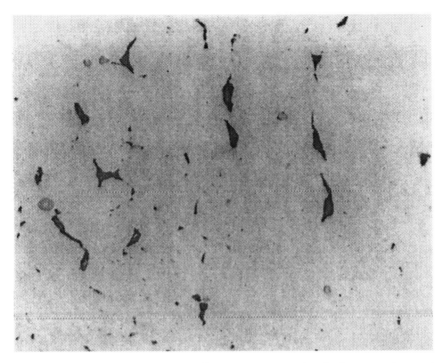

FIGURE 4.7.4. Microstructure near fracture edge contains the intergranular cracks associated with creep failures (Magnification 500×, not etched).

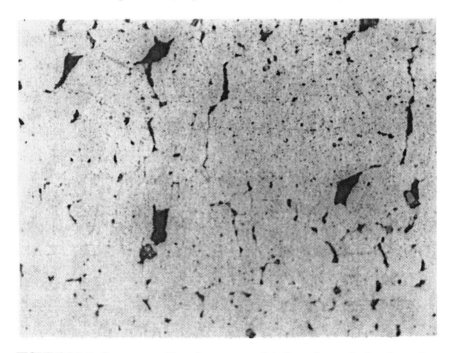

FIGURE 4.7.5. Same general location as preceding figure but etched to show sigma-phase (Magnification 500×, etched).

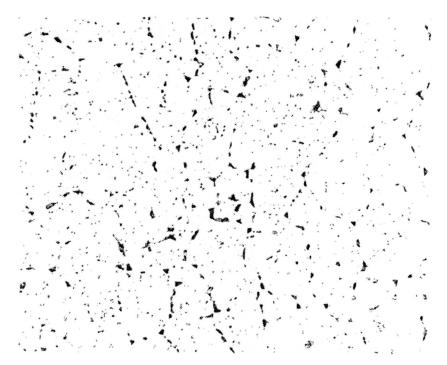

FIGURE 4.7.6. Midwall microstructure some 5 in from end of the opening, in line with the rupture, contains both creep damage and sigma-phase (Magnification 500×, etched).

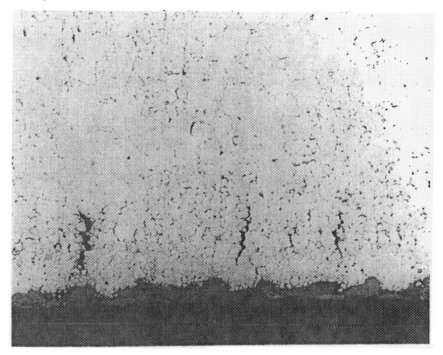

FIGURE 4.7.7. The ID surface at the same location contains extensive intergranular cracking (Magnification 100×, etched).

without confusion with sigma-phase. The microstructure near the fracture edge, in the etched condition, is presented in Figs. 4.7.4 and 4.7.5.

The microstructures at the end of the tube sample, but in line with the failure, are presented in Figs. 4.7.6 and 4.7.7. The microstructure contains creep voids and sigma-phase, and there is extensive ID cracking.

In summary, the overall microstructural features of creep damage in austenitic stainless steel are similar to those in ferritic steels—that is, grain-boundary cracks and voids. However, instead of spheroidization of any carbides present in ferrite steel, sigma-phase formation leaves the microstructure with creep voids and the sigma-phase present, principally at grain boundaries.

REFERENCES

The metallurgy of stainless steels was taken from:

1. J. G. Parr and A. Hanson, *An Introduction to Stainless Steel,* ASM International, Metals Park, Ohio, 1965.
2. *Source Book on Stainless Steels,* ASM International, Metals Park, Ohio, 1976.
3. *ASM Metals Handbook,* 8th ed., Vol. 8, "Metallography Structures, and Phase Diagrams," ASM International, Metals Park, Ohio, 1973.
4. *Boiler and Pressure Vessel Code, Section I,* ASME, New York, 1981.
5. *ASM Metals Handbooks,* 8th ed., Vol. 2, "Heat Treating Cleaning and Finishing," ASM International, Metals Park, Ohio, 1964.

CHAPTER FIVE

FAILURES CAUSED BY GAS—METAL REACTIONS

This chapter will discuss the reactions between a gas and the metal surface that alter the properties of the metal at the surface and change the behavior of the metal so as to cause its degradation and/or failure. Reactions on the steam side between water and steel will form magnetite, Fe_3O_4, alter the heat transfer, and raise metal temperatures. Under certain conditions, steam will also decarburize the surface. On the fire side, reactions between flue-gas constituents and steel will form oxides and sulfides, and will carburize or decarburize the surface.

STEAM-SIDE REACTIONS

Iron will react with steam to form Fe_3O_4 and hydrogen in accordance with the net chemical reaction

$$3Fe + 4H_2O = Fe_3O_4 + 4H_2$$

All ferritic alloys will have a thin layer of magnetite on the steam side of the tube. The formation of this oxide layer is essential to the protection of the steel. The rate at which the oxide-film thickness X increases with time t is parabolic,[1] that is, $X \alpha \sqrt{t}$, Fig. 5.1.[2] A thin film forms almost immediately, but because the diffusion of oxygen and iron through the oxide film is sluggish, the rate at which the film thickens decreases with increasing time. The rate of this reaction increases with increasing temperature; that is, a high temperature favors the reaction. The danger of having too thick a magnetite layer is that it impedes heat transfer between the tube metal and the steam on

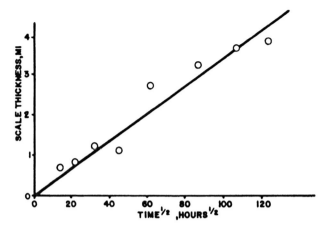

FIGURE 5.1. A plot of scale thickness versus the square root of time for times up to 16,000 hr for the oxidation of Cr–Mo steels in 1100°F (593°C) steam. (Data from Ref. 2.)

the inside of the tube. This results in overheating on the inside of the tube and shortens tube life. Ferritic alloys form oxide layers that are quite uniform in thickness. The oxides usually grow inward and outward from the original surface of the metal to form a double-layer structure. Magnetite can form either by the inward diffusion of oxygen or by the outward diffusion of iron. Since oxygen, a small atom, has a faster diffusion rate in magnetite than in iron, the outer layer has an iron deficiency, more vacant iron lattice sites, and the inner layer has a slight oxygen deficiency, more vacant oxygen lattice sites. The outer surface tends to be more porous since the vacant iron lattice sites coalesce to form visible pores.[3,4] Figure 5.2 shows such an oxide scale. For reasons that are not clear, the two-layer structure repeats itself in multiple layers, like growth rings on a tree; see Fig. 5.3.

Iron oxide has a much lower thermal conductivity than the tube metal and, thus, acts as an insulating blanket. As the magnetite film thickness increases, so does the tube-metal temperature; and higher tube-metal temperatures for a given operating stress level decrease the life expectancy of the material. Compounding the problem is that as the metal temperature increases so does the rate of reaction between steam and iron to form magnetite; thus, the magnetite layer further increases in thickness. Superheaters and reheaters are more prone to this type of failure than are economizers or furnace tubes.

The morphology of austenitic stainless-steel oxide scales is somewhat different from that of the ferritic alloys. At the same temperature and time, the thickness is much less and quite variable.[3] In general, inward diffusion of oxygen is made up of deep penetrations; see Fig. 5.4. The outer scale is also porous, but thinner and usually more uniform in thickness. Samples of T-22 and 321H steels, separated by less than 2 in, were taken from a high-temper-

FIGURE 5.2. The oxide on the ID surface of grade SA-213 T-22 steel. The oxide is formed by the reaction between steam and the tube material. Note the double-layer appearance of the oxide scale (Magnification 250×, unetched).

ature superheater that had been in service 14 years, with a finishing steam temperature of 1005°F (540°C). Careful measurements of these samples showed that the T-22 scale was 0.013 in thick, and that the deepest penetration of the 321H scale was 0.005 in.[5]

EXFOLIATION

Steam-side magnetite layers under certain operational conditions, usually during start-up or shut-down, will spall. The hard particles of magnetite become entrained in the steam and are carried to the turbine to cause erosion of turbine blades. The detailed causes of exfoliation are not well understood;[6] for instance, there are examples of thin films causing turbine erosion and of thick films remaining intact. The principal cause appears to be differences in coefficients of expansion between oxide and metal, so that temperature changes cause the magnetite film to spall off. The problem is more prevalent in cycling boilers and in boilers that have been operated for long periods of time without a chemical cleaning to remove the magnetite scale.

Exfoliation is more of a problem in reheaters than in superheaters, and may be related to greater temperature excursions. During start-up, no steam

FIGURE 5.3. For reasons that are not clear, the two-layer structure of Fig. 5.3 repeats itself in multiple layers (Magnifications 100× above and 250× below, unetched).

FIGURE 5.4. The structure of scale formed on a grade SA-213 321H stainless steel. Compare this structure with Fig. 5.2, and note that the scales formed on stainless steel show an irregularity to the interface between the scale and the steel (Magnification 100×).

flows through the reheater, raising the metal temperature to 1000 or 1100°F (540 or 590°C). The first reheat steam to arrive after the intercept valve opens may be 300°F (150°C) or so at the inlet, which effectively quenches the reheater. Differences in the coefficients of thermal expansion between oxide and metal cause the scale to spall. In superheaters, the temperature rises more smoothly without the quenching action, because steam formed in the walls is vented through the superheater during start-up.

Figure 5.5a shows the inside of a superheater tube that shows the effects of exfoliation. Note the small, pockmarked appearance of the ID surface where the scale has spalled. In Fig. 5.5b from another tube sample, the scale shows large blisters, some of which have broken. It may be that exfoliation is related in some way to the reactions that form oxide and hydrogen. At the reaction site, atomic hydrogen is formed and diffuses away, both through the scale and into the steel. If, however, two atoms of hydrogen join to form molecular hydrogen at the interface between oxide and metal, the molecule is too large to diffuse. The hydrogen gas pressure separates the oxide layer from the steel. A blister will then be created that is more easily broken than tightly adhering scale.

There are two reasons why periodic inspections should be made of all the tubing in order to prevent excessive scale buildup. First, the scale acts as an

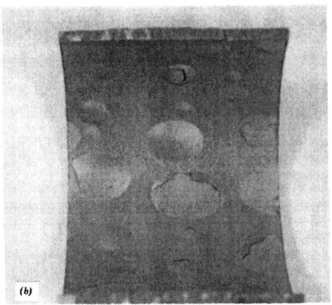

FIGURE 5.5. The inside of superheater tubes that show the exfoliation: (*a*) ID surface is pockmarked where pieces of scale have spalled; (*b*) another tube sample, showing some blisters still intact and others which have broken.

insulating barrier to increase metal temperature and cause premature failure; second, exfoliation of thick scale films leads to erosion of turbine blades downstream of the boiler.

Figure 5.6 presents the ID surface of a reheater tube with serious exfoliation. In cross section, Fig. 5.7, the scale thickness is variable, and there are long, circumferential cracks in the oxide.

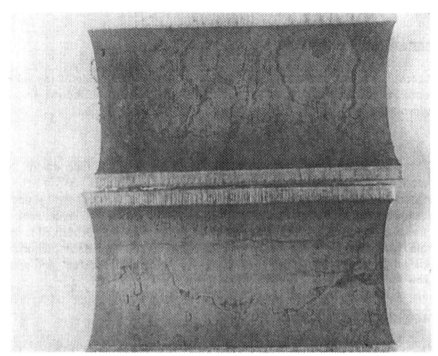

FIGURE 5.6. ID surface of a reheater tube presents overall view of the serious exfoliation (Magnification 0.7×).

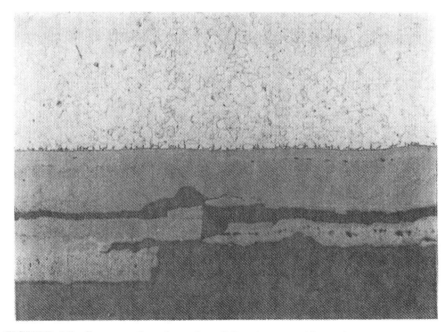

FIGURE 5.7. Cross section through exfoliated steam-side scale shows variable thickness and extensive cracking (Magnification 100×, etched).

284 CHAPTER 5 FAILURES CAUSED BY GAS–METAL REACTIONS

DECARBURIZATION

Another gas–metal reaction that can occur on the steam side of the tubes is the reaction between carbon and steam to form carbon monoxide and hydrogen:

$$H_2O + C = H_2 + CO$$

Carbon diffuses from the interior of the metal to the surface where the reaction occurs, depleting the microstructure of carbon and leaving behind essentially pure ferrite. As the material is decarburized, mechanical properties are reduced; except in extreme cases, these decarburized layers remain thin and do not affect tube life. Figure 5.8 shows such a decarburized layer; note that grain growth has occurred in the ferrite grains. Some decarburization may be present on new boiler tubing, but from the extent and grain growth evident, Fig. 5.8 represents an in-service condition.

FIRE-SIDE REACTIONS

The oxidation of iron is the most obvious fire-side gas–metal reaction. Too high a temperature or the presence of other corrosive species in the flue gas can lead to very rapid oxidation, metal wastage, and failure. Figure 5.9

FIGURE 5.8. A decarburized layer on OD of an SA-210 A-1 steel. Note the excessive grain growth of the ferrite grains at the surface. Some of the pearlite colonies are still visible in the lower portion of the figure (Magnification 100×, nital etch).

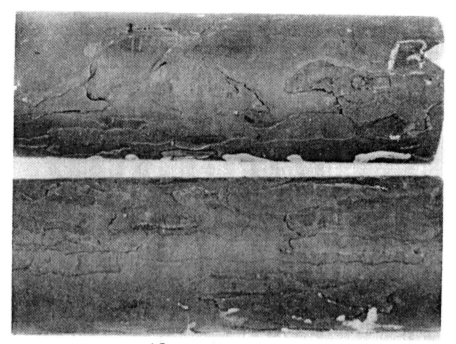

FIGURE 5.9. The outside surface of a grade SA-213 T-22 steel superheater tube from a gas-fired boiler in service about 15 years. The OD scale is more than 0.050 in thick in places.

shows the outside surface of a T-22 superheater tube from a gas-fired boiler in service about 15 years; the OD scale is more than 0.050 in thick in places.

The formation of oxide scales on iron and steel has been extensively studied for many decades. There is general agreement that the growth of thick scales on iron follows a parabolic-rate law for the increase in film thickness.[1] Three forms of iron oxide are stable at different temperatures and oxygen compositions: wüstite, corresponding to the compound FeO; magnetite; corresponding to Fe_3O_4; and hematite, corresponding to Fe_2O_3. Figure 5.10 shows the iron–oxygen phase diagram.[7]

Oxidation first occurs at the surface of the metal, and the resulting scale forms a barrier to the further growth of the oxide film. Growth of oxide scales occurs by both the diffusion of iron from the steel to the surface of the scale to react with oxygen, and by oxygen diffusion through the oxide film to the metal interface to react with iron at the scale–metal interface.[3] In order for the scale to form a protective layer, the volume of the oxide must occupy a greater volume than the metal from which it is formed, and the scale thus formed must be adherent or tightly bound to the metal surface.

Fortunately, most of the steels used in boiler construction develop reasonably adherent protective oxide films. Figure 5.11 shows the oxide layer on grade T-22 steel tubing in service for 16 years in a coal-fired boiler with a

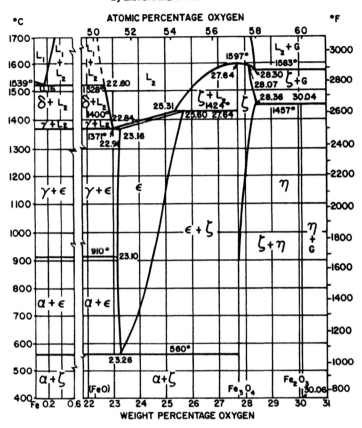

FIGURE 5.10. The iron-oxygen phase diagram. From Darken, L. S. and Gurry, R. W., Fe–O diagram, "Constitution of Alloys," *Metals Handbook,* Editor, Lyman, Taylor, American Society for Metals, 1948, p. 1212.

finishing steam temperature of 1005°F (540°C). Note the grain-boundary penetration of the oxide.

The addition of chromium to ferritic steels imparts an improved corrosion resistance. The selective oxidation of chromium produces a tightly adhering, protective oxide scale. In the case of stainless steels, the formation of both chromium and nickle oxide produces a protective film that gives these alloys their exceptionally high-temperature-oxidation resistance.

CARBURIZATION

The diffusion of carbon to the surface of a steel is called carburization. This is the foundation of the case hardening of steel for use in gears, shafts, and

CARBURIZATION   287

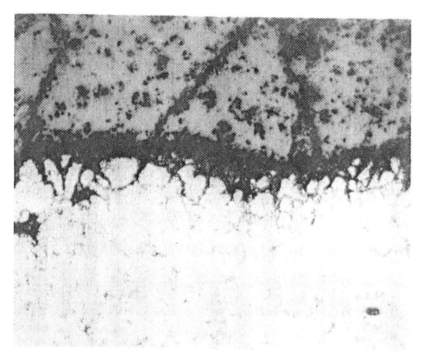

FIGURE 5.11. The oxide layer on a grade SA-213 T-22 steel tube in service 16 years in a coal-fired boiler. The oxides formed by reaction with flue gas are slightly different in character from those shown previously, formed by reaction with steam. Note the oxide penetration along grain boundaries and the more porous appearance to the scale (Magnification 500×, nital etch).

other applications where a hard surface layer is desirable. However, carburization also is one of the mechanisms of corrosion of stainless steel.[8] The diffusion of carbon into the surface of a stainless steel results in the formation of chromium carbide, depleting the surface of chromium to the extent that at the surface the material it is no longer a stainless steel. Rapid oxidation or corrosion can then waste the material to the point of failure. For the most part, the operating temperatures of the pressure parts of a boiler are below the temperature at which carbon diffusion can occur rapidly enough to cause problems. However, in the high-temperature support systems for superheaters and reheaters, or at burner nozzles, coal spreaders, and other portions of the boiler where temperatures can reach 1500°F (820°C) or more, failures by carburization can occur. Carbon, carbon monoxide, and methane (natural gas) all can act as carbon carriers to increase the carbon content of steel surfaces.

Figures 5.12 through 5.15 present the microstructures of a 304 stainless-steel coal-burning component. At the edge, Fig. 5.12, the wastage has occurred by carburization. The carbon content at the lip is about 0.24%. This

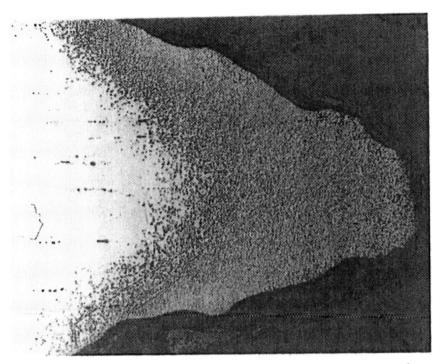

FIGURE 5.12. Failure lip of a 304 stainless-steel fuel-burning component from a coal-burning unit. The lip shows extensive carburization (Magnification 100×, etched).

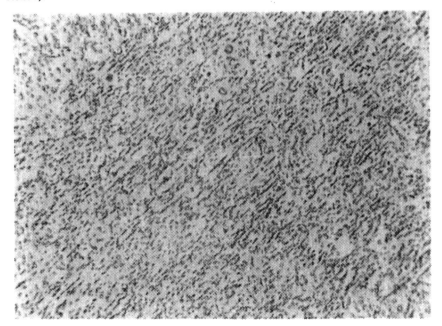

FIGURE 5.13. Microstructure is seen to be small carbide particles within the austenite; %C is 0.24 (Magnification 500×, etched).

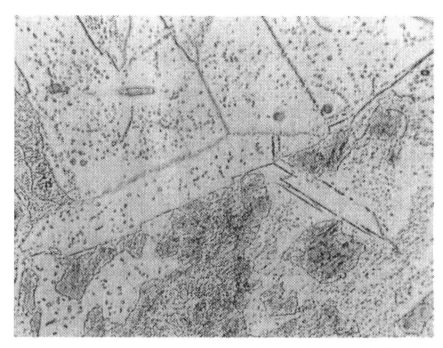

FIGURE 5.14. Microstructure at another location is similar to pearlite in a plain-carbon steel (Magnification 500×, etched).

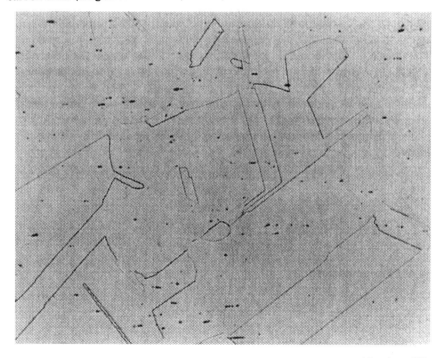

FIGURE 5.15. Normal microstructure, 304 stainless steel (Magnification 500×, etched).

290   CHAPTER 5 FAILURES CAUSED BY GAS–METAL REACTIONS

fuel-burning component suffered failure by the effects of carburization of a 304 stainless steel. Figure 5.12 shows the failure edge, and the microstructure at the tip is dramatically different. At higher magnifications, Fig. 5.13, the microstructure may be seen to be composed of fine carbide particles within the austenite. Some distance away, the surface shows fully developed carbides, and the structure is reminiscent of pearlite in a plain-carbon steel, Fig. 5.14. Away from the surface that has carburized, the microstructure is normal austenite; see Fig. 5.15.

EROSION

While not really a gas–metal reaction, the problems of erosion of superheater and reheater tubing caused by fly ash cannot be ignored. Even though fly ash is in very small particles, high-velocity-gas streams containing fly ash wear away, or erode, the leading tubes. In severe cases, tube shields, similar to those shown in Fig. 5.16, are necessary to protect the pressurized parts. Sootblowers will also cause erosion to boiler tubes in the neighborhood of their operation. Figure 5.17 shows such an example. The high-temperature superheater-tube section is nearly 3 ft long, and a cross-sectional view, Fig. 5.18, displays the damage. Note that the wall thickness has been reduced from 0.248 to 0.050 in. The basic mechanism is the removal of the protective oxide film. The exposed clean tube metal in the high-temperature furnace-gas atmosphere is reoxidized, consuming a little tube metal at each cycle. By alternate scale formation and removal, failure occurs when the tube-metal thickness is reduced to the point where it can no longer contain the internal steam pressure.

Similar failures occur in tubes adjacent to a ruptured tube. Steam escaping from the rupture will wash neighboring tubes, causing them to fail. Usually the failure of steam-eroded tubes is more trouble than the original tube rupture.

FIGURE 5.16. An example of the type of shield necessary to protect pressurized parts in areas where severe erosion may occur.

FIGURE 5.17. A high-temperature superheater tube of SA-213 T-22 1¾-in OD × 0.220-in-minimum wall eroded on one side by excessive use of a sootblower.

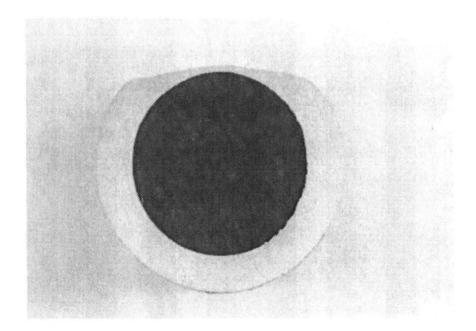

FIGURE 5.18. The cross-sectional view of the tube shown in Fig. 5.17; note the erosion-caused flat is nearly 1½ in wide. The thinnest wall measured 0.050 in, reduced from 0.248 in.

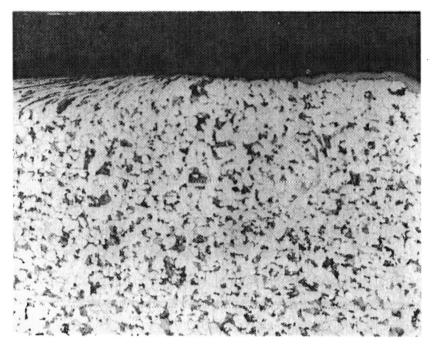

FIGURE 5.19. OD surface of an SA-178 C superheater tube removed from a steam-washed area. The center of the photograph is the edge of the washed region. The OD oxide is intact on the right, and the cold-worked surface is to the left (Magnification 100×, etched).

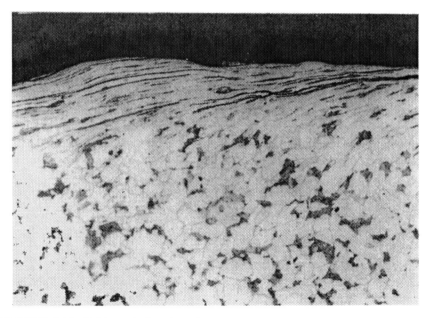

FIGURE 5.20. Cold-worked surface from fly-ash impingement in a case of steam washing (Magnification 200×, etched).

Depending on the amount of fly ash in the neighborhood of the rupture, the steam erosion of the nearby tubes can alter the surface microstructure. Figure 5.19 presents the edge of a washed OD surface. To the right, the oxide scale is still intact; to the left is the cold-worked structure caused by fly-ash impingement. At higher magnification, the distortion to the surface is more evident; see Fig. 5.20.

CASE HISTORIES

Case History 5.1

Microstructures SA-213 T-5 Material

In some supercritical units, the overall efficiency is improved by the addition of a second reheat stage. The second reheaters operate at design pressures in the neighborhood of 500 psi. During start-up, these components see high flue-gas temperatures for a period of time before steam flow is fully established. Thus, tube-metal temperatures can be well above the design condition for this start-up circumstance. The saving grace is that pressure is low, and failures do not often occur.

In this example, the second reheater is of a pendant design. The two samples were removed from the leading tube in the pendant close to the bottom. Tubes A and B are separated by five pendants, but are from the same relative position, near the bottom of the pendant from the front tube. No debris or exfoliated scale was noted in either bend. The tubes are $2\frac{1}{2}$ in OD × 0.220 in MWT, SA-213 T-5 material.

Figure 5.1.1 shows the ring section removed from tube A. Figure 5.1.2 presents the ring sections from tube B through the bulge and at the end of the tube sample. Dimensional measurements of OD and wall thickness were made in four equally spaced positions around the circumference. These data are given in Table 5.1.1.

There has been some wall thinning from tube A, and that is reflected in the OD dimensions and wall thickness. However, the greatest extent of the reduction in the wall was only 5 mils below the specified minimum of 0.220 in. The case for tube B is different. The tube had swollen 2.4% at the end, and the wall thickness had been reduced to 0.140 in. At the bulge, the swelling was 11.5%, and the minimum wall thickness measured was 0.075 in. Also noted were substantial differences in the amount of steam-side scale. Tube A had a maximum of 10 mils; and tube B, a maximum of 79 mils at the thinnest wall measured.

The microstructures are given in Figs. 5.1.3 through 5.1.6. Tube A has a microstructure that is ferrite and spheroidized carbides, but the carbide particle size is too fine to be readily resolved at 500×. The hardness is Rockwell B 75. The microstructure through the bulge, Fig. 5.1.4, shows

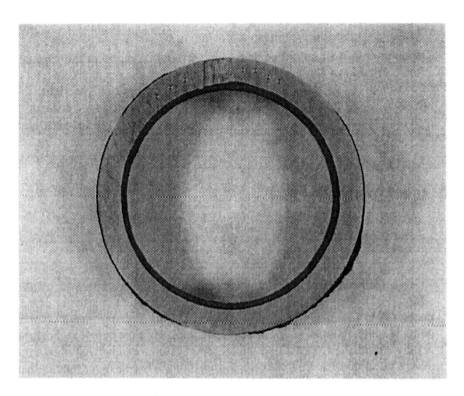

FIGURE 5.1.1. Ring section from tube A (Magnification 1.2×).

FIGURE 5.1.2. Ring sections from tube B, through the bulge and at the end of the sample (Magnification 0.8×).

CASE HISTORIES 295

TABLE 5.1.1. Dimensional Measurements

Location	Position	OD, in	Wall, in	R_B Hardness
Tube A	12:00	2.470	0.230	75
	3:00	2.475	0.225	
	6:00		0.220	
	9:00		0.215	
Next to bulge, tube B	12:00	2.560	0.140	62
	3:00	2.525	0.175	
	6:00		0.160	
	9:00		0.185	
	Tube Swelling 2.4%			
Bulge, tube B	12:00	2.790	0.075	62
	3:00	2.720	0.135	
	6:00		0.175	
	9:00		0.125	
	Tube Swelling 11.5%			

FIGURE 5.1.3. Microstructure through tube A at the thinnest measured wall. R_B hardness is 75, and ID scale is 10 mils (0.010 in) (Magnification 500×, etched).

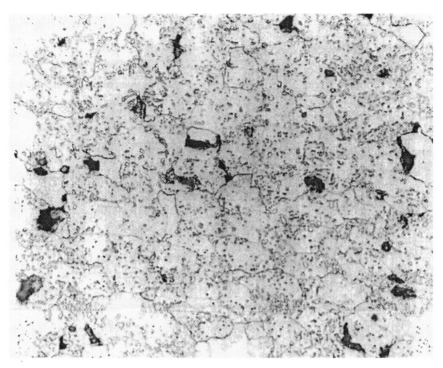

FIGURE 5.1.4. Microstructure through tube B at the thinnest measured wall. R_B hardness is 57, and ID scale is 79 mils (0.079 in) (Magnification 500×, etched).

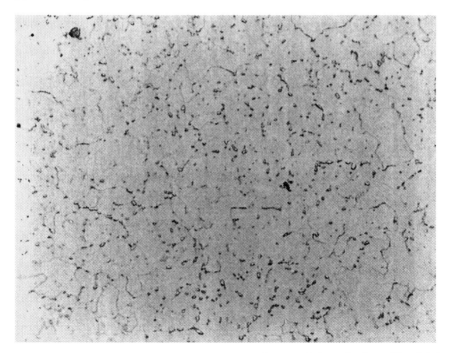

FIGURE 5.1.5. Microstructure 180° around perimeter of the tube from Fig. 5.1.4. R_B hardness is 62, and ID scale is 42 mils (0.042 in) (Magnification 500×, etched).

FIGURE 5.1.6. Microstructure from the end of the sample, tube B, in line with the microstructure in Fig. 5.1.4. R_B hardness is 62, and ID scale is 48 mils (0.048 in) but with some exfoliation (Magnification 500×, etched).

large particles of carbide and extensive creep damage; R_B hardness is 62. The location 180° around the tube perimeter from the thinnest wall measured in Tube B and at the end of the tube sample are given in Figs. 5.1.5 and 5.1.6. The carbide particle sizes here are large enough to be readily resolved, but the amount of creep damage is minimal, if any.

The likely explanation for these widely differing microstructures between two tubes is unbalanced steam flow that does not maintain tube-metal temperature within the desired range.

To establish actual steam-temperature conditions, one should make thermocouple measurements at the outlet of the second reheater. This temperature measurement will provide information on the start-up condition when steam flow has not been fully established and on the steady-state conditions during normal operation. Depending on the tube-to-tube steam-temperature variation, a redesign, or certainly a readjustment, of the steam flow from tube to tube will be necessary to ensure proper tube-metal temperatures during operation.

Case History 5.2

Microstructure of 15-Year-Old Superheater Tubes

BOILER STATISTICS

Size	650,000 lb steam/hr (295,000 kg/hr)
Steam temperature	950°/950° (510°C/510°C)
Steam pressure	1675 psig (120 kg/cm^2)
Fuel	Natural gas

This example deals with the investigation of several superheater sections taken from the high-temperature superheater of a boiler that had been in service for approximately 15 years (Fig. 5.2.1). No failures were observed in any of the samples, and all tubes are 2-in OD × 0.300-in-thick wall, SA-213 T-22 alloy steel.

The visual examination of these particular samples revealed the following:

A. All contained a thick, tightly adhering scale on the OD surface, which measured 0.011–0.030 in thick. The ID surface also exhibited a thick scale, which measured 0.019 in.

FIGURE 5.2.1. Microstructure typical of all superheater sections examined. The structure is one of spheroidized carbides in a ferrite matrix, indicating exposure above the design temperature for extended period of time (Magnification 500×, nital etch).

B. No wall thinning was noted in any of the tubes.
C. No swelling or bulging was observed on any of the tubes.
D. No significant corrosion from outside sources was noted on either the ID or the OD.

The microstructure of all samples revealed complete spheroidization of the carbide in ferrite, indicating exposure to high metal temperatures for prolonged periods of time (Fig. 5.2.1). The microstructure can be attributed to the nearly 15 years of service, and the pearlite would be expected to have spheroidized completely. However, there is further danger in the insulating effect of the nearly 20 mils of scale observed on the ID surface. The thermal conductivity of magnetite is about 2% that of steel.

Using the thermal analysis given in Chapter 2, we can estimate the tube-metal-temperature increase resulting from the 0.019-in scale. For this example the following data are used:

$U_0 = 16$ Btu/hr · ft^2
$Q/A_0 = 15,000$ Btu/hr · ft^2
$T_0 = 2000°$F (538°C)
$T_3 = 950°$F (510°C)
$h_i = 485$ Btu/hr · ft^2

Tubing is 2-in OD × 0.300-in-thick wall, grade T-22 steel.

$k_1 = 0.342$ Btu/hr · ft · °F
$k_2 = 16.7$ Btu/hr · ft · °F

From these data and Eq. (2.2.2), h_0 is found to be 17.4 Btu/hr · ft^2. With 0.019 in of scale,

$r_3 = 1.0$ in (0.0833 ft)
$r_2 = 0.710$ in (0.0592 ft)
$r_1 = 0.691$ in (0.0576 ft)

Equation (2.2.3) is used to calculate a new Q/A_0, 13,700 Btu/hr · ft^2, and from Eq. (2.2.7), T_3 is 1115° (600°C) or about 75°F (40°C) above the no-scale condition.

The Larson–Miller parameter $P = T(20 + \log t)$, where T is absolute temperature in °R (°F + 460°) and t is time in hours, has been used to describe changes in steel that are diffusion controlled. Larson and Miller[9] originally used it to estimate stress-rupture performance at 100,000 hr from high-temperature tensile data and short-time stress-rupture data. It has also been used to relate microstructural changes, such as spheroidization and graphitization, which are measured at high temperatures where these changes occur fairly rapidly, with expected changes that occur at lower temperatures over longer periods of time.

These superheater tubes are now operating at 75°F (42°C) above the no-scale condition. For purposes of calculation, if the design assumes a 1% creep in 100,000 hr at 1040°F (560°C), then

$$P = 37,500 = (1040 + 460)(20 + \log 100,000)$$

At 1115°F, P is still 37,500 but t drops to 6400 hr, which is less than one year.

Unless the thick ID scale is removed by suitable chemical-cleaning procedures, these higher operating temperatures will cause creep failures with increasing frequency. Further complications arise from these higher metal temperatures and heavy scale:

1. The rate of OD oxidation increases with temperature, so increased metal wastage will occur.
2. The thick ID scale tends to spall off, and tiny particles of iron oxide can lead to turbine-blade erosion.
3. The decrease in heat absorption by the superheater leads to loss of thermal efficiency.

Case History 5.3

Creep Failure in a High-Temperature Superheater

BOILER STATISTICS

Size	1,310,000 lb steam/hr (595 kg/hr)
Steam temperature	1005°F/1005°F (540°C/540°C)
Steam pressure	1980 psig (140 kg/cm^2)
Fuel	Pulverized coal

The high-temperature superheater tube is grade SA-213 T-22 steel, 1¾-in OD × 0.260-in-thick wall. The visual examination of the tube revealed the following:

A. The failure was a 1½-in-long longitudinal split, showing virtually no reduction in wall thickness; see Fig. 5.3.1.
B. The wall thickness 1 in away from the end of the failure and in the same plane measured 0.230 in.
C. The wall thickness 180° from the failure measured 0.300 in.
D. The OD scale and deposits measured 0.052 in; the ID scale, 0.032 in.

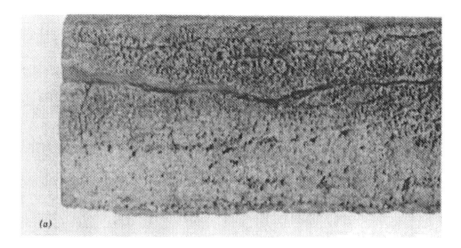

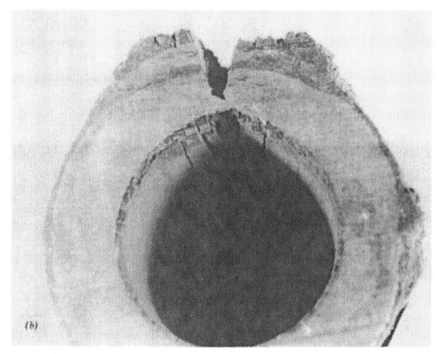

FIGURE 5.3.1. The as-received superheater tube showing the longitudinal split. Note that there is virtually no expansion of the tube at the failure area.

The microstructure of the failed area shows a completely spheroidized structure; see Fig. 5.3.2. The surface of the tube shows an intergranular penetration of the OD oxide; see Fig. 5.3.3. Figure 5.3.4 shows creep voids in the vicinity of the failure.

From Eq. (2.2.4), the temperature increase is calculated to be 80°F (44°C). At the time of failure, the outer skin temperature was about 1200°F

FIGURE 5.3.2. The microstructure of the failed area showing a complete spheroidization of the carbides (Magnification 500×, nital etch).

FIGURE 5.3.3. The OD surface of the as-received tube sample, showing intergranular penetration of the oxide (Magnification 500×, nital etch).

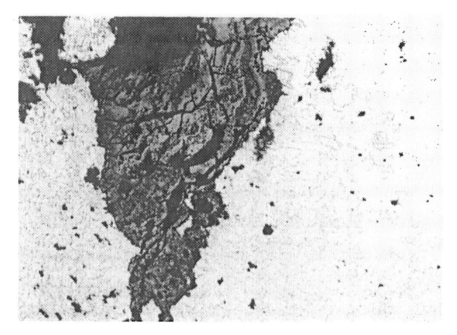

FIGURE 5.3.4. The microstructure at the tip of the failure crack. Note the creep void in the vicinity of the crack tip. Creep voids may be distinguished from graphite particles because the graphite dissolves in a 1700°F (925°C) heat treatment, whereas creep voids remain (Magnification 500×, unetched).

(605°C); the safe design oxidation limit for grade T-22 steel was 1125°F (605°C) at the time this boiler was designed in the early 1960s.

The failure appearance is a classic example of creep failure; many voids precede the failure crack. At high temperatures, the grain boundaries are a source of weakness. Where planes of atoms are in disarray, the binding energy between atoms is a little less than for the regular atomic arrangement within crystals. Adjacent grains slide, and where three grains come together a void will start. Grain-boundary sliding is accompanied by dislocation movement; dislocations are defects of atomic size that make up the grain boundary. At the intersection of three grains, these atomic-size defects collect or pile up and form the void.

The cause of failure was creep. Thick ID scale increased the overall thermal resistance of the tube so that tube-metal temperatures increased about 80°F (44°C). Tube-metal temperatures above 1200°F (650°C) for grade T-22 steel for an extended period of time led to the failure of this tube.

Case History 5.4

Decarburization of a Waterwall Tube

BOILER STATISTICS

Size	165,000 lb steam/hr (75,000 kg/hr)
Steam temperature	370°F (120°C)
Steam pressure	160 psig (11 kg/cm^2)
Fuel	Oil

This waterwall tube sample came from approximately the quarter point of the front wall opposite and at about the same elevation as the burners.

The material is grade SA-178 A steel, 2½-in OD × 0.105-in-thick wall. The visual observations revealed the following:

A. The tube did not contain any ruptures, but did show a multitude of large blisters; see Fig. 5.4.1.
B. The blisters occurred only on the fire-side surface of the tube.
C. The fire side of the tube was coated with a black deposit from the products of combustion.
D. The inside surface of the tube was coated with a brownish gray deposit, which measured 0.010 in thick on the entire internal surface of the tube.
E. The wall thickness of the tube averaged 0.113 in, except, of course, at the blisters where it was slightly thinner.

FIGURE 5.4.1. The as-received tube section showing the large blisters on the fire-side surface.

Samples were taken for microstructural analysis from the fire side at and between the blisters, and from the backside of the tube in a plane with blisters. The structure between the blisters shows an *in situ* breakdown of the pearlite; see Fig. 5.4.2. Here the pearlite colonies are clearly defined in a matrix of ferrite, but the carbide within the pearlite has spheroidized. The microstructure at the blisters themselves shows a nearly complete decarburization. Figure 5.4.3 shows a microstructure of ferrite with only a few spherical carbide particles present. The specimens taken 180° away from the blister show no exposure to elevated temperatures and are the normal microstructure for this low-carbon steel; see Fig. 5.4.4. Here the microstructure of pearlite and ferrite is clearly discernible with no high-temperature deterioration of the pearlite.

The failures appeared to be caused by the steam-side deposit that impedes heat transfer and caused localized overheating to a greater or lesser extent. Decarburization is caused by the carbon dioxide in the flue gas reacting with carbon in the steel to form carbon monoxide, $CO_2 + C = 2CO$, thus depleting the OD surface of carbon. On the steam side, the reaction between carbon and steam to form hydrogen and carbon monoxide, $H_2O + C = H_2 + CO$, can deplete the ID surface of carbon. In either

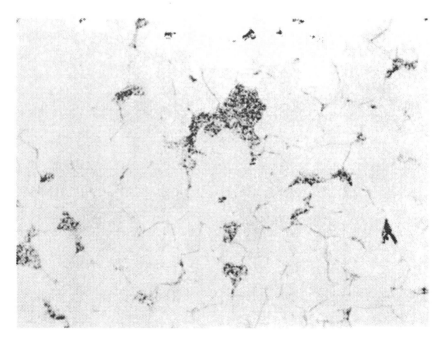

FIGURE 5.4.2. Microstructure between the blisters near the OD surface shows an *in situ* breakdown of the pearlite. The pearlite colonies are still clearly defined; the cementite is completely spheroidized within the colony (Magnification 500×, nital etch).

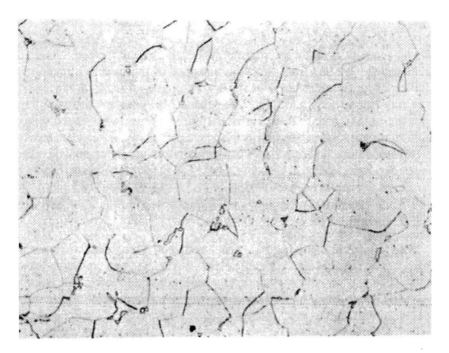

FIGURE 5.4.3. Microstructure near the OD surface through the blisters on the fire side of the tube. Note the absence of many carbide particles, indicating nearly complete decarburization of the tube at these blisters (Magnification 500×, nital etch).

FIGURE 5.4.4. Microstructure of the tube taken 180° away from the blister shows a completely normal ferrite and pearlite microstructure (Magnification 500×, nital etch).

case, the localized high temperatures aid the depletion of carbon, causing a decrease in the high-temperature strength of the material, which leads to the formation of these blisters. The blisters were caused by overheating and by the loss of strength as a result of the reduction in carbon content.

Case History 5.5

Carburization of Type 304 Stainless-Steel Gas Burner

BOILER STATISTICS

Size	1,800,000 lb steam/hr (820,000 kg/hr)
Steam temperature	950°F/950°F (510°C/510°C)
Steam pressure	1765 psig (125 kg/cm^2)
Fuel	Natural gas

The gas ring was a 4-in Schedule 40, welded AISI Type 304 stainless-steel pipe. While no photographs exist of the failure, visual examination revealed the following:

A. A large irregularly shaped hole, about 3 in in diameter, existed on the fire side, that is, the hottest side, of the gas ring.
B. Both the OD and ID were covered with the usual oxide scale.
C. Wall thickness at the edge of the hole was less than $\frac{1}{16}$ in thick; no more precise measurement could be made because of the scale.

Specimens were taken from the failed area, at the edge of the hole, and 6 in away from it and were prepared for microstructural analysis. The microstructure at the failure revealed some minor intergranular penetration by oxide and very coarse grains, indicating exposure to high metal temperatures, about 1800–2000°F (980–1090°C); see Figs. 5.5.1 and 5.5.2.

The structure away from the failure area, Fig. 5.5.3, shows the same small amount of grain-boundary attack. The grain size is not quite as large as at the failure area, and there appears to be quite a bit of surface carburization extending for a few grain diameters into the material.

The cause of the failure is attributed to carburization of the stainless steel. In methane (CH_4) atmospheres, at suitably high temperatures, the surface of the stainless steel is carburized. Carbon enters the surface, $CH_4 = C + 2H_2$, and at these high temperatures diffuses inward to form chromium carbide ($Cr_{23}C_6$). The formation of chromium carbide ties up the chromium, locally reducing the chromium content below 11%; the metal is no longer stainless, and its oxidation and corrosion resistance are severely reduced; the oxidation–corrosion products spall off until the metal thickness is reduced and failure occurs. Reducing atmospheres favor carburization by the decomposition of carbon monoxide into carbon

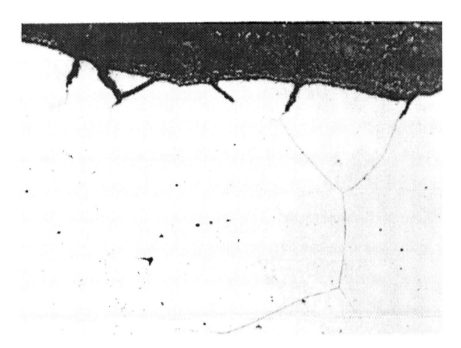

FIGURE 5.5.1. The microstructure in the vicinity of the failure of a Type 304 stainless-steel gas ring. Note the minor amount of intergranular penetration by the oxide and the very coarse austenitic grains, indicating exposure to high metal temperatures, about 1800–2000°F (980–1090°C) (Magnification 250×, electrolytic etch).

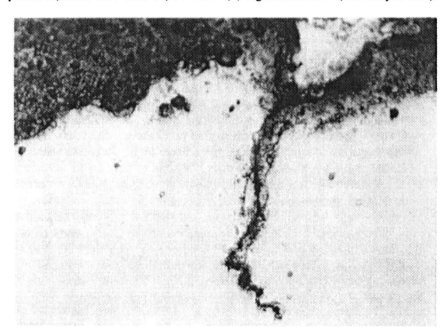

FIGURE 5.5.2. The same area as in the previous figure but at 500× magnification, electrolytic etch.

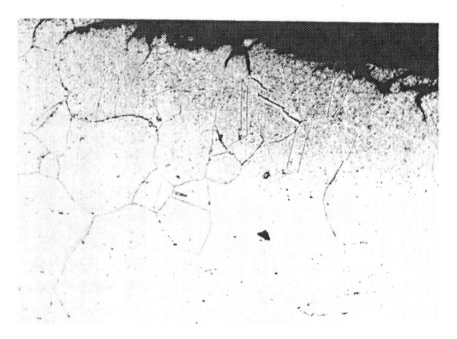

FIGURE 5.5.3. The structure away from the failure area shows the same small amount of oxide attack of the grain boundaries (Fig. 5.5.1); the grain size is not as large, and there appears to be some surface carburization (Magnification 250×, electrolytic etch).

and carbon dioxide and by the incomplete combustion of methane as mentioned before. The free carbon, $2CO = CO_2 + C$, is deposited on the surface to start the carburization-failure mechanism. High temperature, estimated to be around 1800–2000°F (980–1090°C), promotes the rapid diffusion of carbon inward to form $Cr_{23}C_6$.

Case History 5.6

Steam-Washed and Steam-Eroded Superheater Tubes

BOILER STATISTICS

Size	3,000,000 lb steam/hr (1,360,000 kg/hr)
Steam temperature	1005°F/1005°F (540°C/540°C)
Steam pressure	2400 psig (170 kg/cm²)
Fuel	Oil

Tubes in the vicinity of a rupture are often exposed to the high-velocity steam erupting from the point of failure. The high-velocity steam can erode the tube surface. Figure 5.6.1 shows three such examples. In no

310 CHAPTER 5 FAILURES CAUSED BY GAS–METAL REACTIONS

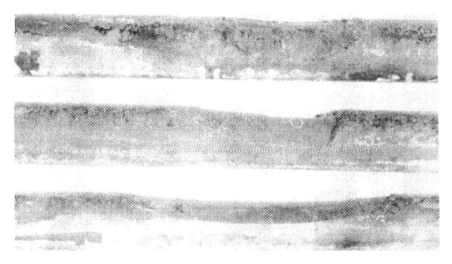

FIGURE 5.6.1. Three examples of erosion caused by escaping steam from a nearby rupture. The tubes are grade SA-213 T-22 steel, with 1.5-in OD × 0.287-in-thick wall and 1.75-in OD × 0.400-in-thick wall.

case was the tube eroded to the point of failure; but, as can be seen, both the surface deposits and some tube metal have been removed. Had the tubes not been replaced, premature failure most certainly would have occurred.

Case History 5.7

Steam-Washed and Steam-Eroded Reheater Tube

BOILER STATISTICS

Size	1,600,000 lb steam/hr (725,000 kg/hr)
Steam temperature	1005°F/1005°F (541°C/541°C)
Steam pressure	1980 psig (140 kg/cm^2)
Fuel	Coal

This example, similar to the previous case history, shows the effects of steam-washed tubes caused by adjacent tube ruptures. Figure 5.7.1 shows the severity of the erosion by steam. In this case, the wall of the tube is eroded to the point of failure, as can be seen in the figure.

FIGURE 5.7.1. A steam-washed reheater tube of SA-213 T-11, 1.75-in OD × 0.134-in-thick wall. This example is similar to the previous one, but, the tube wall is completely eroded.

Case History 5.8

Water-Washed and Water-Eroded Waterwall Tubes

BOILER STATISTICS

Size	2,800,000 lb steam/hr (1,270,000 kg/hr)
Steam temperature	1005°F/1005°F (541°C/541°C)
Steam pressure	3735 psig (260 kg/cm^2)
Fuel	Coal

The tubes involved are from the lower slope of the nose arch and were identified as tubes 98 and 99. The rupture actually occurred in tube 98, while tube 99 failed as a result of erosion from escaping steam emitting from the rupture in tube 98. Figure 5.8.1 shows the two tube samples.

312 CHAPTER 5 FAILURES CAUSED BY GAS–METAL REACTIONS

FIGURE 5.8.1. The as-received waterwall tubes showing the failure and the adjacent steam-washed tube failure. The original failure is in the lower tube; the steam-washed tube is the upper tube in the figure. The tubes are 1⅜ in OD × 0.210-in-thick wall, Type SA-213 T-11 steel.

Case History 5.9

Localized Hot Spots and Failures in Reheater Pendant Tubes

Pendant-style superheaters, especially reheaters, can fail at the very bottom of the pendant. This example deals with failures caused by exfoliation of steam-side scale or other debris that collects at the bottom of the loop. The steam-side scale acts as an insulating layer to heat transfer. The net effect is to raise tube-metal temperatures and to hasten the onset of creep failures. The failure is localized and confined to the region of the debris.

Figure 5.9.1 shows an example. The location and orientation of the failure are at the bottom of the pendant, and the opening is at the very bottom of the loop. Figure 5.9.2 is a close-up of the failure. The rupture is a narrow, fissure type and representative of a creep or stress-rupture failure. The ID surface, Fig. 5.9.3, shows the remnants of considerable debris confined to a fairly narrow region of the tube. Figure 5.9.4 shows another example from a similar pendant but a different boiler. Ring sections of this sample, taken through the failure and at the straight end of the sample, show thinning and debris at the failure, see Figure 5.9.5.

The microstructural evidence shows the failure to be a creep or stress-rupture type, Figs. 5.9.6 and 5.9.7. The microstructure ½ in away from the

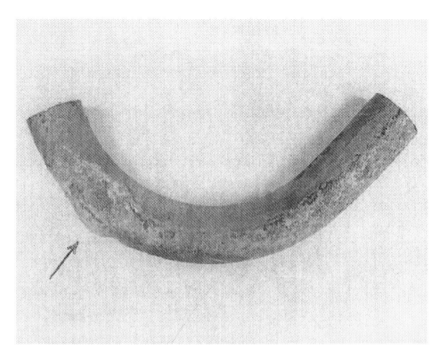

FIGURE 5.9.1. As-received reheater tube. The sample location is at the bottom of a pendant, and the failure is at the very bottom of the loop (Magnification 0.3×).

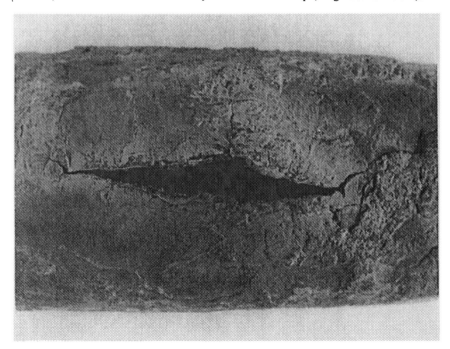

FIGURE 5.9.2. Close-up of the failure. The rupture is a narrow, fissure type, representative of a creep or stress-rupture failure (Magnification 1.3×).

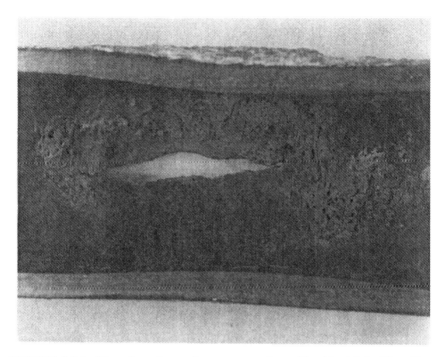

FIGURE 5.9.3. ID surface shows the remnants of considerable debris (Magnification 1×).

FIGURE 5.9.4. Similar failure from another tube from another boiler with same type of problem (Magnification 1×).

FIGURE 5.9.5. Ring sections from straight end of sample and through the failure show the thinning and debris (Magnification 0.6×).

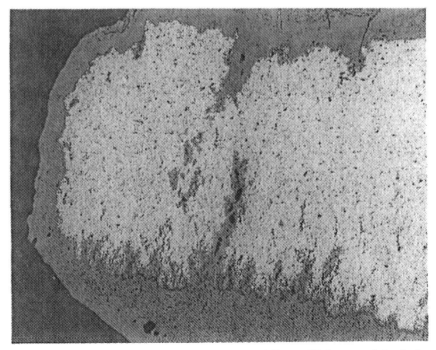

FIGURE 5.9.6. Fracture lip is typical of creep failures. ID is toward the bottom (Magnification $37\frac{1}{2}\times$, etched).

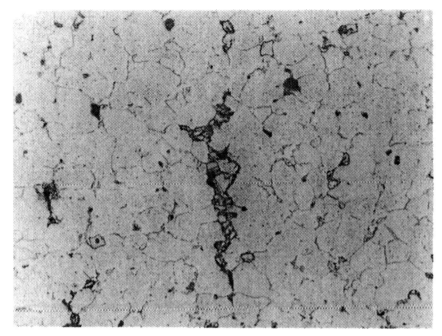

FIGURE 5.9.7. Microstructure is fully spheroidized carbides, ferrite, and the expected creep damage (Magnification 500×, etched).

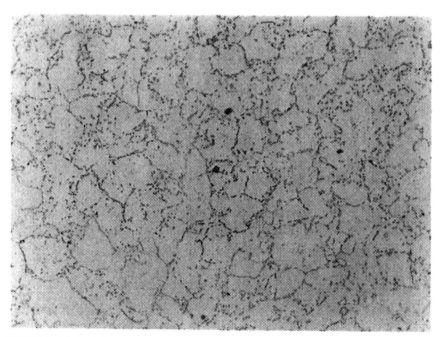

FIGURE 5.9.8. Microstructure ½ in from failure lip is spheroidized carbides, ferrite, and much less creep damage (Magnification 500×, etched).

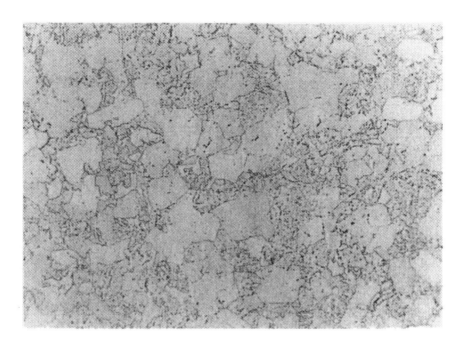

FIGURE 5.9.9. Microstructure 180° around tube perimeter from failure is ferrite and spheroidized carbides, but pearlite colonies are still vaguely visible (Magnification 500×, etched).

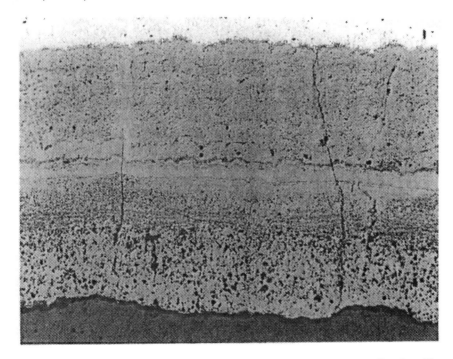

FIGURE 5.9.10. ID scale near failure is 55 mils (0.055 in). (Magnification 50×, etched).

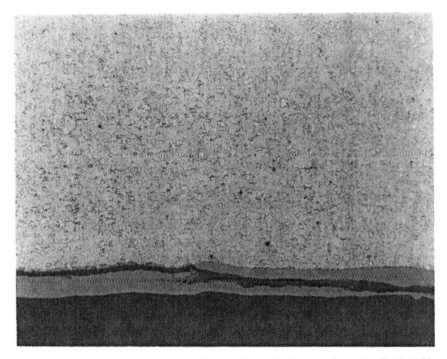

FIGURE 5.9.11. ID scale 6 in from failure in line with the opening is 5 mils (0.005 in) and shows signs of exfoliation (Magnification 50×, etched).

failure, Fig. 5.9.8, shows a microstructure with much less creep damage. For 180° around the tube perimeter from the failure, the microstructure is ferrite and spheroidized carbide, but the pearlite colonies are still vaguely visible (see Fig. 5.9.9). These dramatic changes in microstructure indicate excessive tube temperature is confined to the area of debris only.

The ID scale at the failure and 6 in from the failure but in line with the opening, Figs. 5.9.10 and 5.9.11, displays these differences. At the failure, or near the failure, the scale is 55 mils; in the tube away from the debris, it is only 5 mils, but does show some signs of exfoliation.

This type of failure at the bottom of a pendant occurs because of exfoliation of steam-side scale and collection of the oxide at the bottom of the loop. Excessive scale at the bottom of the pendant can be detected by radiography, but the scale is difficult to remove. In reheaters chemical cleaning is difficult, if not impossible. The use of stainless steel at the toe of these pendants provides a material better able to withstand the higher temperatures, but it requires two dissimilar-metal welds with the attendant problems of those weld types (see Chapter 7).

Case History 5.10

Carburization Corrosion: Stainless-Steel Superheater and Reheater Tubes

The formation of NO_x and NO_x emissions can most easily be reduced by lowering the combustion flame temperature. More nitrogen oxide will form at higher flame temperatures by the direct reaction of the nitrogen and oxygen in the combustion air. Reducing the flame temperature by off-stoichiometric or staged combustion is perhaps the most common technique employed. However, off-stoichiometric firing can lead to unburned carbon in the fly ash and excessive carbon monoxide in the flue gas.

All fossil fuels contain hydrogen and carbon in various amounts and compounds. Natural gas is nearly pure methane (CH_4), 75% carbon and 25% hydrogen by weight. Fuel oils usually have more carbon and less hydrogen. Coals vary widely in their composition, but all contain carbon as well as other hydrocarbons. All give off heat energy when burned, and all behave in a similar fashion when burned with air in a boiler.

The hydrogen (H_2) component burns first and completely, in air or oxygen (O_2), to water vapor (H_2O),

$$2H_2 + O_2 = 2H_2O \qquad (5.10.1)$$

and gives off considerable heat energy. The carbon (C) behaves in a different fashion. There are two common oxides of carbon: carbon monoxide (CO) and carbon dioxide (CO_2). Depending on the relative amount of oxygen, one or both of these oxides of carbon will form; thus,

$$2C + O_2 = 2CO \qquad (5.10.2)$$

$$C + O_2 = CO_2 \qquad (5.10.3)$$

Combinations of hydrogen and carbon in fuel will burn to water vapor, carbon monoxide, carbon dioxide, or, perhaps in the absence of sufficient oxygen, unburned carbon or soot. For simplicity we use methane to illustrate:

$$CH_4 + 2O_2 = CO_2 + 2H_2O \qquad (5.10.4)$$

$$2CH_4 + 3O_2 = 2CO + 4H_2O \qquad (5.10.5)$$

$$CH_4 + O_2 = C + 2H_2O \qquad (5.10.6)$$

Boiler steels develop corrosion resistance by the formation of protective oxide scales in oxidizing environments. Under reducing conditions, porous oxides may form, or, in the presence of sulfur in the fuel, less-protective sulfide films may occur. When there are fuel-ash corrosion

320 CHAPTER 5 FAILURES CAUSED BY GAS–METAL REACTIONS

conditions superimposed on the reducing conditions, the porous oxides or sulfides are more easily dissolved in the liquid-ash species. This is true for stainless steels as well as ferritic steels, even though 304-type stainless steels usually have satisfactory corrosion resistance under fuel-ash corrosion conditions.

In the high-temperature portions of a superheater or reheater, when there are both fuel-ash corrosion and reducing conditions, there is the possibility of carburization corrosion of stainless steel. Figure 5.10.1 shows ring sections removed from reheater and superheater tubes made of 304H stainless steel, in service about six years in a coal-fired boiler. Both tubes have suffered considerable wastage, and the microstructures show the effects of carbon diffusion and absorption in the regions of most serious wastage; refer to Figs. 5.10.2 and 5.10.3.

The microstructural features in Figs. 5.10.2 and 5.10.3 were revealed by a nital etch. The OD surface is readily corroded by this mild acid (5% nitric acid in alcohol), an etchant that is not normally used for stainless steel. Etching of a metallographic sample is, in reality, a controlled corrosion test used to reveal the microstructural details. The more corrosion-resistant 304H stainless steel usually is immune to nital etchants. The contamination of the surface by carbon has decreased the local corrosion

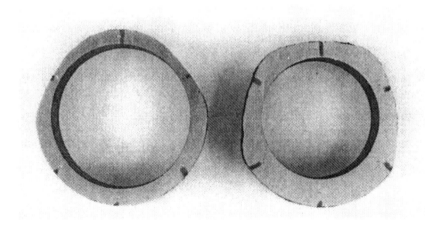

FIGURE 5.10.1. Ring sections from reheater (left) and superheater (right) tubes that suffer carburization corrosion. Note wastage (Magnification ~1×).

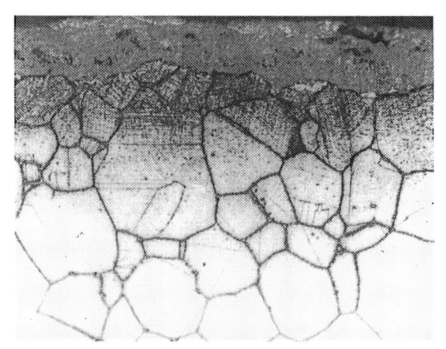

FIGURE 5.10.2. Microstructure of reheater tube in region of severe wastage (Magnification 500×, nital etch).

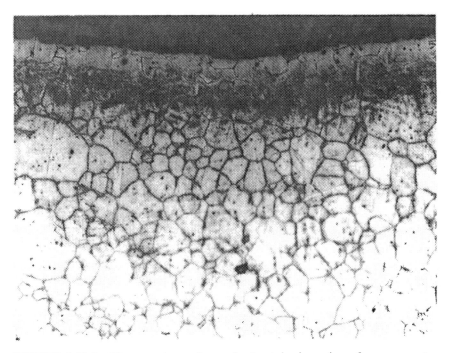

FIGURE 5.10.3. Microstructure of superheater tube in region of severe wastage (Magnification 100×, nital etch).

322 CHAPTER 5 FAILURES CAUSED BY GAS–METAL REACTIONS

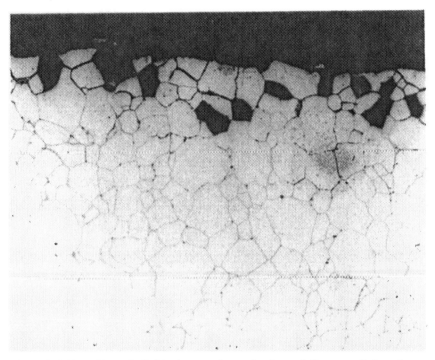

FIGURE 5.10.4. Many individual austenite grains have fallen out of the reheater-tube sample along the OD, proof of intergranular corrosion (Magnification 200×, etched).

resistance to the point where microstructural features are displayed by this mild etch.

One further test was performed to prove the carbon content at the surface. The corrosion debris was wire-brushed from the surface of the superheater tube, and filings were taken of the outer 10 mils (0.010 in) or so. The carbon content was 0.43%. Chemical analysis of the tube shows that it contains a maximum carbon content of 0.06%, well within the specification requirements of SA-213 TP-304H.

The surface region is also severely sensitized. Carbon diffusion is more rapid along grain boundaries than through the austenite grains. Corrosion penetration is also more rapid along these same sensitized grain boundaries. Individual grains surrounded by corrosion debris can fall out during metallographic-sample preparation. The resultant structure shows holes like the irregularly shaped austenite grains; see Fig. 5.10.4.

As carbon is added to the stainless steel, chromium carbides form, which reduce the chromium content of the remaining steel and render the steel less corrosion-resistant—in effect, a sensitized surface. Carbides form preferentially along the grain boundaries, since atomic mobility is greater. Thus, the grain boundaries will be carburized to a greater depth than the bulk austenite grains, as is evident in the two microstructures. The reaction is similar to the sensitization of austenitic stainless steel by the formation of chromium carbide along the austenite grain boundaries.

In summary, these superheater and reheater tubes suffer wastage from a combination of reducing conditions within the flue gas and coal-ash corrosion on the tube surface, which leads to the rapid carburization of the 304H by the unburned carbon or carbon monoxide. The liquid-coal-ash species dissolves the protective oxide scales and allows carburization to occur. As staged-combustion techniques become more prevalent as a means for the reduction in NO_x emissions, alloys more resistant to carburization than 304-type stainless steel will be required. In the absence of improved alloys, better balance in the tube-to-tube steam temperatures will help keep peak metal temperatures low enough to keep carburization and wastage rates at manageable levels.

REFERENCES

1. M. H. Davies, M. T. Simnad, and C. E. Birchenall, "On the Mechanism and Kinetics of the Scaling of Iron," *J. Metals*, Oct. 1951, pp. 889–896.
2. H. L. Solberg, G. A. Hawkins, and A. A. Potter, "Corrosion of Unstressed Steel Specimens and Various Metals by High-Temperature Steam," *Trans ASME*, May 1942, pp. 303–316.
3. F. B. Eberle and J. H. Kitterman, "Scale Formations on Superheater Alloys Exposed to High Temperature Steam," in *Behavior of Superheater Alloys in High Temperature, High Pressure Steam*, ASME, New York, 1968.
4. F. B. Eberle and J. L. McCall, "Electron-Microprobe Study of Scale Formations on Cr–Mo Superheater Tubing after Long-Time Exposure to 2000 PSI Steam of 1100°F and 1200°F," *Trans ASME*, April 1965, pp. 205–300.
5. D. N. French, "High-Nickel Joints Unite Dissimilar Steels," *Welding Design and Fabrication*, May 1981, pp. 92–93.
6. P. J. Grobner, C. C. Clark, P. V. Andreae, and W. R. Sylvester, "Steam Side Oxidation and Exfoliation of Cr–Mo Superheater and Reheater Steels," Presented at Corrosion '80, NACE, Chicago, Illinois, March 3–7, 1981, Paper No. 172.
7. *ASM Metals Handbook*, ASM International, Metals Park, Ohio, 1948.
8. A. J. Sedriks, *Corrosion of Stainless Steels*, Wiley, New York, 1979, pp. 251–254.
9. F. R. Larson and J. Miller, "A Time-Temperature Relationship for Rupture and Creep Stresses," *Trans ASME*, July 1952, pp. 765–771.

CHAPTER SIX

CORROSION-CAUSED FAILURES

This chapter will deal with the degradation of boiler steels by their environment. First, we will deal with the special circumstances of fire-side corrosion by the constituents in fuel ash with special emphasis on coal- and oil-fired boilers. The final portion will deal with steam-side corrosion as a result of boiler-feedwater chemical upsets.

BASIC PRINCIPLES

Regardless of the form that corrosion takes, the underlying principles are essentially the same.[1-3] Corrosion is an electrochemical process; there is a flow of electrons from one part of the metal surface to another through a suitable conducting medium. The term *anode* is used to describe that portion of the surface at which chemical oxidation occurs. For example, an iron anode dissolves to ferrous ions. A cathode is that portion of the surface at which chemical reduction occurs. For example, hydrogen ions become hydrogen gas, or copper ions become copper metal. Anodic reactions lead to corrosion of the metal or oxidation. Cathodic reactions are reduction reactions that can lead to the evolution of hydrogen.

Typical reduction (cathode) reactions are

$$2H^+ + 2e^- = H_2$$
$$Cu^{2+} + 2e^- = Cu$$
$$Fe^{3+} + 3e^- = Fe$$

Typical oxidation (anode) reactions are

$$Zn = Zn^{2+} + 2e^-$$
$$Fe = Fe^{2+} + 2e^-$$

For ease of understanding the reactions at anodes and cathodes, reference is made to the electrochemical action in an electric dry cell (see Fig. 6.1). A dry cell is made up of a zinc metal cup filled with an ammonium chloride (NH_4Cl) solution as the electrolyte and a carbon electrode. The zinc container is the anode and the graphite electrode is the cathode, forming an example of a dissimilar-electrode cell. When the external circuit is closed, electric current is generated by the corrosion of the zinc, and electrons flow from the anode to the cathode through the external circuit. Other examples of dissimilar-electrode cells are a copper pipe connected to an iron pipe, and magnesium anodes used to protect buried steel pipe (here the anode is expected to corrode and is an example of galvanic protection).

Concentration cells have electrodes of the same metal in contact with solutions of differing composition. The differential-aeration cell is most common; it accounts for the rusting of steel and crevice corrosion. The area under a scab of rust or in a crevice has a low concentration of oxygen and is anodic; the edge of the blister of rust has a higher oxygen level and is cathodic. Anodic areas dissolve and cathodic areas are protected.

Differential-temperature cells have electrodes of the same metal that are at different temperatures. The high-temperature corrosion of superheater or reheater tubes in liquid ash may be examples of this phenomenon. Tubes will

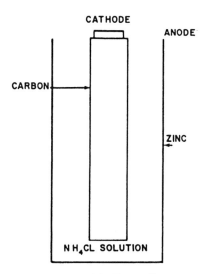

FIGURE 6.1. Dry cell.

show a row of large shallow pits at about the 2 and 10 o'clock positions with a conically shaped ash deposit between them. The insulating effect of the ash leads to high heat transfer at the 2 and 10 o'clock positions, which, in turn, leads to rapid corrosion and a double row of grooves on the tube.

TYPES OF CORROSION

The reaction known as corrosion, between a metal and its environment, may take many forms. Several more common types of corrosion will be described.

Uniform Attack

This includes the formation of a protective magnetite (Fe_3O_4) layer on the inside of a superheater or reheater tube, by the reaction of steam with steel. As the name implies, the thickness of the Fe_3O_4 corrosion product, for all practical purposes, is the same from point to point on the inside surface. Other examples are the familiar tarnishing of silver and the rusting of steel. See Fig. 5.2 for an example of uniform attack.

Pitting

Pitting is a localized attack in which the rate of corrosion is greater in some areas than others. The pits may be either large and shallow or small and deep; and the depth of the pitting is sometimes referred to as the pitting factor, which is the ratio of the deepest metal penetration to the average metal loss.

Figure 6.2 is an example of severe pitting attack from a small industrial boiler. Figure 6.2a shows the general appearance of the tube ID, and Fig. 6.2b displays the cross section through the pits. Two factors led to the severe oxygen pitting; the pH of the boiler feedwater dropped from the normal 8.5–10 range to 6.5–7; during lay-up, vents were opened without first drying the unit. The combination of slightly acidic moisture and oxygen from the air caused the pitting in a boiler that had previously operated more than 20 years without serious problems.

A second example of pitting attack is seen in a return bend of type SA-178 A material. Figure 6.3a shows the ID of the tube; note the row of pits, nearly a complete groove, along the neutral axis of the bend. Figure 6.3b displays a cross section through the pits; note the cold-worked grains along the pit. During bending, a defect on the mandrel severely scored the tube ID. The heavily cold-worked structure of the bend became the anode and was corroded by a dissimilar-metal-galvanic-cell effect when condensed steam filled the tube during storage.

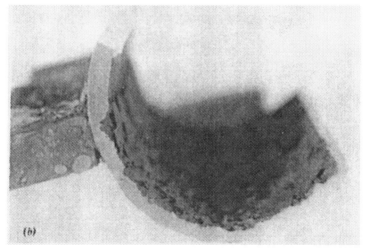

FIGURE 6.2. (*a*) An overall view of the tube ID showing the severe oxygen pitting. (*b*) Cross section through ID pit.

The bottoms of pendant loops or sagged tubes in horizontal reheaters and superheaters are likely locations for oxygen pitting. Reheaters are more susceptible to pitting corrosion because they are more difficult to isolate when the unit is shut down. Condensate collects at the low points, and, during shut-down, with the boiler open to the atmosphere, oxygen and moisture combine to form oxygen pits. Figure 6.4 shows a through-wall hole in a reheater from oxygen pitting during shut-down. The location is from the bottom of a pendant, and the material is SA-213 T-22. Horizontal reheaters and superheaters are also subject to similar pitting attacks at the low points, from the same cause of condensate collection. Figures 6.5 and 6.6 show two

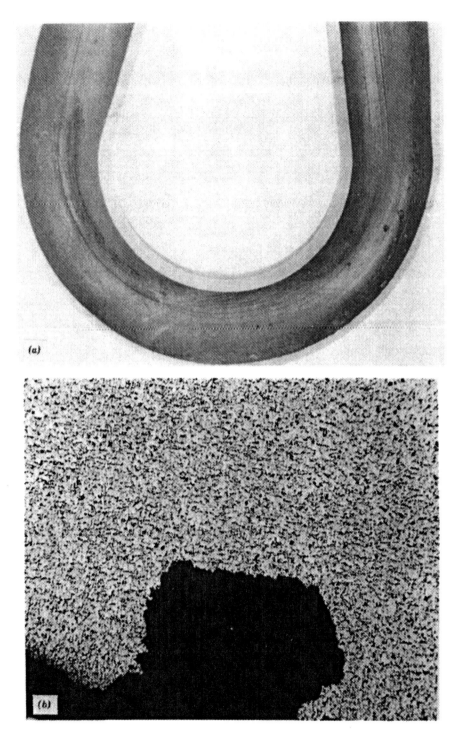

FIGURE 6.3. (*a*) A row of oxygen pits along the ID of a return bend can be seen along the die mark in the left-hand portion of the bend. (*b*) Cross section through the pit (Magnification 100×, nital etch).

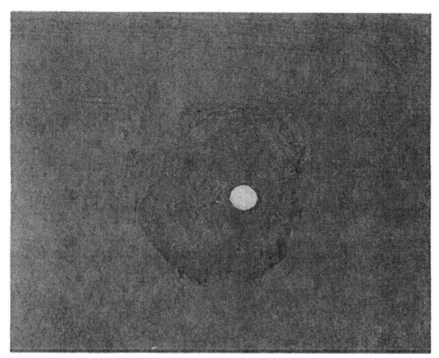

FIGURE 6.4. Through-wall hole from oxygen pitting during shut-down periods. The location is the bottom of a pendant-style reheater of T-22 (Magnification 2×).

FIGURE 6.5. Oxygen pits from a horizontal reheater (Magnification 2×).

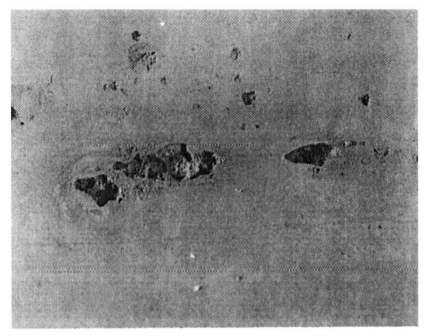

FIGURE 6.6. Oxygen pits from a horizontal superheater (Magnification 4.2×).

examples. In stainless steel, sensitized structures are also likely to develop grain-boundary attack from oxygen corrosion; see Fig. 3.7.

Fretting Corrosion

Fretting corrosion occurs when there is slight relative motion, as caused by vibration, of two metals in contact with each other. The relative motion destroys the protective oxide film so that clean metal is in contact with the corroding media. An example of this may be the more rapid attack of superheater- or reheater-tube ties or supports as they rub.

Figure 6.7 shows the cross section through a region of fretting-corrosion damage. The microstructure at such damage often reveals cold-work features in the region of maximum contact stress and relative motion.

Erosion–Corrosion

Erosion–corrosion results from the removal of the protective oxide film by mechanical means, for example, by abrasive particles in a gas stream, such as from fly ash in a boiler or frequent use of sootblowers. Figure 6.8 shows the cross section of two superheater tubes, one of Type T-22 steel and the other of Inconel 625® from a boiler burning municipal refuse. Pluggage of a

TYPES OF CORROSION 331

FIGURE 6.7. Fretting-corrosion microstructures often show cold-worked microstructures in the region of maximum contact stress (Magnification 200×, etched).

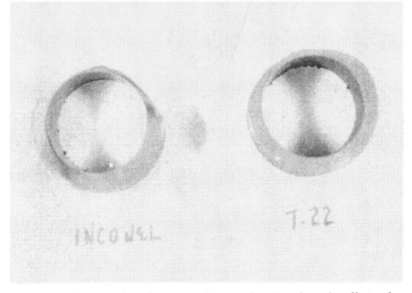

FIGURE 6.8. Cross section of two superheater tubes that show the effects of corrosion–erosion metal loss.

332 CHAPTER 6 CORROSION-CAUSED FAILURES

portion of the superheater raised gas velocities through the open lanes, eroding the tubes as shown.

Intergranular Corrosion

This localized type of attack occurs at the grain boundaries of a metal. Overheated boiler tubes will occasionally show oxide penetration into the grain boundaries of the material and may lead to premature failure. Figure 6.9 shows an example of intergranular oxide penetration taken from a grade T-22 steel superheater tube.

Corrosion Fatigue

As the name implies, corrosion fatigue is failure by repeated or cyclic stresses in a corrosive environment. As described in Chapter 3, there is a stress level for steels, called the endurance limit, below which failure will not occur, regardless of the number of cycles. In a corrosive environment, no true endurance limit or safe working stress may exist. Corrosion-fatigue cracks are typically transgranular; they may be branched, but usually are not; and several minor cracks may be visible at the metal surface in the

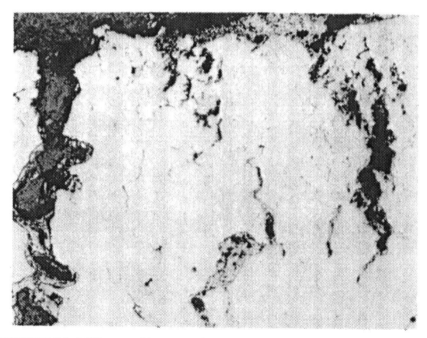

FIGURE 6.9. Mild case of intergranular oxide penetration, normally not a severe problem for ferritic steels below about 1250–1300°F (680–700°C) (Magnification 100×, nital etch).

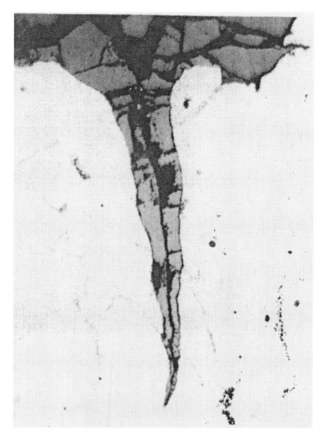

FIGURE 6.10. Corrosion fatigue; a single spear-shaped oxide wedge is typical of this failure (Magnification 500×, nital etch).

vicinity of the major crack that accounted for final failure. Fatigue cracks, on the other hand, while they are similarly transgranular, have rarely more than one major crack. On occasion, the fatigue crack may originate at an oxygen pit, in effect a stress-concentration factor, but oxygen pits are not necessary. Figure 6.10 shows an example of a corrosion-fatigue crack. An oxide wedge forms once a fatigue crack occurs. Since the volume occupied by the oxide is greater than the volume of metal from which it forms, the wedge adds stress to the crack tip and prevents the crack from closing. On the next cycle, the crack moves further and propagates faster than without the formation of the wedge and the corroding media in which it formed.

Stress-Corrosion Cracking

Stress-corrosion cracking is the failure of a metal stressed in tension below the yield point and exposed simultaneously to a specific corroding environ-

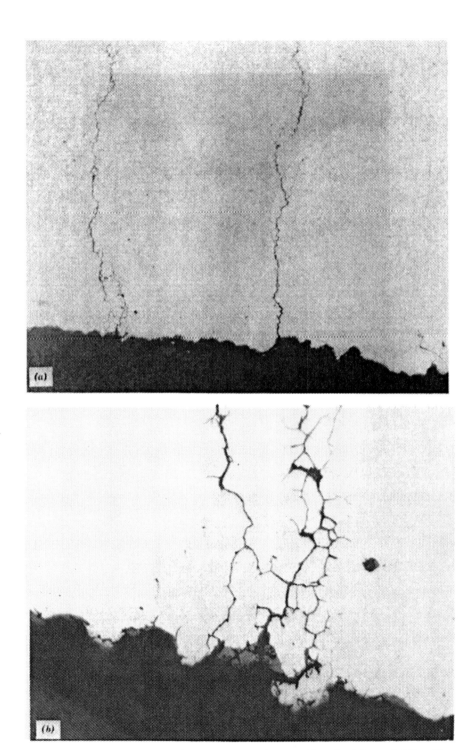

FIGURE 6.11. (*a*) Intergranular stress-corrosion cracking, ferritic steel (Magnification 100×, nital etch). (*b*) Intergranular stress-corrosion cracking, same sample as (*a*). Note the decarburized layer at the ID (Magnification 500×, nital etch).

ment to which the metal is sensitive. The stress may be either applied, as, for example, by pressure force, or residual, as, for example, by the effects of cold work. A net compressive stress prevents stress-corrosion-cracking damage. Carbon steels will fail in concentrated hydroxide solutions by stress-corrosion cracking, also known by the name caustic embrittlement. Riveted steam-drum failures of half a century ago were an early example of caustic embrittlement. Figure 6.11 shows stress-corrosion cracking of a ferritic boiler tube. Stainless steels will fail by stress-corrosion cracking in chloride solutions. Cracks usually are intergranular, but can be transgranular under certain conditions. Figure 6.12 shows stress-corrosion cracking of austenitic stainless steel. In Fig. 6.12a the cracks are predominantly intergranular; in Fig. 6.12b the crack is transgranular.

Hydrogen Damage

Hydrogen damage occurs in high-pressure boilers, usually under heavy scale deposits, on the steam side of a boiler tube.[4,5] The most frequent location is in the highest-heat-release regions of the furnace, the area of the burner elevations. Other locations where hydrogen damage has been found are the nose arch, both top and bottom; along the roof; and, rarely, within the economizer. Local upsets to the uniform fluid flow, often at tube-to-tube butt welds and bent tubes near burners, are primary sites. These locations create turbulence where iron-oxide particles and other corrosion debris can most easily form deposits.

Hydrogen damage is coincident with under-deposit corrosion and may be caused by either strongly basic or strongly acidic conditions. Concentrated sodium hydroxide will remove the protective magnetite film:

$$4NaOH + Fe_3O_4 = 2NaFeO_2 + Na_2FeO_2 + 2H_2O$$

The more frequent boiler-water condition that leads to hydrogen damage is an acid or low pH excursion. However, there is evidence to suggest that hydrogen damage has occurred in boilers with no history of any low pH upset.[6]

Under acid conditions, for example when seawater enters the boiler through a condenser leak, the pH will drop to the acid range, a pH less than 7. With these conditions, iron reacts with hydrochloric acid, for example, to form iron chloride and hydrogen:

$$Fe + 2HCl = FeCl_2 + 2H$$

With the protective oxide destroyed, the reaction of iron with concentrated hydroxide forms sodium ferroate and atomic hydrogen:

$$Fe + 2NaOH = Na_2FeO_2 + 2H$$

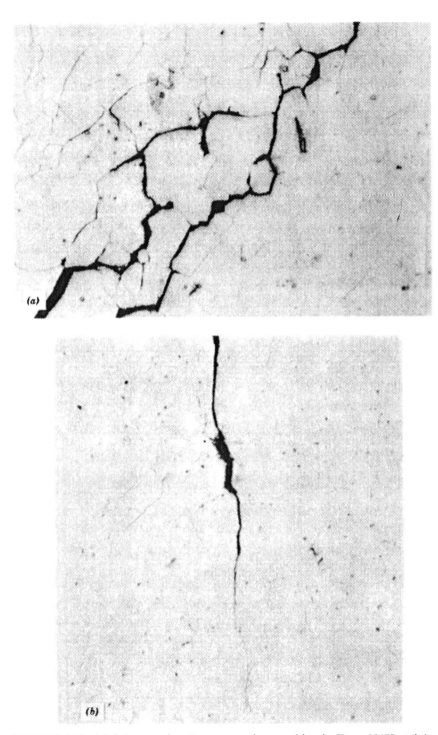

FIGURE 6.12. (*a*) Intergranular stress-corrosion cracking in Type 321H stainless steel (Magnification 250×, electrolytic etch). (*b*) Transgranular stress-corrosion cracking in Type 321H stainless steel (Magnification 100×, electrolytic etch).

The hydrogen is trapped between the scale and the steel, and some hydrogen diffuses into the steel. Since the hydrogen is a small atom, it can easily diffuse into the steel where it reacts with iron carbide to form methane and iron:

$$4H + Fe_3C = CH_4 + 3Fe$$

Methane is a large molecule and cannot easily diffuse and, therefore, collects at grain boundaries within the steel. When sufficient methane collects, a series of intergranular cracks that weaken the steel are formed. Hydrogen-damage-related failures are typically thick-edged, low-ductility failures. The microstructure of hydrogen-damaged steel shows the methane-induced cracks and decarburization. See Figs. 6.13 and 6.14. Often the damage is localized and may be in spots as small as the size of a silver dollar. Figure 6.15 shows the ring section taken through a hydrogen-damaged area and a neighboring ring section from 4 in away. Under-deposit corrosion has reduced the wall thickness from 0.320 to 0.106 in before failure occurred.

Depending on the extent of attack, the microstructures can vary from fully decarburized, Fig. 6.16, to cracks surrounding the remnants of the pearlite colonies; see Figs. 6.13 and 6.17. Sometimes the hydrogen-damaged microstructure contains spheroidized carbides or graphite. The transforma-

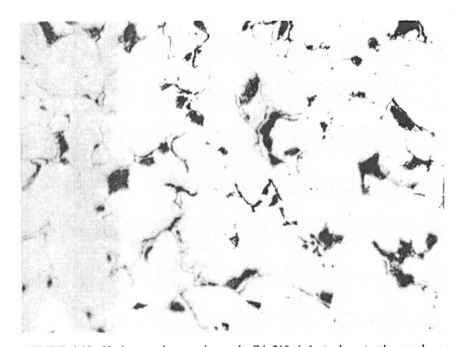

FIGURE 6.13. Hydrogen damage in grade SA-210 A-1 steel; note the cracks or fissures around the pearlite colonies (Magnification 500×, nital etch).

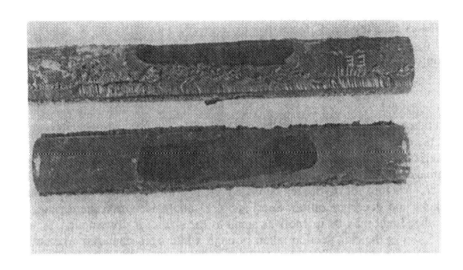

FIGURE 6.14. Hydrogen-damage failures are often thick-lipped, limited-ductility ruptures with a window blown out (Magnification 0.25×).

FIGURE 6.15. The damage may be localized to small corrosion zones an inch or so in diameter. These two rings are separated by a distance of about 4 in (Magnification 0.7×).

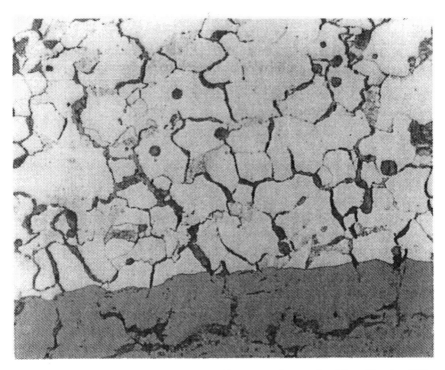

FIGURE 6.16. Total decarburization and intergranular cracks (Magnification 500×, etched).

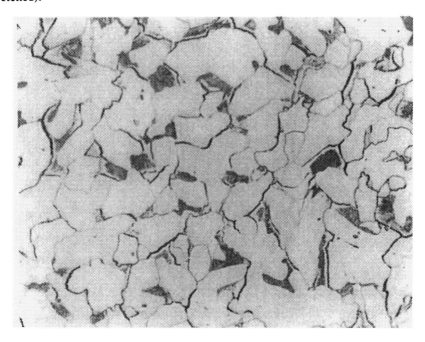

FIGURE 6.17. Partial decarburization and intergranular cracks (Magnification 500×, etched).

340 CHAPTER 6 CORROSION-CAUSED FAILURES

tion of pearlite to graphite (see Fig. 6.18) indicates elevated metal temperatures from the insulating effects of the thick, water-side deposits.

Caustic Attack or Grooving

Caustic attack or grooving occurs as a result of the following:

1. An upset in the boiler-feedback chemistry, and too much caustic (NaOH) is introduced.
2. A steam blanket forms on the fire side of the furnace-wall tube that concentrates the hydroxide so that metal attack can happen. Figure 6.19 shows two views of severe cases of waterwall caustic attack from a feedwater upset.

Copper Deposits

Not strictly a boiler corrosion product, copper deposits are more in the nature of indicator deposits, on the scene of the crime, but not the cause. A leak in the preboiler circuits that use copper alloys, the condenser or boiler-

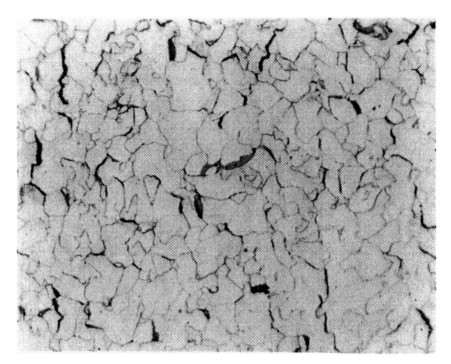

FIGURE 6.18. On occasion, the hydrogen-damaged microstructure includes spheroidized carbides or graphite, indicating elevated temperatures from the water-side deposits (Magnification 500×, etched).

TYPES OF CORROSION 341

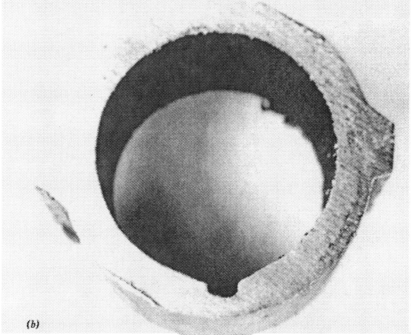

FIGURE 6.19. Two views of a groove on the ID of a waterwall tube; grooving at the crown is characteristic of caustic gouging.

342 CHAPTER 6 CORROSION-CAUSED FAILURES

feedwater heater, for example, will cause a feedwater-chemistry upset that cannot be handled by the water-treatment equipment. Oxygen and corrosion products from these copper alloys dissolve in the boiler feedwater. In the boiler, the oxygen reacts with the steel to form iron oxide, and the copper ions react with the boiler steel

$$3Cu^{2+} + 2Fe = 3Cu + 2Fe^{3+}$$

to leave copper particles mixed with the other deposits. The real cause of the corrosion is the leak in the condenser, and copper becomes a part of the evidence in the boiler. Figure 6.20 shows copper particles—light areas—intermixed with the darker iron oxide and other corrosion deposits. The example is taken from a large utility boiler. It is not unusual for several hundred pounds of copper to be removed during a chemical cleaning of the boiler. In fact, modern cleaning procedures routinely call for a copper-removal step.

Fire-side Corrosion

Fire-side corrosion is caused by the reaction of certain fuel-ash constituents and flue gas with the steel components. Corrosion occurs in superheaters and reheaters at about 1100°F (590°C) and above, and on furnace walls at about 700° to 800°F (370° to 430°C).

FIGURE 6.20. Copper deposits, the light-colored specs within the gray ID oxide and corrosion debris; note copper within the pit (Magnification 100×, unetched).

Liquid-Phase Corrosion, Coal Ash

Detailed chemical analyses of coals mined in the United States show significant variations in the mineral content. Table I of the appendix displays coal analyses from several kinds of U.S. coal. From a boiler-corrosion point of view, the bad actors are sodium, potassium, and sulfur.[7-9] In spite of the fact that all coals contain more or less of all three, not every coal-fired boiler suffers from coal-ash corrosion. More than just oxides of these are necessary to produce liquid-ash attack.

A nearly unique combination of factors is necessary. Tube-metal temperatures must be high enough to allow the liquid-ash phase to form, and tube-metal temperatures are strongly influenced by steam-side scale thicknesses. The way the fuel is burned and the ash is deposited on the tube surface are important. The amount of SO_3 in the deposit determines the stability of both the trisulfates and pyrosulfates. Other constituents in the ash and mineral form, especially calcium and magnesium, can alter the sulfates to those that have higher melting points.[10] Finally, the amount and way that carbon enters the ash deposit are perhaps most important.

Coal is usually burned as fine particles, most are less than 200 mesh in size, well mixed with air, and blown into the furnace in a turbulent flame. Residence time in the flame is long enough for nearly complete combustion, so that only a few percent carbon ends up in the bottom ash. In the high-temperature portion of the flame, above 3000°F (1650°C) or so, the mineral matter is stripped of both sodium and potassium. Volatilized alkali elements react with oxygen to form gaseous sodium oxide and potassium oxide, Na_2O and K_2O; K_2O vaporizes at 2690°F (1480°C) and Na_2O at 3210°F (1770°C).[7]

Sulfur in coal is present either as pyrite (FeS_2) or sulfates (Na_2SO_4, K_2SO_4, or $FeSO_4$). Within the flame, sulfide minerals are burned to metallic oxides, for example, Fe_2O_3 and SO_2; sulfates are unstable at flame temperature and decompose into metallic oxides, for example, Na_2O and SO_3. The amounts of SO_2 and SO_3 present in the flue gas will depend on the temperature. The reaction

$$2SO_2 + O_2 = 2SO_3$$

is reversible; low temperatures favor SO_3, and high temperatures favor SO_2. Formation of SO_3 is enhanced by some oxides that act as catalysts. Once formed, both Na_2O and SO_3, or K_2O and SO_3, combine in different ways to form liquid-ash components.

As the flue gas cools from, approximately, 3000°F (1650°C), the alkali oxides condense to liquids and make the ash quite sticky. Even though the amounts of sodium and potassium are similar for many coals, their amounts within the ash may be different. The volatilization temperature of K_2O is appreciably lower than Na_2O, so more K_2O will deposit from the gas than will Na_2O. These alkali oxides form a host of components with silica (SiO_2), alumina (Al_2O_3), magnesia (MgO), calcia (CaO), and other coal-ash components. Once the partially liquid ash sticks to the boiler tube, reactions with SO_3 and the tube metal can proceed to promote liquid-ash corrosion. The K_2O, Na_2O, and SO_3 migrate through the ash deposit toward the cooler metal surface to react and form the low-melting-point liquids that do the damage.

High-Temperature Corrosion of Superheater and Reheater

At this point, it is important to differentiate between what occurs in superheaters and reheaters at high temperature from what occurs at lower temperatures on furnace walls. In a superheater or reheater, the liquids that form are the trisulfates. Once the trisulfates form, tube wastage occurs by the solution of the protective iron-oxide scale in the liquid and the transport by the liquid of oxygen to the metal surface to form more iron oxide. There are several intermediate steps that may occur, and some of these lead to the formation of metallic sulfides.

The corrosion steps start with the formation of Na_2O and K_2O in the flame, and the reaction with SO_3 to form sodium or potassium sulfate (Na_2SO_4 or K_2SO_4) in the ash deposit. Iron oxide (Fe_2O_3), alkali sulfate (Na_2SO_4 or K_2SO_4), and sulfur trioxide (SO_3), react to form the trisulfates on the cooler tube surface:

$$3K_2SO_4 + Fe_2O_3 + 3SO_3 = 2K_3Fe(SO_4)_3$$

$$3Na_2SO_4 + Fe_2O_3 + 3SO_3 = 2Na_3Fe(SO_4)_3$$

Reid[7] has proposed the following to account for tube wastage:

$$9Fe + 2K_3Fe(SO_4)_3 = 3K_2SO_4 + 4Fe_2O_3 + 3FeS$$

$$4FeS + 7O_2 = 2Fe_2O_3 + 4SO_2$$

$$2SO_2 + O_2 = 2SO_3$$

$$3K_2SO_4 + Fe_2O_3 + 3SO_3 = 2K_3Fe(SO_4)_3$$

The net reaction is

$$4Fe + 3O_2 = 2Fe_2O_3$$

which accounts for the metal loss. Iron sulfide has been observed in these cases and can be detected by a sulfur print. (To make a sulfur print, a piece of photographic paper is dipped in a 2% sulfuric-acid solution, and the excess acid wiped off. The tube sample is rough polished—a 120 grit belt sander is satisfactory—pressed onto the damp photographic paper for a few seconds, and removed. The paper is fixed as usual for photographic paper.

What remains is an image of the sulfide distribution on the tube. Sulfuric acid reacts with metallic sulfides to form hydrogen-sulfide gas, H_2S; the H_2S reacts with the silver salts on the photographic paper to form silver sulfide, which is dark brown or black.)

The following example is typical of many studied at Riley Stoker. Figure 6.21 shows the appearance of a reheater tube that suffers from the early stages of liquid-ash corrosion. Very little wall thinning is noted at this time, but all the other features of liquid-ash attack are present. Figure 6.22 shows

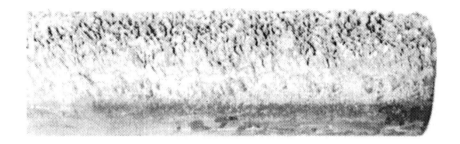

FIGURE 6.21. As-received tube showing OD ash deposit.

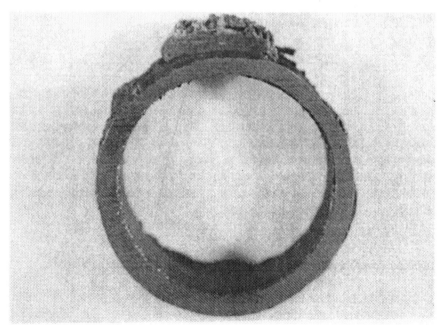

FIGURE 6.22. Cross section through tube shown in Fig. 6.21.

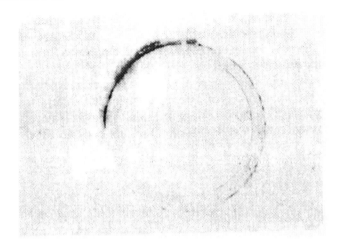

FIGURE 6.23. Sulfur print of tube in Fig. 6.22; sulfide deposits are primarily under the ash deposit.

TABLE 6.1. Ash Analysis—Superheater Tube Corrosion

Sodium as Na$_2$O	0.14%
Silicon as SiO$_2$	21.0%
Aluminum as Al$_2$O$_3$	7.0%
Sulfur as SO$_3$	9.0%
Iron as Fe$_2$O$_3$	51.9%
Calcium as CaO	8.0%
Magnesium as MgO	1.04%
Carbon	1.02%
Potassium as K$_2$O	0.22%

the cross section; note especially the conically shaped deposit on the leading "face" of the tube and the dark glassy appearance to the inner layer of the deposit. Figure 6.23 shows a sulfur print of the tube cross section; note the higher concentration of sulfides under the deposit. Table 6.1 lists the ash-deposit analysis; note the presence of carbon. Tube wastage is the greatest under the ash where the sulfur print shows the most sulfides. In time, the wall thickness will decrease until tube failure occurs, and the tube must be replaced.

MORPHOLOGY SUPERHEATER AND REHEATER CORROSION

The appearance of superheaters and reheaters which suffer coal-ash corrosion is somewhat variable and in large part determined by the flue-gas flow past the corroding tube. Figures 6.24 and 6.25 illustrate this variable. Both

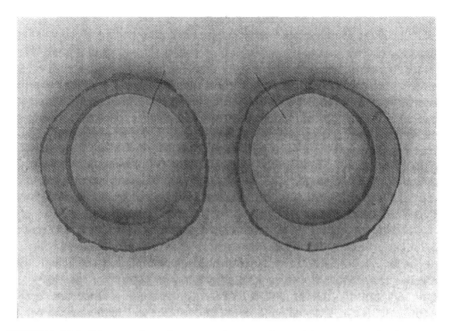

FIGURE 6.24. Ring sections from superheater tubes with severe coal-ash corrosion. Maximum wastage is at 10 and 2 o'clock positions. Lines point in direction of flue-gas flow (Magnification 1.3×).

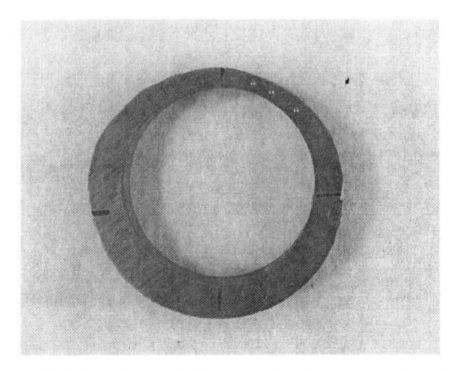

FIGURE 6.25. Another example of a superheater tube with coal-ash corrosion wastage. Here principal wastage is at 12 o'clock position (Magnification 1.3×).

figures show ring sections cut through corroding tubes. In one case, Fig. 6.24, the greatest wastage is at the 10 and 2 o'clock positions (the cross section of the tube is referenced to a clock face with the 12 o'clock position facing directly into the flue-gas flow). The greater wastage along the flanks of the tube is more typical, but not always the case. Figure 6.25 shows a superheater tube with the major wastage at the 12 o'clock position.

The morphology of the wastage is sometimes referred to as "alligatoring" or "alligator hide." Figures 6.26 and 6.27 show this characteristic appearance when all of the inner layer of debris has been removed. In cross section, Figs. 6.28 and 6.29, the shallow grooves give the general impression of thermal-fatigue cracks.

As shown on page 357, circumferential grooving in waterwall tubes is caused by a corrosion-fatigue mechanism. It is also likely that a similar mechanism is the cause of the alligator-hide appearance in superheater and reheater tubes. A liquid-ash species forms on the surface of the tube. At room temperature, these liquids sinter the ash particles and corrosion debris to form the dense, tightly bound, inner layer to the ash deposit, as shown in Fig. 6.22. The liquid layer on the tube can support only so much fly ash. As the ash layer is shed or spalls, the local heat transfer and, thus, tube-metal temperature, increases sharply. When the ash layer re-forms, tube-metal temperatures return to normal. The cycling of metal temperatures as the ash

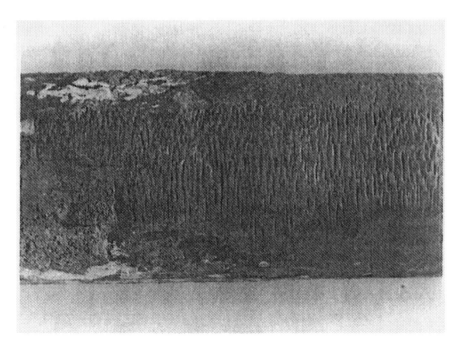

FIGURE 6.26. "Alligator hide" appearance under ash deposit (Magnification 1×).

FIGURE 6.27. "Alligator hide" appearance contains shallow circumferential grooves (Magnification 2.4×).

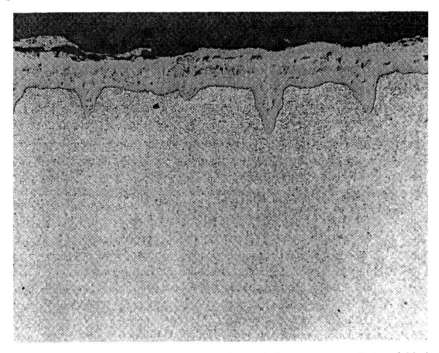

FIGURE 6.28. In cross section, the spacing and shape are somewhat variable but generally appear similar to thermal-fatigue cracks (Magnification 25×).

350 CHAPTER 6 CORROSION-CAUSED FAILURES

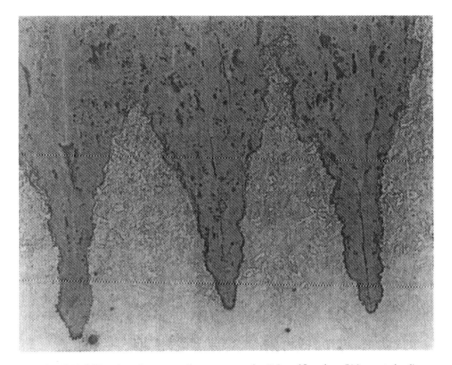

FIGURE 6.29. Another superheater sample (Magnification 500×, etched).

layer forms and sheds, creates the corrosion-fatigue pattern known as alligatoring.

CONTROL OF ASH CORROSION

In severe cases of superheater and reheater corrosion, it has been found that shields added to the tube surface offer good results in preventing further metal loss. The trisulfates decompose at 1250–1300°F (680–700°C) and cease to be liquids. Figure 6.30 shows the corrosion rate, weight loss, as a function of testing temperature for Type 321 stainless steel.[11] The dramatic drop in corrosion rate between 1250 and 1300°F (680 and 700°C) is caused by the decomposition of the trisulfates. Once the liquid phase has been removed, the corrosion rate is that of simple oxidation in contact with flue gas. By the use of shields, the overall heat transfer between flue gas and steam is reduced. The net effect is to increase the outside metal temperatures above the range of stability of the trisulfates. Once the liquid decomposes, rapid liquid-ash corrosion can no longer continue.

The formation of complex sulfates with calcium and magnesium can occur, and the melting points of the resulting compounds are higher than normal superheater and reheater temperatures. Simple calcium or magne-

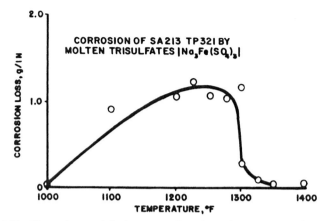

FIGURE 6.30. Corrosion weight loss as a function of temperature for Type 321H stainless steel. Note sharp drop in corrosion at temperatures about 1250–1300°F (680–700°C) (Ref. 7).

sium sulfate, $CaSO_4$ or $MgSO_4$, tie up the SO_3 in an innocuous form. Regardless of the exact chemistry, the presence of particular compounds of calcium and magnesium does reduce the corrosion rate. But, the use of additives has not been a great success in coal-fired boilers because the volume of ash is so great and the chemistry so variable.[10]

FURNACE-WALL CORROSION

The mechanism of furnace-wall corrosion is similar to the corrosion of superheaters and reheaters, but the low-melting-point liquid is different. The problem is more common in supercritical units, but natural- or forced-circulation boilers with drum pressures less than 2800 psig are not immune. Furnace-wall metal temperatures of 750–800°F (400–425°C) are more than 300°F (170°C) lower than the 1100–1150°F (590–620°C) metal temperatures found in the superheater and reheater. The lowest melting point of a mixture of $K_3Fe(SO_4)_3$ and $Na_3Fe(SO_4)_3$ is about 1030°F (550°C), much too high to be a problem for the furnace walls. Sodium and potassium pyrosulfate ($Na_2S_2O_7$ and $K_2S_2O_7$) have been blamed in the past for furnace-wall corrosion.[12-14] Both melt below 800°F (425°C): $Na_2S_2O_7$ at 750°F (400°C) and less than 570°F (300°C) for $K_2S_2O_7$, according to handbook data.[7] Mixtures of the two presumably have lower melting points than either pure substance.

The examples that follow illustrate the features common to all of the samples studied over the past two decades or so. The failures show:

1. Circumferential penetration of the corrosion products on the fire side of the tube; Fig. 6.31 displays the surface of the tube with the ash and

FIGURE 6.31. OD of furnace-wall tube with ash deposit removed to show sulfide-grooving attack morphology.

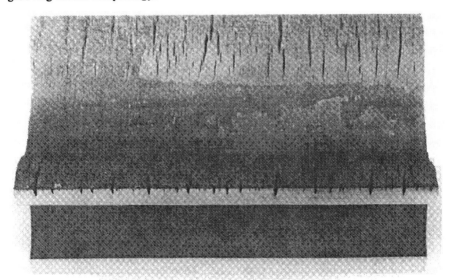

FIGURE 6.32. A section of a pair of waterwall tubes presents the characteristic appearance of circumferential grooving. From a supercritical, coal-fired boiler. Sample courtesy of Mr. John Alice, GPUN, Reading, Pennsylvania (Magnification 0.7×).

some of the deposit removed. Figure 6.32 shows a section of a pair of waterwall tubes. In this example the ash-corrosion debris has been removed by dissolving the deposits in boiling hydrochloric acid. This figure is more characteristic of the appearance of circumferential grooving. Figures 6.33a and 6.33b display longitudinal cross sections of the circumferential cracking and grooves, but also present is similar circumferential grooving from the ID surface, especially at the corners of the internal rifling. Figure 6.34 shows the longitudinal cross section, and these grooves have a characteristic similar to the thermal-fatigue cracks shown in Figs. 3.39 and 3.40. All of the coal-ash-corrosion cross sections, whether it be from a waterwall or a superheater tube, have similar daggerlike appearances.

FIGURE 6.33. (a) A longitudinal cross section clearly shows the OD cracks and grooves; note also the water-side cracks (Magnification 2×).

354 CHAPTER 6 CORROSION-CAUSED FAILURES

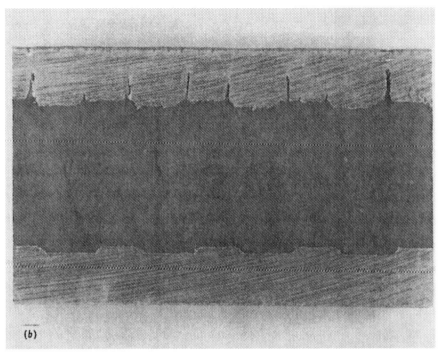

FIGURE 6.33. (*b*) Occasionally, internal cracks also form, especially at the internal rifling. The OD grooving is less severe (Magnification 1.4×).

2. Sulfur prints show the corrosion deposits within the tube-wall penetration on the surface to contain sulfides. Figure 6.35 shows both transverse and longitudinal cross-sectional views of a corroding tube.
3. The ash deposits are in two layers: a reddish brown outer layer and a dark black, glassy, inner layer.
4. The ash deposits contain carbon as well as the usual oxides associated with coal ash; see Table 6.2.
5. Low-melting-point materials are contained in all of the samples. Melting points were measured by a differential scanning calorimeter and range from 635 to 770°F (335 to 410°C).[15]

Corrosion probably proceeds in several stages:

1. Sodium and potassium in the coal become oxides in the flame:

$$4Na + O_2 = 2Na_2O$$

$$4K + O_2 = 2K_2O$$

FURNACE-WALL CORROSION 355

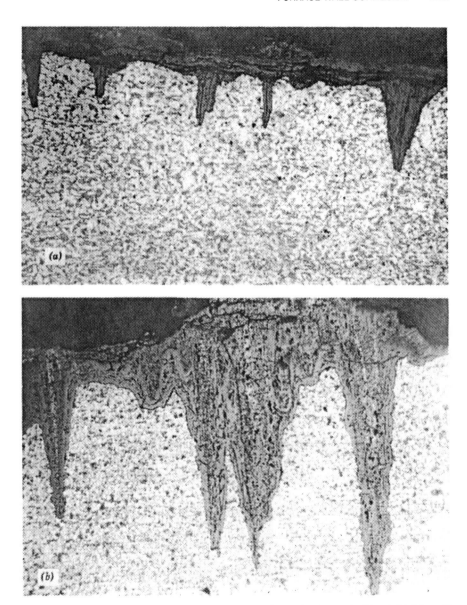

FIGURE 6.34. (*a*) Longitudinal cross section of sample shown in Fig. 6.31 (Magnification 100×, nital etch). (*b*) A more severe case of furnace-wall corrosion, longitudinal cross section (Magnification 100×, nital etch).

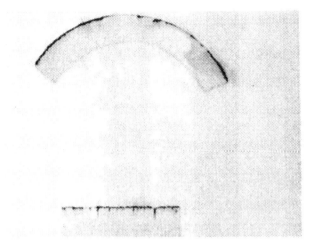

FIGURE 6.35. Circumferential and longitudinal sulfur prints of furnace-wall corrosion.

2. Sulfur in the coal becomes sulfur dioxide in the flame:

$$S + O_2 = SO_2$$

3. Sulfur dioxide becomes the trioxide, catalyzed on the tube surfaces or within the ash deposit in the low-temperature region near the furnace wall:

$$2SO_2 + O_2 = 2SO_3$$

4. Pyrosulfates form in the ash deposit close to the metal surface:

$$Na_2O + 2SO_3 = Na_2S_2O_7$$

$$K_2O + 2SO_3 = K_2S_2O_7$$

TABLE 6.2. Ash Analysis—Furnace Wall Corrosion

Sodium as Na_2O	0.4%	0.4%
Silicon as SiO_2	33.6%	8.6%
Aluminum as Al_2O_3	19.0%	11.8%
Sulfur as SO_3	1.9%	2.3%
Calcium and magnesium as $CaO + MgO$	1.1%	0.4%
Iron as Fe_2O_3	36.2%	67.6%
Potassium as K_2O	1.6%	0.4%
Carbon	0.64%	1.23%
Melting point	770°F (410°C)	635° and 730°F (335° and 388°C)

5. Carbon deposits in the ash by either unburned coal or by reduction

$$2CO = CO_2 + C$$

6. Tube wastage occurs by either

$$SO_3 \text{ (from pyrosulfate)} + 3C + Fe = FeS + 3CO$$
$$2SO_3 \text{ (from pyrosulfate)} + 9C + Fe_2O_3 = 2FeS + 9CO$$

The role of carbon in both 1100°F (590°C) and 800°F (430°C) liquid-ash corrosion is somewhat unclear. Deposits studied contain both sulfides and carbon. It appears likely that sulfur may be transported to the metallic interface as SO_3 in the sulfate where it is reduced by the carbon to form a metallic sulfide. Carbon is necessary to maintain, at least on a microenvironmental scale, the reducing conditions needed for sulfide stability. This also suggests that the liquid-phase sulfate, in whatever form, acts as the vehicle for sulfur transport.

There is meager data on the gas composition in the vicinity of waterwalls that suffer circumferential grooving. However, some data suggest that there is a higher carbon-monoxide content to the flue gas. This higher concentration of carbon monoxide in the flue gas is consistent with the chemical analysis of the ash deposits that show free carbon; see, for example, Table 6.2. Sulfur prints show the presence of sulfides within the corrosion deposit; see, for example, Fig. 6.35. All of this information suggests that reducing atmospheres are an essential part of the grooving-corrosion mechanism.

Chordal-thermocouple measurements of waterwall metal temperatures show unusual circumstances.[16] These circumstances can explain the thermal-fatigue-like appearance of the circumferential grooving. Figure 6.36 shows these chordal-thermocouple measurements. During sootblower operation and slag falls, the fire-side surface temperature jumps nearly 100°F (60°C). As the ash layer reforms, the metal temperature returns to "normal." This varying metal temperature leads to a varying temperature-induced stress. The varying stress, in turn, leads to the thermal-fatigue component to the corrosion and wastage of the tube metal.

While no similar chordal-thermocouple measurements have been made on superheaters and reheaters, the alligatoring observed has similar cross sections (see Figs. 6.28 and 6.29) and may also be explained by slag shedding. However, the temperature spike is apt to be much less severe, since the heat flux is substantially less. Thus, frequent but mild temperature spikes would lead to a difference in details of the appearance, but overall would have the same general corrosion-fatigue appearance.

These observations suggest the following circumstances as the cause of circumferential grooving within waterwalls and the alligator-hide appearance to superheater and reheater corrosion. There are three essential features:

358 CHAPTER 6 CORROSION-CAUSED FAILURES

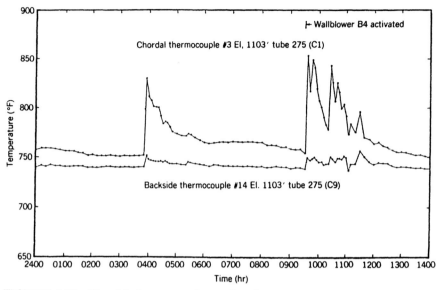

FIGURE 6.36. Chordal thermocouple reading from a supercritical boiler at Keystone Station. Unit No. 2 steady-state operation 5/15/84, front wall. Courtesy of Mr. James LaFontaine, Penelec Division, GPUN.

1. *A Liquid Species in the Ash.* Depending on the temperature, the liquid species is, of course, different. In superheaters and reheaters the mixtures of sodium- and potassium-iron trisulfates are the liquid species; while waterwall tubes may have mixtures of sodium or potassium pyrosulfates. Differential scanning calorimetry does show melting points within the normal reach of waterwall tubes; see Table 6.2.
2. *Reducing Atmospheres.* At least on a microenvironmental scale, the atmosphere at the waterwall-tube surface where corrosion occurs is reducing. The evidence for this is sulfides within the corrosion deposits, free carbon within the ash deposits, and the reported higher carbon-monoxide content measured within the flue gas.
3. *Variable Stress That Leads to Corrosion-Fatigue-Like Grooves or Cracks.* The evidence for this is the chordal-thermocouple measurements on waterwall tubes; by implication, a similar circumstance occurs in superheaters and reheaters.

These three conditions are interconnected. The liquid species in the ash weaken the bond between the slag or ash cover and the tube-metal surface. When the slag layer builds to a sufficient thickness and weight, the reduced strength at the liquid film allows the slag to shed or fall more easily and completely. The radiant heat in contact with the bare tube, from the fireball in a waterwall, or the hot flue gases in a superheater or reheater, raises the

FURNACE-WALL CORROSION 359

tube-metal temperatures on the surface, as shown in Fig. 6.36. Reducing atmospheres led to the formation of, and stability to, iron-sulfide scales along the tube surfaces. Sulfides are less protective against further corrosion or wastage than are oxides, and thus more rapid wastage at the tube occurs.

In some circumstances, the temperature spikes are severe enough that water-side cracking is also present; see Figs. 6.33 and 6.34.

The use of a corrosion-resistant weld overlay will not be totally effective. While the corrosion wastage may be prevented, the fatigue component still leads to the formation of circumferential cracks. Figures 6.37–6.39 present three views of a stainless-steel (304) overlay that was ineffective in the prevention of circumferential grooving on a waterwall of a supercritical boiler. Figure 6.37 shows the end view of a three-tube panel with the stainless steel on the fire side. The circumferential grooves are obvious in Fig. 6.38, and, in cross section, are presented in Fig. 6.39.

In the absence of variable stress, or in a boiler where heat flux does not lead to high metal temperatures, the appearance of corrosion wastage is expected to be different. Metal loss would be severe, but the appearance would be relatively smooth (see Fig. 6.40). Figure 6.40 is a waterwall tube from a refuse-fired boiler. The fuel is burned on a stoker, and thus the heat flux is likely to be substantially less when slag is shed. Therefore, tempera-

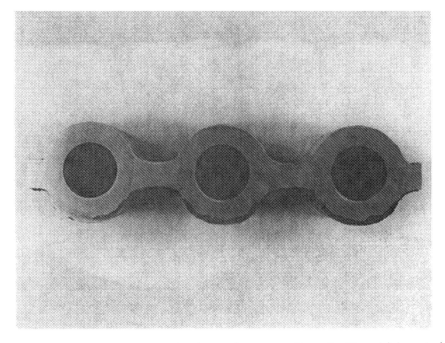

FIGURE 6.37. Cross section of a three-tube waterwall panel with a stainless-steel weld overlay on the fire side (Magnification 1×). Courtesy of GPUN, Reading, Pennsylvania.

FIGURE 6.38. Circumferential grooves still form even with the corrosion-resistant overlay (Magnification 3×).

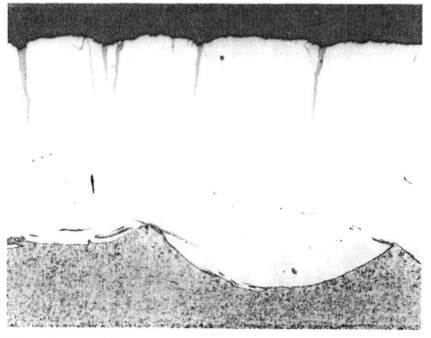

FIGURE 6.39. Longitudinal cross section through circumferential grooves. At this time, the corrosion-fatigue cracks have not fully penetrated the stainless steel (Magnification 25×, nital etch).

FIGURE 6.40. Severe wastage in a refuse-fueled boiler waterwall tube (Magnification ~½×).

ture spikes do not exist to the same degree as in a pulverized-coal-fired unit. Circumferential grooving does not occur, but severe wastage can lead to steam leaks.

OIL-ASH CORROSION

The amount of ash in fuel oils rarely exceeds about 0.1%. Table XII of the appendix[17] lists the metallic elements found in several crude oils from around the world, and Table III[7] gives ash analyses from several fuel oils. As can be seen from the table, compositions are quite variable and contain a wide variety of metals. As far as corrosion is concerned, the culprits are vanadium, sodium (and potassium if present), and sulfur. Vanadium is usually present in the oil as an organometallic compound, sulfur as complex organic sulfides or sulfates, and sodium and potassium as the sulfate or chloride. Sodium as the chloride causes the most trouble because it is readily volatilized when burned. Unfortunately, much of the fuel oil burned in power stations is moved by barge or tanker over sea routes and seawater is unavoidably entrapped, so some NaCl is inevitable.

The combustion of fuel oil produces metallic oxides and sulfur oxides. Similar to the problems in coal combustion, SO_2 and SO_3 form, the relative amounts depending primarily on the temperature. In excess air, vanadium forms the pentoxide V_2O_5, and sodium forms Na_2O. Together V_2O_5 and Na_2O form a whole range of compounds that melt at temperatures down to less than 1000°F (540°C). Figure 6.41 shows the phase diagram for this system. Similarly, Na_2O and SO_3 can form Na_2SO_4 within the ash deposit. Na_2SO_4 and V_2O_5 also form low-melting-point mixtures (see Fig. 6.42[6]), the lowest melting point being around 1000°F (540°C). The problems of oil-ash corrosion are usually in the superheaters and reheaters, because the melting points in the V_2O_5–Na_2O and V_2O_5–Na_2SO_4 systems are too high to cause furnace-wall problems, but are right for operating temperatures found in superheaters and reheaters. In principle, there is no reason why pyrosulfates cannot form on furnace walls; however, the Riley Stoker Corporation has had no example of furnace corrosion in an oil-fired boiler caused by liquid-ash components.

The oil-ash-corrosion mechanism is similar to the mechanism with coal ash; a low-melting-point liquid forms that dissolves the iron-oxide protective film, and facilitates transport of oxygen from the flue gas to the metal surface. The several steps are:

1. The vanadium compounds are oxidized to V_2O_5, and sodium to Na_2O, in the flame.
2. Ash particles stick to the metallic surfaces; Na_2O acts as the binder.
3. On the tube surface, V_2O_5 and Na_2O react to form the low-melting-point liquid. Alternatively, Na_2O reacts with SO_3 to form Na_2SO_4, which then combines with V_2O_5 to form a low-melting-point liquid. In either case, the low-melting-point liquid attacks the tube surface.

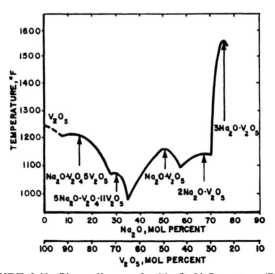

FIGURE 6.41. Phase diagram for Na_2O–V_2O_5 system (Ref. 7).

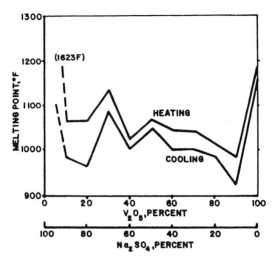

FIGURE 6.42. Melting points in system V_2O_5–Na_2SO_4 (Ref. 7).

Figure 6.43 shows an Inconel 601 support lug from a large oil-fired boiler. The original thickness of the lug was 0.405 in, but less than 0.25 in remains at the zone of greatest ash attack. That this form of liquid-ash attack increases with temperature may be deduced from this example. The support clip was welded to a reheater tube along the right-hand edge and passes through a

FIGURE 6.43. Inconel 601 support corroded by oil ash.

division wall. To prevent its sliding out, a bar was welded along the left side; it is still intact on the sample. Most metal loss is from a region between the reheater tube and the division wall where the lug would be hottest, not cooled by the tube, nor protected by the division water-cooled wall tubes.

CORROSION IN REFUSE-FIRED BOILERS

In the past decade, an increasing number of boilers using municipal refuse as a fuel have been installed. For the most part these boilers burn the fuel on a stoker or perhaps in a rotary kiln furnace. In either case, the heat fluxes are considerably smaller than for pulverized-coal- or oil-fired utility units. Heat fluxes are less than about 50,000 Btu (hr·ft^2), more typical of stoker-fired, coal-burning, industrial steam generators.

There are two principal differences in burning municipal refuse on a stoker:

1. Municipal refuse contains appreciable amounts of chlorine. The precise amount is hard to measure and varies from site to site, but estimates ranging between 0.5 and 2% chlorine have been reported.[18]
2. The method of adding fuel to the stoker tends to leave a nonuniform thickness on the grate. Since some of the combustion air comes from under the stoker, variable bed thicknesses lead to alternating reducing and oxidizing conditions.

Two corrosion mechanisms are possible:

1. Chlorides of zinc and iron (among other possibilities) form low-melting-point species. These chloride mixtures act as a flux and dissolve the protective iron oxide. With the scale removed, wastage of the steel may proceed. Thus the wastage is, in effect, caused by a liquid-ash-corrosion mechanism.[19]
3. Under reducing conditions, less-protective sulfide scales form. Without suitable protection, hydrogen chloride in the flue gas attacks the steel as a chlorination wastage.[20]

Erosion from high-velocity combustion air would, of course, exacerbate tube wastage regardless of the corrosion mechanism.

The appearance of the ash deposits within the boiler shows two distinct and separate layers: an outer, soft, friable layer and an inner, dark, or black one. Table 6.3 presents the analysis of these two layers. One other observation on these ash samples is that the outer, tan-colored deposit often smells strongly of hydrogen sulfide.

The principal features of these analyses are the presence of carbon, sulfur, and chlorine within the outer ash, and a much higher value of carbon and

TABLE 6.3. Ash and Corrosion Deposit Analysis (%)[a]

Element	Outer, Friable Ash	Inner, Black, Corrosion Layer
Carbon	0.06	4.83
Sulfur	13.8[b]	2.29[c]
Chlorine	0.66	17.22
Iron	3.83	26.07
Silicon	7.40	0.78
Calcium	6.21	0.89
Sodium	3.64	8.54
Aluminum	5.95	0.40
Zinc	1.70	1.87
Lead	0.72	3.44
Plus 11 other elements, less than 0.75%		

[a] Average of three different boilers burning RDF.
[b] Smells of H_2S.
[c] Tests positive for sulfide.

chlorine within the inner corrosion layer. This suggests a concentration of those elements along the steel surface that promote the corrosion and wastage. Note also the presence of sodium, zinc, and lead. The inner layer is dark and easily scraped off the tube with a pocketknife, indicative of less-protective scales than pure iron oxide. A small amount dissolved in water leads to an acid pH. This implies that hydrogen chloride may also be present. The presence of carbon and sulfides within both the outer and inner layers confirms the presence of reducing conditions. Under oxidizing conditions, hydrogen sulfide is burned to sulfur dioxide and perhaps sulfur trioxide. The amount of carbon is considerably less or none at all.

The implications of the chemical analyses given in Table 6.3 are as follows:

1. The high percentage of carbon along the tube surface ensures, at least at the corrosion site, a strongly reducing atmosphere.
2. Chlorine migrates to the tube surface and forms an iron chloride as a corrosion product; thus, the wastage is a chlorination reaction. Note also the increase in iron concentration in the inner layer from the wastage of the steel tube.
3. The presence of iron, sodium, zinc, and perhaps lead, suggests the formation of low-melting-point chlorides. In effect, this would support the liquid-ash-corrosion mechanism of tube wastage.
4. The anion mix of the leachable chlorides taken from the inner layer shows about 90% sodium, 10% iron, and no zinc. Lead chloride is not very soluble in water, so its presence would not show up in this test. Zinc chloride, if present, would be soluble in water, but none was found.

There are two ways to view the corrosion mechanism that results in the rapid wastage of carbon-steel tubes:

1. The presence of elemental carbon and hydrogen sulfide within the ash deposits indicates strongly reducing conditions and implies the presence of carbon monoxide in the flue gas. With reducing conditions and hydrogen sulfide present in the flue gas, iron-sulfide scales form rather than iron oxide. These iron-sulfide scales are inherently more porous and less protective than oxides. Hydrogen chloride more easily attacks the sulfides and forms iron chloride as a corrosion product. For this mechanism, reducing conditions are essential for the wastage to proceed by the formation of iron chloride. In effect, wastage is a chlorination-corrosion mechanism.

2. Reducing conditions, while present, are not a necessary condition for the corrosion. The presence of chlorides leads to the formation of low-melting-point liquids along the tube surface that may contain zinc, lead, iron, and sodium chlorides, among others. These chloride species would be expected to have a low melting point and would dissolve the protective iron oxide, leaving bare steel to be corroded by hydrogen chloride. This mechanism is a liquid-ash-corrosion attack. The liquid chlorides, both from the products of combustion and corrosion of the steel furnace tubes, behave as a soldering or brazing flux and remove oxide protection. The ultimate wastage, however, is still by a chlorination-corrosion mechanism.

EFFECTS OF CHLORINE[21,22]

Within the United States, the damaging effects of chlorine compounds on liquid-ash corrosion have become a serious problem only recently as more and more refuse-derived fuels (RDF) are burned. Very little chlorine, if any at all, is found in U.S. coals; of the commercially important production areas, only Illinois coals contain any chlorine. Oil-fired boilers often receive fuel from an oceangoing tanker or barge, so some sodium chloride from seawater is present; the amount is small, but is a significant part of the ash. However, it is in RDF-fired incinerators where chlorine compounds, mostly polyvinyl chloride (PVC) plastics, contribute to corrosion problems.

Regardless of the form of the chloride compound in the fuel, at flame temperatures it is molecular chlorine; on cooling, the chlorine may combine with either hydrogen or water vapor to form hydrogen chloride (HCl) or it may remain as molecular chlorine. On the surface of steel tubing, one of the three reactions occurs to cause tube-metal loss:

$$2Fe + 3Cl_2 = 2FeCl_3$$
$$2Fe + 6HCl = 2FeCl_3 + 3H_2$$
$$Fe_2O_3 + 6HCl = 2FeCl_3 + 3H_2O$$

Aside from the direct corrosion of steel, ferric chloride has a melting point of 540°F (280°C), and so contributes to the low-melting-point liquids that cause liquid-ash attack. Once FeCl₃ forms, it may act as a flux to promote the formation of liquids within the ash deposit. As noted previously, mixtures of ash constituents may form even lower-melting-point compounds by combining with other ash constituents.

EFFECTS OF CARBON[23]

The role of carbon is somewhat unclear. In some cases, the corrosion products contain sulfides and carbon. It appears likely that the V_2O_5–Na_2SO_4 liquid transports both carbon and sulfur trioxide to the metal surface, where the carbon reduces sulfur trioxide and iron oxide and forms iron sulfide:

$$Fe_2O_3 + 2SO_3 + 9C = 2FeS + 9CO$$

The other likely reaction assumes the Fe_2O_3 is dissolved by the liquid, and the bare iron reacts with SO_3 and carbon to form FeS:

$$Fe + SO_3 + 3C = FeS + 3CO$$

CORROSION MORPHOLOGY

Aerodynamic considerations will, to a large extent, determine the features of the corrosion pattern that develops on the surface of the superheater or reheater tube. Ash builds up on the leading portion of the tube, at the flanks of the ash deposit; the highest-heat-transfer zones lead to the melting of a portion of the ash, and rapid deterioration of the tube follows. Figure 6.44 shows three views of a section of a stainless-steel, grade SA-213 TP-321H, reheater tube that failed in just a few months of operation. Note the deep pits at roughly the 10 and 2 o'clock positions, the general pitting attack in between, and a virtual attack-free side that extends from the 3 o'clock to the 9 o'clock position, opposite the ash buildup.

ADDITIVES

Use of additives for corrosion control of superheaters and reheaters has had the greatest success in oil-fired boilers.[24-26] With only a small amount of ash to treat, fuel-oil additives have proven economically feasible. Most additives are magnesium compounds of either the oxide or hydroxide. The particle shape, size, and size distribution vary from supplier to supplier. The corrosion-suppression reactions involve the formation of magnesium sulfate

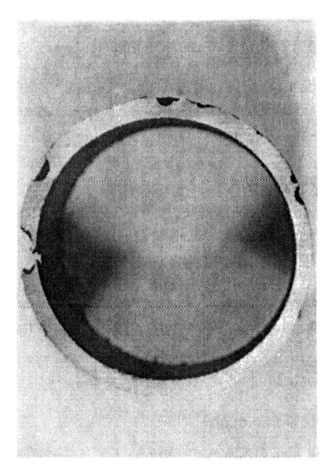

FIGURE 6.44. Three views of a Type 321H stainless-steel reheater tube severe corroded by liquid oil-ash components.

FIGURE 6.44. *(Continued)*

($MgSO_4$), which forms higher-melting-point mixtures than does Na_2SO_4. Thus, liquids do not form at superheater and reheater metal temperatures.

Another way to help suppress the ash components that form liquids is to fire the boiler with low excess air. If V_2O_5 formation can be prevented and V_2O_4 formed instead, the melting points of complex compounds are high enough to be above those encountered on tube surfaces.

Shields are not an effective remedy for preventing liquid oil-ash corrosion. Mixtures of V_2O_5–Na_2O or of V_2O_5–Na_2SO_4 do not thermally decompose; thus, raising tube-metal temperatures will not be helpful. As can be deduced from Fig. 6.43, the higher the metal temperatures, the more rapid the corrosion.

CORROSION PREVENTION

In refuse-fired boilers, more corrosion-resistant materials, for example, stainless steel or other nickel–chromium alloys, are expected to perform exceptionally well. Chromium oxide is stable even under reducing conditions. These materials would not be attacked by sodium chloride or hydrochloric acid. Thus, use of Inconel 625 weld overlays has proven successful in several years of service. Of course, the weld overlay is only as corrosion-resistant as the overlay is complete. Any holidays or voids are subject to rapid corrosion attack and lead to pinhole steam leaks.

Weld overlays in pulverized-coal-fired boilers would not be expected to perform as well as in refuse boilers. The wastage has a fatigue component that could lead to the formation of circumferential grooves within the waterwall tubes even in the absence of corrosion. In the case of superheaters and reheaters, where the fatigue component is substantially less, stainless-steel superheater and reheater tubes do have substantially better corrosion resistance than T-22.

EFFECTS OF ID SCALE

As discussed in Chapter 2, internal scale or corrosion deposits are an effective insulating barrier to the transfer of heat from flame to steam. Figure 6.45 plots the metal-temperature increase over the clean-tube condition, ΔT, versus the scale thickness for waterwalls assuming nucleate boiling using the thermal analysis of Chapter 2.

For high-pressure utility units, the steam–water temperature in the furnace is about 680°F (360°C) and the crown temperature is around 730°F (390°C). The oxidation limit for the carbon steel usually used for furnace construction is 850°F (450°C), and the stress allowed under the ASME Boiler Code drops sharply above 750°F (400°C) from 13,000 to 7800 psi at 850°F (450°C). From Fig. 6.45 it can be seen that a thin internal deposit will raise the tube-metal temperature into the ash-corrosion range, into the creep-failure range, or into the rapid-oxidation range, any one of which will lead to serious furnace-tube problems.

For superheaters and reheaters, the situation is similar; Fig. 6.46 plots the crown temperature increase over the clean-metal temperature condition versus the ID scale thickness. The curves are similar to Fig. 6.45 except that Q/A is smaller and the temperature increase is less for a given scale thickness. However, references to Figs. 6.41 and 6.42 shows that for oil-fired

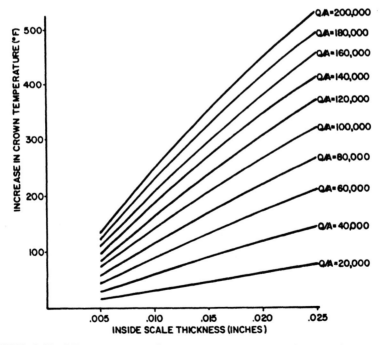

FIGURE 6.45. ΔT, temperature increase, versus scale thickness for waterwall tubes, assuming nucleate boiling.

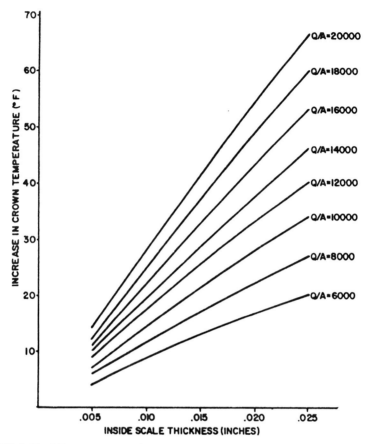

FIGURE 6.46. ΔT, temperature increase, versus scale thickness for steam-cooled tubes (superheaters and reheaters).

boilers, as the tube-metal temperature increases, the range of composition of $Na_2O-V_2O_5$ or $Na_2SO_4-V_2O_5$ that will form liquids expands. The situation is similar for coal-fired boilers; Fig. 6.47[7] shows the melting points of mixtures of the trisulfates, $Na_3Fe(SO_4)_3$ and $K_3Fe(SO_4)_3$. Increasing the metal temperatures expands the range of ash compositions that will lead to severe corrosion; thus, it is not unusual for boilers to operate 8–10 years without any hint of ash-corrosion problems, and "all of a sudden" rapid metal wastage increases the number of forced outages from tube failures to an unacceptable level.[27]

Two comments are in order at this time. The actual or real-life expectancy of the hot ends of superheaters and reheaters is 12–18 years because of ash-corrosion problems caused by internal scale. Design requirements at the time a boiler is purchased should include drainable superheaters and reheaters to facilitate chemical cleaning at regular intervals if it is desired to extend the life expectancy of the unit.

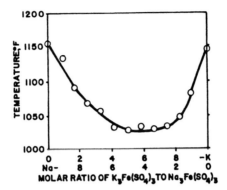

FIGURE 6.47. Melting points in system $Na_3Fe(SO_4)_3$–$K_3Fe(SO_4)_3$ (Ref. 6).

CHEMICAL-CLEANING DECISIONS

The chemical cleaning of a furnace-wall tube is required for the correction of two problems:

1. Steam-side scale and deposits interfere with heat transfer. The net effect of these insulating layers is to raise tube-metal temperatures. In the extreme, furnace-wall tubes can fail by a creep or stress-rupture mechanism due to the overheating.
2. Underdeposit corrosion leads to hydrogen damage. Hydrogen damage can lead to premature tube failure and unreliable boiler operation.

Atwood and Hale[28] have presented guidelines for the definition of the need for chemical cleaning based on internal deposit loading given in g/ft^2. Table 6.4 presents these guidelines. The ASTM has a formalized procedure for the techniques to be used for the determination of water-side deposit loading.[32]

We now discuss the location of the tube sample and the location, within the tube, of the sample used to determine the deposit loading. Tube samples should be removed from the highest-heat-flux regions of the furnace. The reason for this location is clear. Whatever the water-side scale effect, hydrogen damage, or creep failure, it will be the greatest in these hottest locations. Tube-metal temperatures will be the highest, and therefore corrosion rates will be the fastest. Tube-metal temperature increases are a function of the heat flux, and therefore the insulating effect of the steam-side scale will raise tube-metal temperatures the most in these high-heat-flux regions.

In addition to the effect of furnace elevation on heat flux and tube-metal temperature, there is a second consideration for any tube. The metal temperature varies around the perimeter. The highest metal temperature is along the centerline of the fire side of the tube, midway between the membranes. For corner-fired boilers, the highest heat flux is biased toward the burners.

TABLE 6.4. Relationship of Analyzed Deposit Quantity to Unit Cleanliness

Boiler Type	Internal Deposit Quantity Limits[a]		
	Clean Surfaces, mg/cm^2	Moderately Dirty Surfaces, mg/cm^2	Very Dirty Surfaces, mg/cm^2
Supercritical units	<15	15–25	>25
Subcritical units (1800 psig and higher)	<15	15–40	>40

[a] All values are as measured on the furnace side of tube samples and include soft and hard deposits.
Note: For all practical purposes, 1 mg/cm^2 = ~1 g/ft^2.

When caustic gouging or hydrogen damage occurs, the location will be an accurate reflection of the hottest metal temperatures, which are usually centered midway between the furnace membranes. See, for example, Fig. 6.19 to see that the groove caused by caustic gouging is centered between the membranes and reflects the location of the highest tube-metal temperature.

Because the metal temperature varies around the perimeter, the amount of deposit will also vary around the perimeter. The thickest deposits form at the highest metal temperatures. The cold or casing side of the waterwall tube will have the least deposit. Figures 6.48 to 6.50 show the variation in deposit thickness from the centerline of the fire side, 30° off the centerline, and 180° around the perimeter from the centerline of the fire side. The scale thickness varies in these photographs from 3.8 to 1.2 mils.

The procedure that will give the best, most consistent results in deposit loading is as follows (refer to ASTM D3483):[32]

1. Remove the tube sample from the highest-heat-flux regions of the furnace. At least two, and preferably four, samples should be taken.
2. Remove all OD ash deposits and oxides from the tube samples for a distance of about 6 in.
3. Dry-saw-cut the tube samples to a length of 6 in.
4. Dry-saw-cut the test coupon from the centerline of the fire side so that the coupon covers the middle 60° centered on the membrane. A similar coupon is removed from the cold or casing side. Any saw-cut chips should be carefully removed to prevent spurious readings.
5. Weigh each coupon.

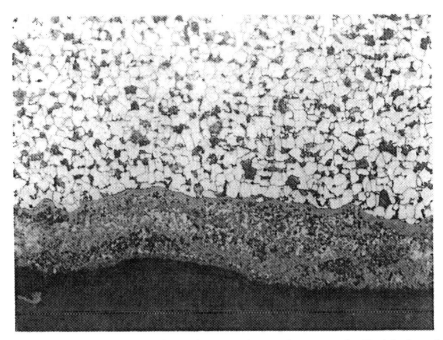

FIGURE 6.48. Centerline, midway between the membranes, at the 12 o'clock position, has an ID scale thickness of 3.8 mils (0.0038 in) (Magnification 200×, etched).

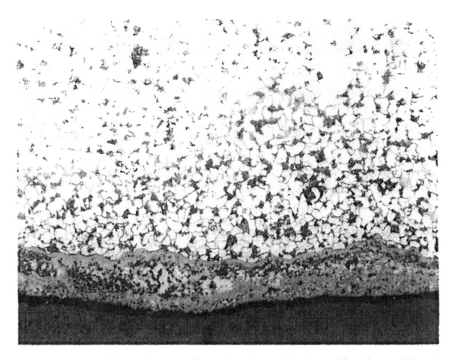

FIGURE 6.49. 30° from the centerline, about the 1 o'clock position, has an ID scale thickness of 2.5 mils (Magnification 200×, etched).

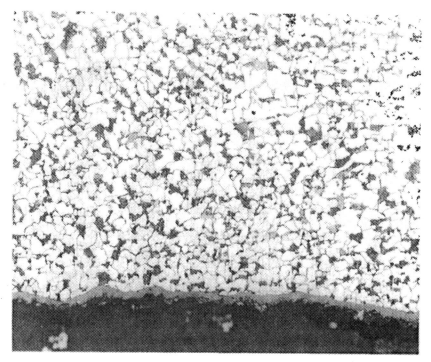

FIGURE 6.50. 180° around the perimeter from the centerline, the 6 o'clock position, has an ID scale thickness of 1.2 mils (Magnification 200×, etched).

6. Remove the water-side deposits in inhibited acid. There may be two steps required, since the chemicals used to remove iron oxide will not remove copper, and vice versa. When the deposits have been removed, reweigh the coupon.
7. The difference in weight is the amount of water-side deposit removed.
8. Calculate the area of the coupon, and convert the deposit loading to g/ft^2 determination.

In addition, metallographic samples should be removed from the fire side and the cold, or casing, side. The metallographic analysis will show both the scale thickness, as in Figs. 6.48 to 6.50, and whether there is any underdeposit corrosion, hydrogen damage, or microstructural changes that would indicate overheating.

SUPERHEATERS AND REHEATERS

In coal-fired boilers, metal wastage of superheaters and reheaters is caused by two corrosion mechanisms. At low temperatures, metal loss occurs by simple oxidation and at a low rate. At higher temperatures, complex liquid-

376 CHAPTER 6 CORROSION-CAUSED FAILURES

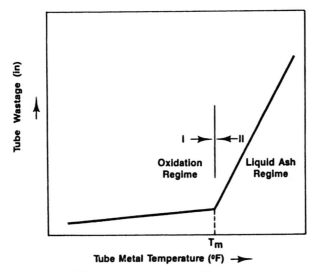

FIGURE 6.51. Schematic presentation of wastage vs. temperature.

ash attack occurs and often at a high rate. Low and high temperatures are relative terms, but both temperature regimes are prevalent in a boiler. Schematically these wastage rates are plotted in Fig. 6.51.[30]

At the low-temperature end, region I of Fig. 6.51, tube wastage follows from the oxidation of steel. At the high-temperature end, region II of Fig. 6.51, rapid wastage is caused by liquid-ash corrosion. A superheater or reheater starts its life in region I of Fig. 6.51. Nothing unusual occurs for many years—no rapid metal loss, no forced outages due to tube failures. When the tube-metal temperature increases enough, due to the steam-side scale, it rises to the melting point of the liquid-ash species. Figures 6.52 and

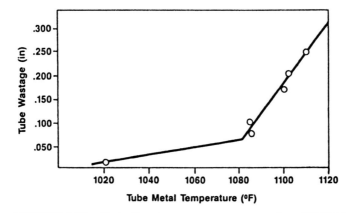

FIGURE 6.52. Plot of wastage vs. temperature after 125,000 hr.

SUPERHEATERS AND REHEATERS 377

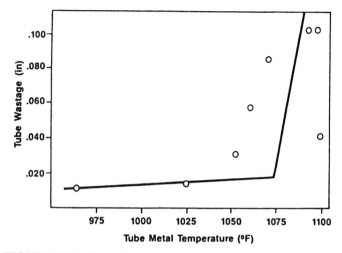

FIGURE 6.53. Plot of wastage vs. temperature after 130,000 hr.

6.53 support the shape of a hockey stick to the corrosion of the tube wastage versus the tube-metal temperature plot.

By a periodic removal of the steam-side scale, tube-metal temperatures effectively cycle between lower temperatures, and life is substantially extended. Figures 6.54 and 6.55 from Ref. 31 show the effects of chemical cleaning on the expected failure time for two superheater examples.

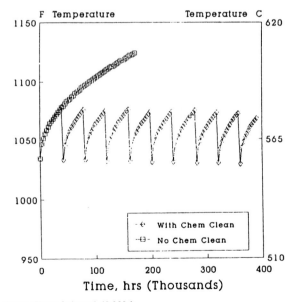

Chem Clean Interval 40,000 h

FIGURE 6.54. Effect of chemical cleaning case no. 1.

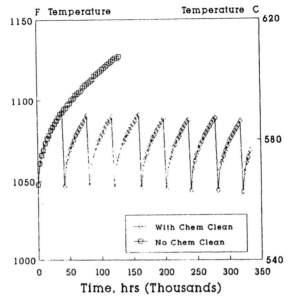

Chem Clean Interval 40,000 h

FIGURE 6.55. Effect of chemical cleaning case no. 3.

LOW-TEMPERATURE CORROSION

The combusion products of coal include SO_2 and SO_3, as mentioned, and water vapor, H_2O, among many others. When the temperatures fall below the acid dew point, sulfuric acid will form:

$$H_2O + SO_3 = H_2SO_4$$

This occurs at the cold end of the back of the boiler. For this reason, the exhaust temperature into the stack is kept above about 300°F (149°C). An example is a bare-tube economizer that failed from the OD by sulfuric-acid corrosion. Apparently, the feedwater heater malfunctioned, and the economizer-inlet water temperatures dropped to below 250°F (120°C), which allowed H_2SO_4 to coat the tube surface. The acid gave the complete tube perforation shown in Fig. 6.56.

STRESS-ASSISTED CORROSION

Stress-assisted corrosion[32] is most likely to occur in those locations that have a high and localized stress concentration. In corrosive environments, more rapid attack occurs in those regions of highest stress. For example, Fig. 6.57 displays the ID surface of a waterwall tube at a buckstay weld. A

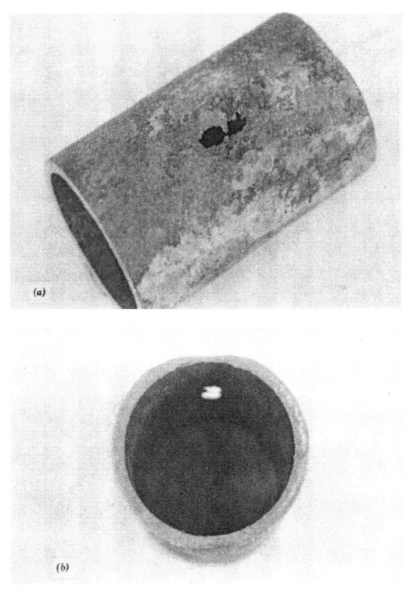

FIGURE 6.56. (a) View of an economizer tube perforated by condensed sulfuric acid. (b) End view of same tube to display more clearly the complete perforation.

close-up of these cracks shows them to be a series of interconnected pits, suggestive of oxygen-pitting corrosion; see Fig. 6.58. A cross section through this attack, Fig. 6.59, shows the line of pits to be about 55 mils deep.

The corrosion is caused by a combination of factors: high local stress, either residual or applied, and a corrosive environment. In this case oxygen contamination of the boiler water or perhaps improper control during chemi-

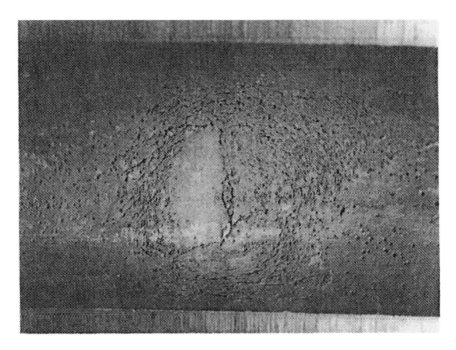

FIGURE 6.57. ID surface of a waterwall tube at a buckstay-attachment weld. Most severe pitting surrounds the weld (Magnification 1.3×).

FIGURE 6.58. A close-up shows the cracks to be a series of interconnected pits (Magnification 6.4×).

FIGURE 6.59. In cross section, the pits are about 55 mils (0.055 in) deep (Magnification 25×, etched).

cal cleaning was responsible. Upsets in boiler operation that would aggravate the local stress (for example, a rapid start-up or fan cooling during shutdown) will exacerbate the problem.

BOILER-FEEDWATER CONTROL

The function of the chemical additives for boiler feedwater is to control the purity and level of impurities so that boiler operation can continue without corrosion or scale buildup on the inside of the tubes.

In any boiler-feedwater-treatment program several steps are required, not all of which may be necessary depending on the quality of the feedstock:[33,34]

1. Remove suspended solids; add a coagulant to agglomerate the contaminants and facilitate settling.
2. Filter to remove coarse material; filter beds are made up of graded gravel, and may or may not include crushed anthracite.
3. Aerate to remove carbon-dioxide (CO_2) and hydrogen-sulfide (H_2S) gases.

4. Mechanically deaerate to remove oxygen, the most damaging impurity. Final oxygen levels are in the low parts-per-billion (ppb) range.
5. Chemically soften to remove "hard"-water contaminants, usually calcium and magnesium compounds and silica. Zeolite softening or ion-exchange processes reduce hardness concentration to 2 ppm, or so. Demineralization, also an ion-exchange process, produces the purest water available. Cations [calcium (Ca^{2+}), magnesium (Mg^{2+}), sodium (Na^+)], anions [carbonate (CO_3^{-2}), bicarbonate (HCO_3^-), sulfate (SO_4^{-2}), and chloride (Cl^-)], and silica are removed by special resins.
6. Physical means; evaporation and/or distillation processes provide pure water by condensing steam vapor. The major drawback is expense.
7. Newest mechanical method: reverse osmosis uses pressure to force water through a semipermeable membrane. Since impurities cannot penetrate the membrane, the water produced is quite pure. Again, the major drawback is expense.

Even with a good pretreatment program, residual oxygen must be removed from the water entering the boiler. Several techniques for preventing oxygen corrosion are discussed below.

Sodium Sulfite Treatment

The most common oxygen scavenger in lower-pressure industrial units is sodium sulfite. Sodium sulfite reacts with oxygen to form sodium sulfate.

$$2Na_2SO_3 + O_2 = 2Na_2SO_4$$

In high-pressure utility units, the sodium sulfite will be broken down by heat to sodium hydroxide and sulfur-dioxide gas:

$$Na_2SO_3 + H_2O + heat = 2NaOH + SO_2$$

These problems occur at about 900–1000 psi. Also, sodium sulfite cannot be used with the coordinated phosphate treatment because the sodium ion alters the balance of the sodium and phosphate that is so important in a coordinated phosphate water treatment.

Hydrazine

In high-pressure units, hydrazine is a common additive for oxygen scavenging. The products are inert; hydrazine reacts with oxygen to form water and nitrogen gas:

$$N_2H_4 + O_2 = 2H_2O + N_2$$

It will also react with ferric and cupric oxides to form ferrous and cuprous oxides.

$$6Fe_2O_3 + N_2H_4 = 4Fe_3O_4 + 2H_2O + N_2$$
$$4CuO + N_2H_4 = 2Cu_2O + 2H_2O + N_2$$

Coordinated Phosphate Control

The coordinated phosphate control depends on a mixture of phosphate or hydroxide and disodium phosphate to control pH. The pH is controlled by the ratio of one phosphate to another.

Phosphates hydrolyze with water to form hydroxide:

$$PO_4^{3-} + H_2O = HPO_4^{2-} + OH^-$$
$$HPO_4^{2-} + H_2O = H_2PO_4^- + OH^-$$
$$H_2PO_4^- + H_2O = H_3PO_4 + OH^-$$

If all phosphate is trisodium phosphate, the ratio of sodium to phosphate is 3. At this or slightly lower ratios, no free hydroxide is present. Free hydroxide is defined as the amount of hydroxide over the equilibrium established between trisodium phosphate and disodium phosphate.

Congruent Control

The coordinated phosphate procedure outlined above did not work as well as planned. Some deposits were not pure trisodium phosphate (Na_3PO_4), but contained disodium phosphate (Na_2HPO_4) or monosodium phosphate (NaH_2PO_4), leaving an excess of sodium hydroxide to corrode the boiler by caustic attack. Within the deposits, the sodium-to-phosphate ratio was as low as 2.6. This corresponds with the ratio of sodium-to-phosphate found in most deposits. Thus, for the congruent control, the ratio of sodium to phosphate is about 2.6. The temporary deposition of soluble chemicals is called "hide-out." Hence, congruent control fixes the sodium—phosphate ratio at 2.6–2.8, so no free hydroxide will form even if some hide-out occurs.

Volatile Treatment

An all-volatile treatment, sometimes called zero-solids control, prevents the formation of hydroxides, since no sodium chemicals are used. Ammonia and/or morpholine or cyclohexylamine are used for pH control, and hydrazine is used for oxygen scavaging. This treatment is used primarily for supercritical units and 2800-psi drum units.

384 CHAPTER 6 CORROSION-CAUSED FAILURES

CASE HISTORIES

The next several pages present examples of corrosion-related failures, both steam–water-side-corrosion problems and liquid-ash attack.

Case History 6.1

Hydrogen Damage in Waterwall Tubes

BOILER STATISTICS

Size	1,550,000 lb steam/hr (705,000 kg/hr)
Steam temperature	1005°F/1005°F (540°C/540°C)
Steam pressure	1950 psig (135 kg/cm^2)
Fuel	Oil/gas

The tubes were taken from various locations in the furnace walls of an oil-fired boiler. Figure 6.1.1 shows the general appearance of a typical as-received tube sample. The tubes are grade SA-192 steel 3.25-in OD × 0.340-in-thick wall. None of the sections contained any failures. The internal surface showed some pitting but no large accumulation of deposits. The internal scale measured 0.015 in thick.

Before any metallographic examination was performed, a ring section was cut from the as-received tube sample and flattened in a vice. During the flattening test, the orientation of the flattening was such that the vice squeezes directly on the fire side of the tube. The ID at the fire side then becomes deformed in tension and, if hydrogen damage exists, will easily crack at that point. Figure 6.1.2 shows the results of such a test. Some grooving on the fire side of the tube is noted in the undeformed ring section adjacent to the flattened test in Fig. 6.1.2.

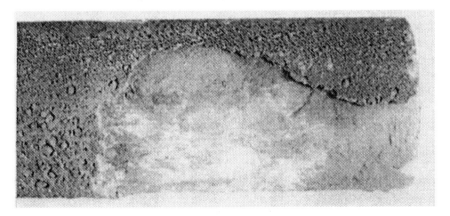

FIGURE 6.1.1. As-received tube sample; nothing is visually unusual about this furnace tube.

CASE HISTORIES 385

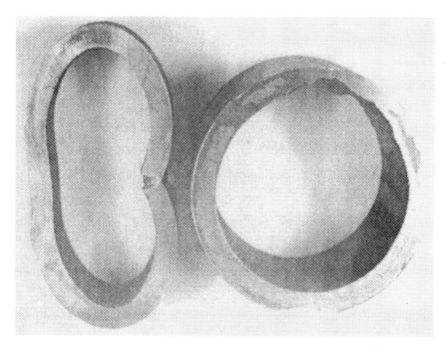

FIGURE 6.1.2. Ring tests for ductility show that hydrogen damage is suspected.

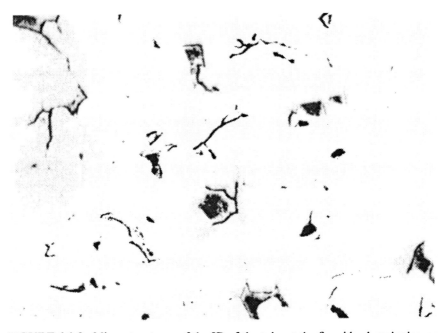

FIGURE 6.1.3. Microstructures of the ID of the tube at the fire side show hydrogen cracks surrounding the pearlite colonies (Magnification 500×, nital etch).

FIGURE 6.1.4. Hydrogen damage from another tube shows complete decarburization, and the resultant structure is left as ferrite with hydrogen-induced cracks (Magnification 500×, nital etch).

In high-pressure boilers, the hydrogen that is generated by the corrosion of the boiler tube is trapped between the scale and the steel. Some of the hydrogen diffuses into the steel, where it reacts with the iron carbide in the pearlite to form methane and iron. Since the methane molecule is too large to diffuse easily, the methane collects at ferrite grain boundaries and leads to the formation of internal cracks or fissures. The microstructure of such attack will show cracks around the individual pearlite colonies, as seen in Fig. 6.1.3. In the case where all of the iron carbide has been converted to methane and ferrite, all that is left is ferrite and the methane-induced cracks, as shown in Fig. 6.1.4.

Another feature of the microstructure is the lack of any evidence of overheating. The pearlite colonies are still well defined, and the platelets of iron carbide show no spheroidization.

Case History 6.2

Oxygen Pitting of Superheater Tubes

BOILER STATISTICS

Size	640,000 lb steam/hr (290,000 kg/hr)
Steam temperature	905°F/900°F (480°C/480°C)
Steam pressure	1600 psig (110 kg/cm^2)
Fuel	Pulverized coal

The following example is of two tubes from a horizontal superheater from a coal-fired boiler. One tube contained a massive rupture with the entire bottom-half section of the tube blown away. The edges of the failure were blunt and exhibited very little ductility. See Fig. 6.2.1. The other tube contained no failure. The tubes were reported as SA-209 T-1, 2-in OD × 0.148-in-thick wall.

Visual observation of both tubes revealed gross pitting on the ID confined to the bottom surface. The top surface had some pits, but to a much lesser extent. There was no wall thinning noted except at the pits themselves, and no swelling of the circumference was measured. Figure 6.2.2

FIGURE 6.2.1. As-received superheater tube with a window blown out; note the thick-edged fracture mode.

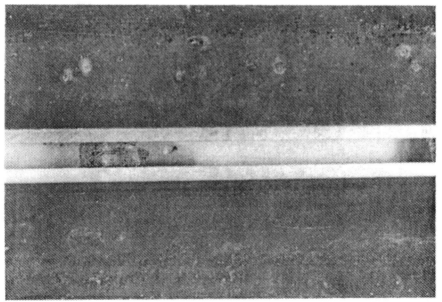

FIGURE 6.2.2. The ID surface that displays the extensive ID pitting.

shows the inside surface of the unfailed tube section. Figures 6.2.3 and 6.2.4 show cross sections of the ID pitting observed on both samples. Figure 6.2.5 shows the deepest pit observed from the tube that contained the failure; note that the pit extends more than halfway through the tube wall. Figure 6.2.6 shows the microstructure of the tube material at approximately the midwall of the unfailed tube. No evidence of overheating is observed, the pearlite is still intact, and the iron carbide shows no evidence of spheroidization.

The boiler in question had been in service for more than 10 years and had been a base-loaded unit for the first 9 years of service. Its most recent service, and the service for the year or so prior to the failure, was one of cyclic duty, with the boiler off the line and cold each weekend. During the idle periods, the boiler was vented to the atmosphere, and air and condensate would collect at the low points of the horizontal superheater. Here we have the ideal conditions for oxygen-pitting attack to occur—a moist environment and ready access to air. During the shut-down, the pitting would occur; and over the year plus of cyclic duty, pitting progressed to the point where failures occurred during operation.

The failure due to oxygen pitting of the superheater tubes in this coal-fired cyclic-duty boiler illustrates a problem that can be created by improper off-duty storage. When a boiler is to be laid up for an extended

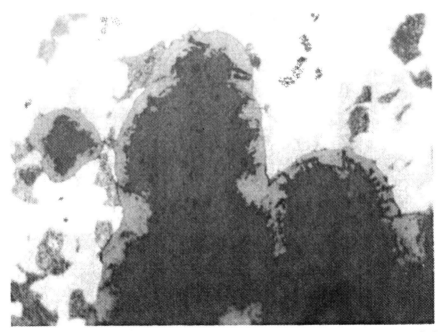

FIGURE 6.2.3. Cross section of the ID oxygen pits (Magnification 500×, nital etched).

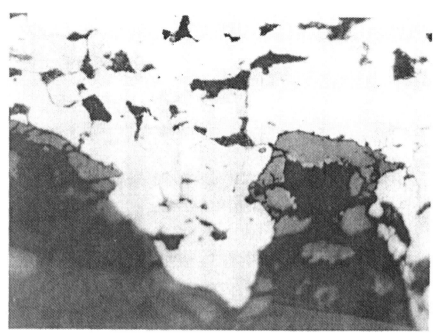

FIGURE 6.2.4. Another set of ID pit cross sections (Magnification 500×, nital etched).

FIGURE 6.2.5. Cross section of the whole tube wall; note that at least one pit nearly penetrates the wall (Magnification 4×, nital etched).

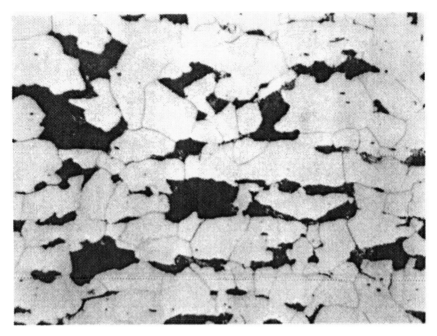

FIGURE 6.2.6. Microstructure of normal ferrite and pearlite and no evidence of any overheating (Magnification 500×, nital etched).

time, extreme caution must be exercised to prevent the leakage of air. Units to be shut down for only a short time should be kept pressurized. For longer storage duration, units should be carefully drained, dried, and back-filled with dry nitrogen to prevent internal corrosion.

Case History 6.3

Corrosion Fatigue Failure of an Economizer Tube

BOILER STATISTICS

Size	30,000 lb steam/hr (1360 kg/hr)
Steam temperature	Saturated
Steam pressure	250 psig (18 kg/cm^2)
Fuel	Municipal refuse

The following example is an economizer-tube failure from a municipal incinerator that failed in service. The tube was grade SA-423 steel, having a 3-in OD swaged to 2-in OD as it entered the drum. Wall thickness was 0.188 in. The failure was located in the swaged section about $\frac{1}{2}$ in from the entrance to the drum. The crack was $\frac{5}{8}$ in long and proceeded in a circumferential manner just outside the drum.

CASE HISTORIES   391

Visual examination of the failed section revealed no unusual scale or deposits on the inside or outside of the tube. The crack length was much larger on the ID (about 1 in) than on the OD (about ⅜ in). Microscopic examination through the failure crack showed the crack propagating from the ID toward the OD. The crack itself showed some debris within and was predominantly transgranular. Figure 6.3.1 shows the crack tip at

FIGURE 6.3.1. Crack tip of thermal-fatigue crack (Magnification 500×, nital etch).

FIGURE 6.3.2. ID pit cross section (Magnification 500×, nital etch).

FIGURE 6.3.3. Cross section of the whole tube wall; note the concentric markings typical of a fatigue failure (Magnification 4×).

500×. Figure 6.3.2 displays some of the numerous ID pits that were found adjacent to the failure area. Figure 6.3.3 shows the fracture face at 4× after cleaning in a solution of 2% nitric acid in alcohol (nital). The figure exhibits concentric curved zones typical of a fatigue failure.

The failure was caused by a corrosion-fatigue crack that originated at an oxygen pit. Oxygen pits are caused by improper boiler-feedwater treatment. Cyclic stress from any of several causes, such as improper and frequent start-up, vibration, unstable inlet temperatures of the feedwater to the economizer that thermally shock these tubes, combined with a design limitation of too stiff a tube, could have caused this failure.

Case History 6.4

Liquid-Ash Corrosion in Coal-Fired Boiler Superheater

BOILER STATISTICS

Size	1,700,000 lb steam/hr (770,000 kg/hr)
Steam temperature	1005°F/1005°F (540°C/540°C)
Steam pressure	1980 psig (140 kg/cm^2)
Fuel	Pulverized coal

The following example is a failed superheater tube of grade SA-213 T-22 steel, having a 1¾-in OD × 0.260-in-thick wall, taken from a boiler that had operated for 12 years without trouble. Visual examination revealed a 1½-in-long longitudinal split; see Fig. 6.4.1. The OD was covered with ash deposits 0.040 in thick, and the ID scale measured 0.012 in thick. The tube-wall thickness was 0.230 in in line with the split and 0.300 in 180° away from it.

FIGURE 6.4.1. The as-received superheater-tube failure; note the small fissure and limited tube swelling characteristic of creep failures.

394 CHAPTER 6 CORROSION-CAUSED FAILURES

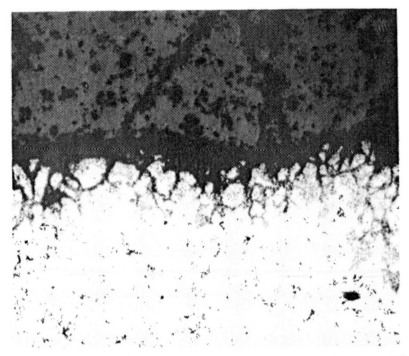

FIGURE 6.4.2. Minor grain-boundary attack on the OD; note the spheroidized microstructure (Magnification 500×, nital etch).

Microstructural analysis at the point of failure revealed a complete spheroidization of the carbides, indicating long-time exposure to elevated temperatures. Some grain-boundary attack was noted on the OD; see Fig. 6.4.2. A sulfur print was made on the failed tube; noted were sulfides indicative of liquid-ash corrosion.

The failure, a narrow split without excessive swelling of the tube or wall thinning at the opening, is typical of creep failures. Overheated microstructures, usually associated with long-term operation at temperatures that are 50–100°F (28–56°C) above the normal expected range, are good examples of the troubles caused by internal scale buildup. From Fig. 6.45, metal temperatures of 50–70°F (28–39°C) above design or actual crown temperatures above 1125°F (607°C) can be expected. Thus, the combination of high tube temperatures and wall thinning from ash corrosion led to tube failures. For all practical purposes, the high-temperature legs of this superheater are "worn out" and will need replacement if an excessive number of failures are to be avoided.

Case History 6.5

Furnace-Wall ID Corrosion

BOILER STATISTICS

Size	2,000,000 lb steam/hr (910,000 kg/hr)
Steam temperature	1005°F/1005°F (540°C/540°C)
Steam pressure	2700 psig (190 kg/cm^2)
Fuel	Pulverized coal

The waterwall tubes, grade SA-210-1A steel, having a 2½-in OD × 0.260-in-thick wall, did not contain any ruptures, but did exhibit swelling and wall thinning on the fire-side surface. The tube OD had increased from 2½ to 2$\frac{18}{32}$ in and 2$\frac{17}{32}$ in, respectively. The fire side of both tubes appeared to have a groove at the 12 o'clock position. Wall thickness on the fire side ranged from 0.220 to 0.248 in and on the rear, 180° from fire side, to 0.280 in. The ID scale and deposits measured 0.011 in. A sulfur print, Fig. 6.5.1, showed the initial stages of ash-corrosion attack on the fire side. Microstructural examination showed spheroidized carbides on the OD crown. Figures 6.5.1–6.5.3 display the essential features.

From Fig. 6.42, an ID scale of 0.011 in can raise crown temperatures by 200°F (111°C) or more, depending on the location in the furnace. Water-treatment upsets and ID corrosion were the cause of the overheated microstructure and the OD ash corrosion.

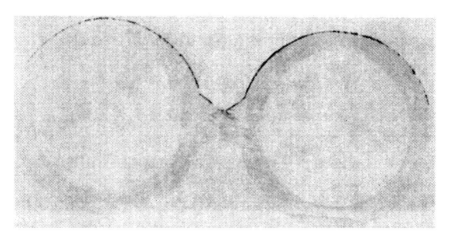

FIGURE 6.5.1. Sulfur print showing the initial stages of liquid-ash attack.

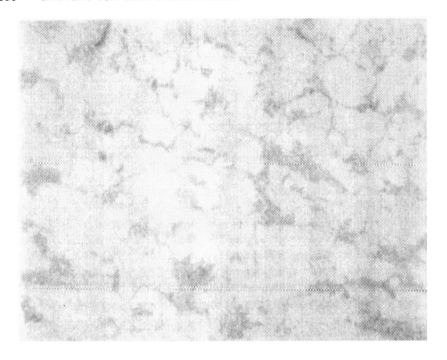

FIGURE 6.5.2. Microstructure of the fire-side crown of the furnace-wall tube; note the spheroidized-carbide particles indicative of metal temperatures over 900–1000°F (480°–540°C) for some time (Magnification 500×, nital etch).

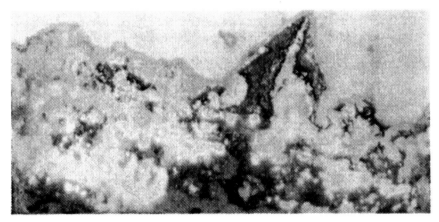

FIGURE 6.5.3. Structure of the ID scale and deposit, 0.011 in thick. The white areas in the scale are copper (Magnification 500×, nital etched).

Case History 6.6

Furnace-Waterwall ID Deposits

BOILER STATISTICS

Size	Unknown
Steam temperature	350°F (177°C)
Steam pressure	125 psig (9 kg/cm²)
Fuel	Gas

This example is from a small, low-pressure industrial unit and shows the interesting microstructures that develop from long-time operation at temperatures around 900–1000°F (482–538°C). Figures 6.6.1 and 6.6.2 show the condition of the as-received tubes; observe the virtual pluggage of the tubes with feedwater chemicals.

The three tube sections contained no failures, but were part of a tube either above or below a point of failure. The tubes were 2-in OD × 0.134-in-thick wall grade SA-178 A steel. The boiler had operated at 125 psi for more than 10 years and was on a phosphate–sulfite water-treatment program.

A simple calculation of the hoop stress,

$$S = \frac{P(D - W)}{2W}$$

where P is pressure (psi), S is tube stress (psi), D is tube OD (in), and W is wall thickness (in), gives a stress value of around 900 psi. The safe operating stress for SA-178 A steel at 1000°F (538°C) is, per Table 1A of the Code, 1300 psi. It is impossible to estimate how long these tubes have

FIGURE 6.6.1. Cross section of the as-received tube; note the complete pluggage.

FIGURE 6.6.2. Another tube section that is virtually plugged with water-treatment chemicals.

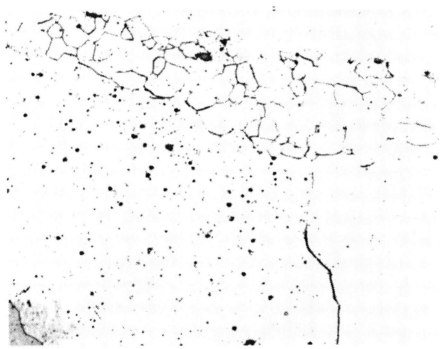

FIGURE 6.6.3. Microstructure at the thinned portion of the tube; note that the tube is completely decarburized and the ferrite grain size is very large (Magnification 250×, nital etch).

been operating at temperatures near 1000°F (538°C), but it was a long enough period for complete decarburization and grain growth to have occurred; see Figs. 6.6.3–6.6.5.

These microstructures are an unusual feature of low-pressure industrial boilers, where hydrogen damage does not occur, but decarburization does. Typically, failure occurs before complete decarburization occurs, and grain growth is excessive.

A look at Fig. 6.45 for a Q/A of 40,000–60,000 Btu/hr · ft^2 shows our present case to be off-scale. But certainly, wall temperatures around 1000°F (538°C) are to be expected from the virtual pluggage of these tubes.

FIGURE 6.6.4. Opposite side of tube shown in Fig. 6.6.3 (Magnification 100×, nital etch).

400 CHAPTER 6 CORROSION-CAUSED FAILURES

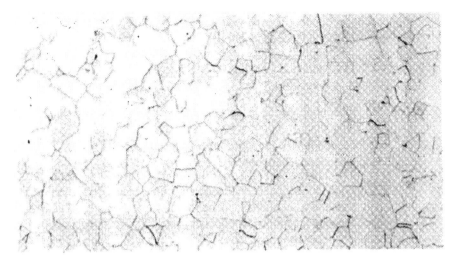

FIGURE 6.6.5. Another section of the tube in Fig. 6.6.3 showing completely decarburized structures and excessive ferrite-grain growth (Magnification 250×, nital etch).

Case History 6.7

Oil-Ash Corrosion of Reheater Tube

BOILER STATISTICS

Size	550,000 lb steam/hr (250,000 kg/hr)
Steam temperature	1005°F/1005°F (540°C/540°C)
Steam pressure	1875 psig (130 kg/cm^2)
Fuel	No. 5 fuel oil

This example deals with the failure of a reheater tube from a small municipal power plant. The reheater tube is SA-213 T-22 steel, having a 2-in OD × 0.148-in-thick wall. Visual examination of the tube revealed the following:

A. The tube was heavily coated with a thick deposit and scale.
B. The inside of the tube contained a tightly adhering magnetite scale, 0.015 in thick.
C. The tube was thinned for a considerable distance; see Fig. 6.7.1.
D. Micrometer measurements showed the wall thickness to be 0.020 in at the thinnest portion of the tube.
E. The wall thickness just under the lug measured 0.152 in.
F. Microstructure showed complete spheroidization of the pearlite, and a carbide network around the ferrite grains; see Fig. 6.7.2.

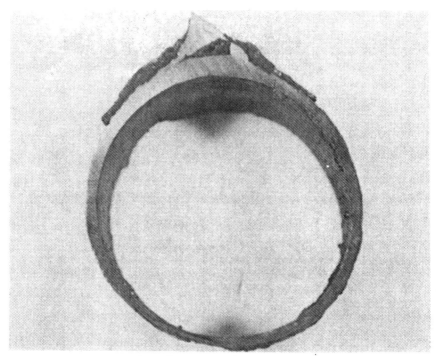

FIGURE 6.7.1. Cross section of the as-received tube section; note wall thinning over most of the tube opposite the lug.

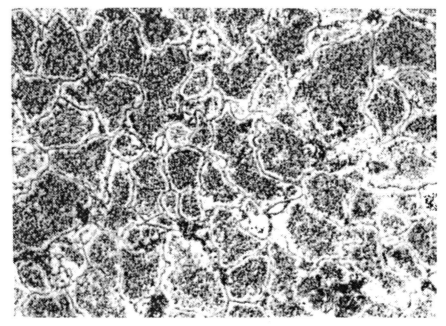

FIGURE 6.7.2. Microstructure shows complete spheroidization of the pearlite, and a carbide network surrounds the ferrite grains (Magnification 500×, nital etch).

The extensive wall thinning indicated a severe OD corrosion problem. Chemical analysis of the ash deposit was as follows:

> Sulfur as SO_3 13.9%
> Vanadium as V_2O_5 39.2%
> Sodium as Na_2O 12.2%
> Iron as Fe_2O_3 31.2%
> Carbon   0.4%

The quantities of these elements suggested that the tube wastage was from liquid-ash corrosion by mixtures of either V_2O_5 and Na_2O or V_2O_5 and Na_2SO_4.

Case History 6.8

Oil-Ash Corrosion of High-Temperature Superheater

BOILER STATISTICS

Size	786,000 lb steam/hr (360,000 kg/hr)
Steam temperature	1005°F/1005°F (540°C/540°C)
Steam pressure	1890 psig (135 kg/cm²)
Fuel	No. 6 fuel gas

This example discusses the effects of carbon on the liquid-ash corrosion in oil-fired boilers. Figure 6.8.1 shows a portion of SA-213 T-22 steel superheater tube that displays the liquid-ash attack. While this particular

FIGURE 6.8.1. As-received tube sample.

tube contained no failure, it was submitted as part of a study of the corrosion problem. Chemical analysis of the fire-side deposits yields:

Vanadium as V_2O_5	64.2%
Sulfur as SO_3	1.5% (average deposit composition)
Sulfur as SO_3	6.5% (at metal–deposit interface)
Sodium as Na_2O	16.9%
Calcium as CaO	0.1%
Magnesium as MgO	4.0%
Chlorine	20 ppm
Carbon	0.01 and 0.25% (two samples)

Note the differences in sulfur content between the average deposit composition and the concentration at or near the tube-metal–deposit interface.

A sulfur print, Fig. 6.8.2, confirmed the presence of metallic sulfides. X-ray diffraction identified the following compounds in the deposit:

Iron sulfide	FeS
Iron oxide	Fe_2O_3, Fe_3O_4
Sodium sulfate	Na_2SO_4
Magnesium silicate	$MgSiO_3$ or Mg_2SiO_4
Iron chloride	$FeCl_2$ or $FeCl_3$
Vanadium pentoxide	V_2O_5

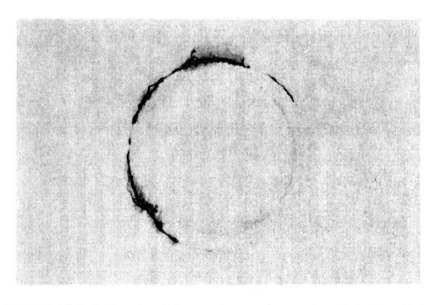

FIGURE 6.8.2. Sulfur print of the superheater-tube sample suffering from V_2O_5–$NaSO_4$ ash attack.

Here, the compounds in the deposit necessary for a typical oil-ash corrosion attack were Na_2SO_4 and V_2O_5. Also, iron sulfide and carbon were identified in the ash deposit. From this, it appears likely that SO_3 and Na_2O from the fuel oil form Na_2SO_4 in the outer portion of the deposit and migrate to the metal surface; carbon, SO_3, and iron then react to form iron sulfide and cause tube wastage:

$$3C + SO_3 + Fe = FeS + 3CO$$

Thus, it appears necessary that carbon be present to reduce SO_3 and maintain the reducing conditions, at least on a microenvironmental scale, required for iron-sulfide stability.

Case History 6.9

Water Washing

Oil-fired boilers are often washed down several times a year. The final rinse should be made with a basic solution of sodium carbonate to neutralize the acid salts in the oil ash. A solution of oil ash in water has an acid pH. As the wash water evaporates, the acid droplets that form can pit the boiler tubes. Figure 6.9.1 shows this effect. While a single water washing

FIGURE 6.9.1. Pits on waterwall tubes from acid salts in oil ash. During water washing, as moisture finally evaporates, acid droplets form and pit the steel.

is unlikely to lead to a pitted condition, repeated washings over several years certainly will.

Case History 6.10

Stress-Corrosion Cracking Inconel 625®

For years the ASME code has precluded the use of austenitic stainless steel from service in water-wetted circuits. Thus the excellent corrosion resistance of these 300-series stainless steels can be used only as weld overlays or bimetallic tubing in waterwall service. In recent years, the use of higher-alloy, nickel-based austenitic materials has been allowed for use in waterwall tubes. Alloys of 35-nickel–20-chrome (Incoloy 800®) and 60-nickel–20-chrome–10-molybdenum (Inconel 625®) have been allowed in this type of service.

Laboratory tests and early operating experience have shown these alloys to resist chloride-induced stress-corrosion cracking to a satisfactory degree to make them safe for boiler operation. However, under certain stress circumstances, a residual stress can be the site of stress-corrosion cracking. Figures 6.10.1 and 6.10.2 show one such case of stress-corrosion cracking of a waterwall tube of Inconel 625®. The failure

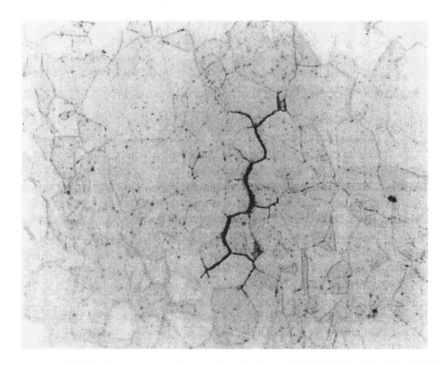

FIGURE 6.10.1. Stress corrosion cracks in Inconel 625® (Magnification 500×, etched).

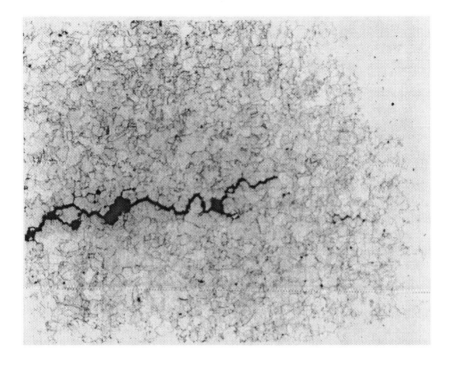

FIGURE 6.10.2. Crack tip (Magnification 100×, etched).

occurred at a dent, and the damage was confined to the area of the dent alone. Both figures display intergranular cracks that are characteristic of stress-corrosion cracking.

Case History 6.11

Chelant Corrosion

Chelants are organic acids that will complex iron oxide to keep it in solution. Therefore these chemicals are often used in chemical-cleaning solutions and may be added to boiler water, especially low-pressure, low-temperature (<600°F) industrial units to keep iron oxide in solution and prevent deposit buildup. The most common chelants are EDTA (ethylenediaminetetraacetic acid) and NTA (nitrilotriacetic acid). NTA is used at 900 psig, and EDTA is used at boiler pressures up to 1200 psig.[35]

These chelants will cause water-side corrosion when the chelant concentrations are excessive which suggests corrosion will occur where departure from nucleate boiling occurs and the chelant can reach high local concentrations. Upsets in the pH will also promote chelant attack. The presence of dissolved oxygen will accelerate corrosion rates as will high turbulence and high fluid velocities. There have been cases of increased corrosion without the usual protective iron-oxide film.[36]

The general appearance of chelant corrosion is a smooth, glassy appearance with a very thin iron-oxide surface film. Figures 6.11.1 and 6.11.2 illustrate this feature. Depending on the fluid dynamics, turbulence and other local conditions, the form of corrosion can be irregular, that is, small regions that have virtually no wastage adjacent to areas of severe metal loss.[37]

The cross section through the wastage will show the effects of turbulence and high local velocities. Figure 6.11.3 is a longitudinal cross section through one of the pits shown in Fig. 6.11.2. The leading edge has a fairly sharp corner with a smooth trailing edge, almost half a teardrop in cross-sectional appearance. Other circumstances will lead to some undercutting or reentrant angles, again, influenced by the fluid turbulence. A transverse cross section, Fig. 6.11.4, shows the appearance to be a segment of a circle.

While high heat fluxes and high metal temperatures will promote more rapid corrosion, elevated temperatures are not a necessary condition. Figure 6.11.5 shows the microstructure to be normal ferrite and pearlite with no evidence of excessive operating temperatures. Energy-dispersive x-ray analysis (EDX) indicates that only iron is present along the surface of the shallow pits (see Fig. 6.11.6).

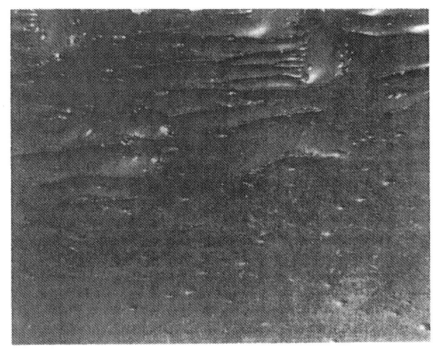

FIGURE 6.11.1. Water-side chelant corrosion (Magnification 6.4×).

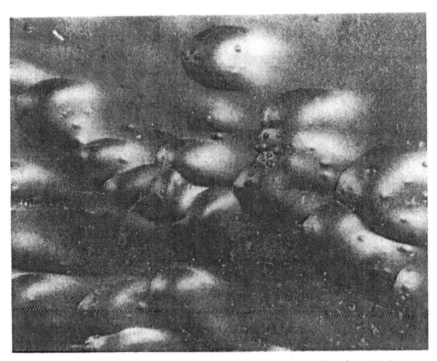

FIGURE 6.11.2. Water-side chelant corrosion (Magnification 6.4×).

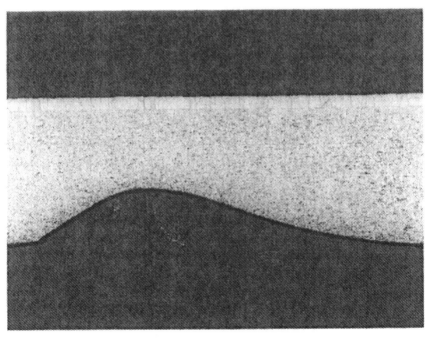

FIGURE 6.11.3. Longitudinal cross section through one shallow pit shown in Fig. 6.11.2. The shape is half a teardrop with smooth transition at the trailing edge (Magnification 25×, etched).

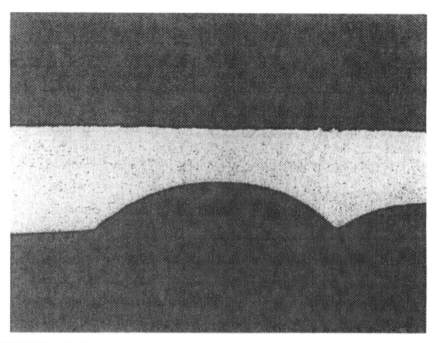

FIGURE 6.11.4. Transverse cross section shows an arc of a circle (Magnification 25×, etched).

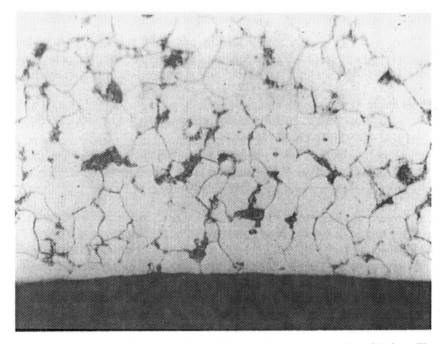

FIGURE 6.11.5. ID surface at the base of the shallow pit is virtually oxide-free. The structure is normal ferrite and pearlite with no overheating (Magnification 500×, etched).

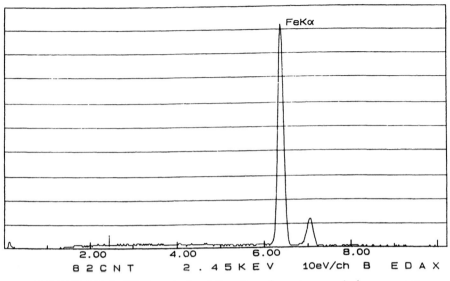

FIGURE 6.11.6. EDX analysis of the pit surface shows only iron present.

Case History 6.12

Waterwall Tubes—Hydrogen Damage and Creep Failures

Thick deposits on the water side of waterwall tubes in high-pressure utility boilers can lead to hydrogen damage and creep failures. These deposits have two adverse effects on the metallurgical soundness:

1. One of the by-products of underdeposit corrosion is elemental hydrogen, which diffuses into the steel and leads to hydrogen damage by the formation of methane.
2. Thick deposits also effectively insulate the tube from the cooling effects of steam formation. The net effect is to raise tube-metal temperature and lead to creep failures.

While the usual circumstance in a high-pressure boiler is the formation of hydrogen damage, occasionally these heavy deposits will lead to failures by a creep, or stress-rupture, mechanism. The microstructural features are somewhat similar in that hydrogen damage occurs by decarburization and intergranular-crack formation as a result of the reaction of hydrogen with iron carbide to form methane. The resultant microstructure will have intergranular cracks and be decarburized. The hydrogen damage always initiates along the interface between the deposit and the steel at the ID of the tube. Creep damage is by a grain-boundary sliding mechanism, and the resultant microstructure will contain intergranular cracks or voids as a result of the creep deformation. Since there is a temperature gradient across the tube wall, the OD or fire side of the tube is always hotter, so creep damage will always be found to initiate at the OD. Concurrent with creep damage will be spheroidization and/or graphitization of the carbide phase within the pearlite, so creep damage will be associated with spheroidization and graphitization. Hydrogen-damage failures will sometimes show creep damage along the OD as well as the hydrogen damage at the ID. This is particularly true if there is a decarburized layer. Carbide-free ferrite is considerably weaker, and intergranular cracking will develop along the ferrite grain boundaries.

The next three examples highlight the similarities and differences between hydrogen damage and creep failures in waterwall tubes. All are removed from high-pressure (>2000 psig) utility boilers.

Figure 6.12.1 presents a typical hydrogen-damage failure. There is intergranular cracking, and the cracking initiates at the ID. Figure 6.12.2 shows the intergranular cracking close to the failure edge. Along the OD surface, Fig. 6.12.3, are the early stages of intergranular creep damage that have formed within the decarburized layer. At higher magnifications, slightly removed from the surface, the structure is seen to be ferrite and spheroidized carbides; see Fig. 6.12.4. For 180° around the perimeter of the tube, the microstructure is normal ferrite and pearlite; refer to Fig. 6.12.5.

FIGURE 6.12.1. Cross section through a hydrogen-damage failure. Cracks initiate and grow from the ID surface (Magnification 25×, etched).

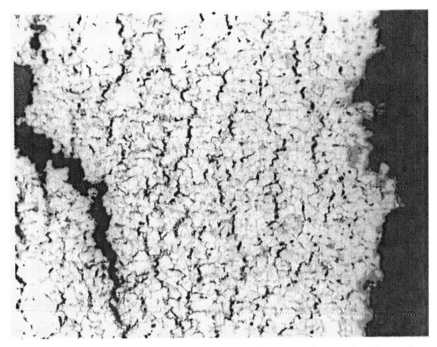

FIGURE 6.12.2. Hydrogen damage appears as intergranular cracks. Depending on the stress and temperature, the cracks tend to be aligned perpendicular to the hoop stress (Magnification 100×, etched).

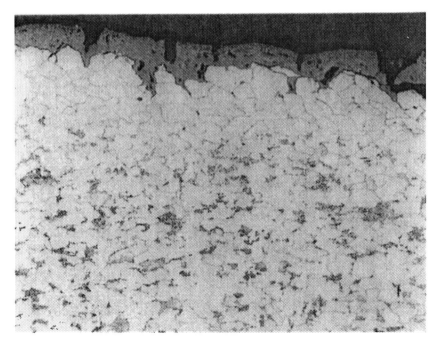

FIGURE 6.12.3. Along the OD surface, early stages of creep crack formation may be evident, especially if there is a decarburized zone (Magnification 100×, etched).

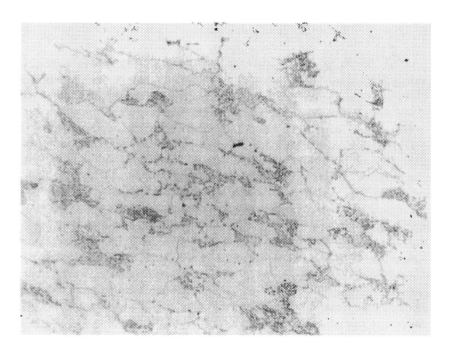

FIGURE 6.12.4. The microstructure toward the OD surface may also show the effects of elevated temperature. Here the pearlite has begun to spheroidize (Magnification 500×, etched).

FIGURE 6.12.5. For comparison, the microstructure 180° around the tube perimeter is normal ferrite and pearlite (Magnification 500×, etched).

While this failure is clearly hydrogen-damage-induced, there are creep features along the OD surface. The microstructure is spheroidized, indicative of operating temperatures higher than the design anticipated, and creep damage has appeared in the decarburized layer.

Figure 6.12.6 displays a roof tube that failed by a creep mechanism. The failures are on either side of a weld, and the excessive water-side deposits built up along the root pass upset. The fracture edge, Fig. 6.12.7, indicates the damage is from the OD, inward. The microstructure near the fracture edge, Fig. 6.12.8, is ferrite, spheroidized carbides, and creep damage. Along the OD surface near, but not at, the fracture edge, are longitudinal grooves; see Fig. 6.12.9. Figure 6.12.6 shows that the OD surface contains many longitudinal cracks in the oxide surface. As the tube expands by a creep-deformation mechanism, the brittle oxide scale cannot follow the tube swelling, and longitudinal cracks develop. These cracks short-circuit the diffusion path between oxygen in the flue gas and steel, and longitudinal cracks form within the steel; see Fig. 6.12.9. Creep voids may be seen near the crack tip. The ID surface, Fig. 6.12.10, shows spheroidized carbides, some creep damage, but no hydrogen damage. Thus, this waterwall-tube failure is by a creep, or stress-rupture, mechanism. The steam-side deposits impede the heat transfer. The net effect is to raise tube-metal tempratures into the creep range, and failure occurs.

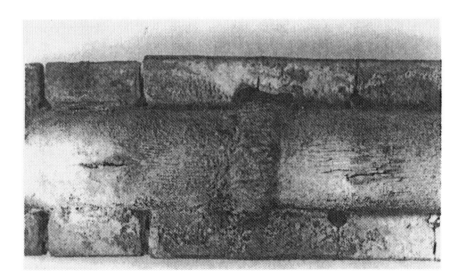

FIGURE 6.12.6. Creep failure in a roof tube (Magnification 0.7×).

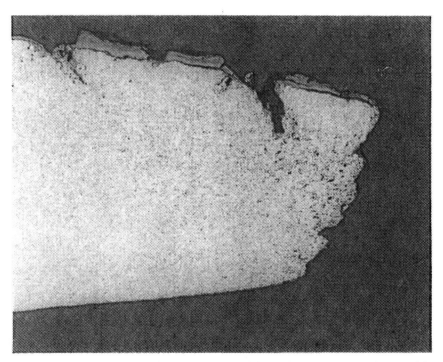

FIGURE 6.12.7. Cross section through the failure lip shows extensive cracking from the OD (Magnification 25×, etched).

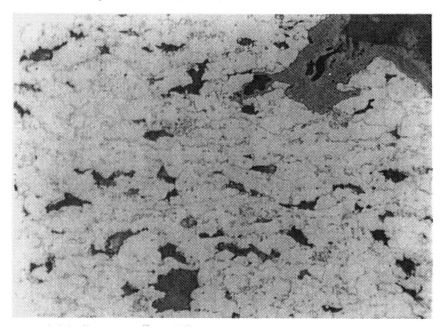

FIGURE 6.12.8. Microstructure near the fracture edge is ferrite, spheroidized carbides, and creep voids and cracks. Usually the creep cracks are not as long as the hydrogen-damage cracks. Compare with Fig. 6.12.2 (Magnification 500×, etched).

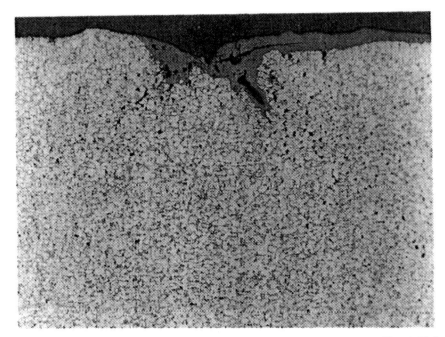

FIGURE 6.12.9. The longitudinal cracks in the OD oxide scale, see Fig. 6.12.6, extend into the steel and grow as creep cracks (Magnification 500×, etched).

FIGURE 6.12.10. Along the ID, the microstructure reflects the high temperature and creep, but there is no hydrogen damage. Note the longitudinal cracks in the scale and the small cracks in the steel (Magnification 500×, etched).

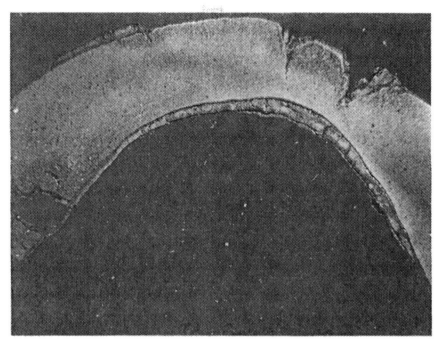

FIGURE 6.12.11. Cross section through a waterwall platen-tube blister. There is considerable creep damage on the OD of the blister (Magnification 6.4×, etched).

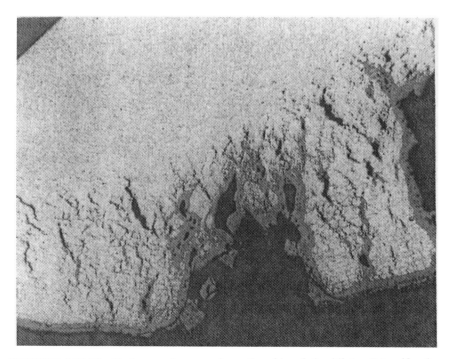

FIGURE 6.12.12. Hydrogen damage along the side of the blister (Magnification 25×, etched).

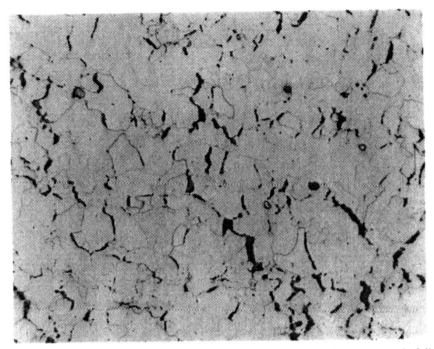

FIGURE 6.12.13. Same ID location as previous figure. The microstructure is fully decarburized with the attendant intergranular cracks formed by methane pressure (Magnification 500×, etched).

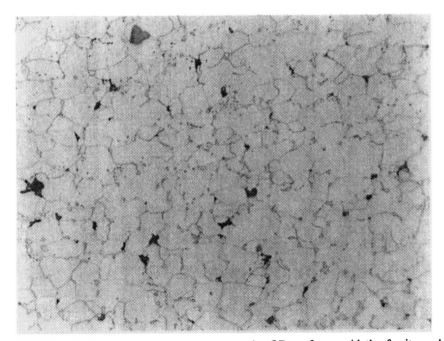

FIGURE 6.12.14. Creep cracks are seen at the OD surface amid the ferrite and spheroidized carbides (Magnification 500×, etched).

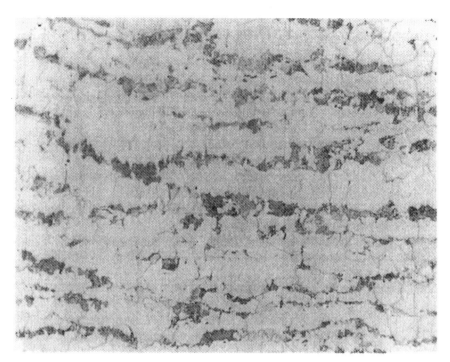

FIGURE 6.12.15. Microstructure on the cold side of the tube is normal ferrite and pearlite, albeit a banded structure (Magnification 500×, etched).

The final example shows a little of both failure types in the same waterwall rupture. Figure 6.12.11 is a photograph of a metallographic mount through a waterwall-tube blister. There is creep damage from the OD, and along the side of the blister is hydrogen damage. Figure 6.12.12 shows the hydrogen damage; at 500×, the microstructure is seen to be fully decarburized, as a result of the hydrogen damage, Fig. 6.12.13. Along the OD, in the region of the creep cracks, the structure is fully spheroidized carbides, ferrite, and creep damage, Fig. 6.12.14. For 180° around the tube perimeter, the structure is normal ferrite and pearlite, albeit a banded structure, Fig. 6.12.15.

The effects of thick, steam-side deposits are twofold: the underdeposit corrosion leads to hydrogen damage, and the insulating effects raise tube-metal temperatures until creep failure occurs.

REFERENCES

1. H. H. Uhlig, *Corrosion and Corrosion Control*, 2nd ed., Wiley, New York, 1971.
2. *NACE Basic Corrosion Course*, National Association of Corrosion Engineers, Houston, Texas, 1970.
3. M. G. Fontana and N. D. Greene, *Corrosion Engineering*, McGraw-Hill, New York, 1967.

4. G. A. Nelson and R. T. Effinger, "Blistering and Embrittlement of Pressure Vessel Steels by Hydrogen," *Welding Journal Research Supplement*, Jan. 1955, pp. 1–11.
5. E. P. Partridge, "Hydrogen Damage in Power Boilers," *Trans. ASME Journal of Engineering for Power*, July 1964, pp. 311–324.
6. A. G. Howell and P. E. Wigglesworth, "Chemistry and Operational Influence on Hydrogen Damage," Presented at International Conference on Cycle Chemistry in Fossil Plants, June 4–6, 1991.
7. W. T. Reid. *External Corrosion and Deposits: Boilers and Gas Turbines*, American Elsevier, New York, 1971.
8. A. J. B. Cutler, T. Flatley, and K. A. Hay, "Fire-Side Corrosion in Power Station Boilers," *CEGB Research*, Oct. 1978, pp. 13–26.
9. J. J. Demo, "Hot Ash Corrosion of High Temperature Equipment," *Materials Performance*, March 1980, pp. 9–15.
10. F. J. Oschell, R. L. Pall, and B. L. Libotti, "Modifying Deposition in Solid Fuel-Fired Boilers," Presented at Corrosion '80, NACE, Chicago, Illinois, March 3–7, 1980, Paper No. 61.
11. C. Cain, Jr. and W. Nelson, "Corrosion of Superheaters and Reheaters of Pulverized-Coal-Fired Boilers, II," *Trans. ASME*, Oct. 1961, pp. 468–474.
12. W. T. Reid, R. C. Corey, and B. J. Cross, "External Corrosion of Furnace-Wall Tubes—I, History and Occurrence," *Trans. ASME*, May 1945, pp. 279–288.
13. R. C. Corey, B. J. Cross, and W. T. Reid, "External Corrosion of Furnace-Wall Tubes—II, Significance of Sulfate Deposits and Sulfur Trioxide in Corrosion Mechanism," *Trans. ASME*, May 1945, pp. 289–302.
14. R. C. Corey, H. A. Grabowski, and B. J. Cross, "External Corrosion of Furnace Wall Tubes—III, Further Data on Sulfate Deposits and the Significance of Iron Sulfide Deposits," *Trans. ASME*, Nov. 1949, pp. 951–963.
15. D. N. French, "Furnace Wall Corrosion Problems in Coal Fired Boilers," Presented at Corrosion '82, NACE, Houston, Texas, March 22–26, 1982, Paper No. 289.
16. D. N. French, "Circumferential Cracking and Thermal Fatigue in Fossil Fired Boilers," Presented at NACE Corrosion '88, March 21–25, 1988, Paper No. 133.
17. R. H. Filby, K. R. Shah, and F. Yaghmaie, "The Nature of Metals in Petroleum Fuels and Coal-Derived Synfuels," in *Ash Deposits and Corrosion Due to Impurities in Combustion Gases*, R. W. Bryers, ed., McGraw-Hill, New York, 1978, pp. 51–64.
18. E. S. Domalski, A. E. Ledford, S. S. Bruce, and K. L. Churney, "The Chlorine Content of Municipal Solid Waste from Baltimore County Maryland and Brooklyn N.Y.," Presented at 1986 National Waste Processing Conference, Denver, Colorado, June 1986.
19. P. L. Daniel, J. L. Barna, and J. D. Blue, "Furnace Wall Corrosion in Refuse-Fired Boilers," Presented at 1986 National Waste Processing Conference, Denver, CO, June 1986.
20. I. G. Wright, "Hot Corrosion in Coal- and Oil-Fired Boilers," *Metals Handbook*, 9th ed. Vol. 13, ASM International, Metals Park, Ohio, 1987.
21. D. N. French, "Waterwall Corrosion in RDF Fired Boilers," in *Incinerating*

Municipal and Industrial Waste, R. W. Bryers, ed., Hemisphere, Washington, DC, 1989.

22. A. J. B. Cutler, W. D. Halstead, J. W. Laxton, and C. G. Stevens, "The Role of Chloride in the Corrosion Caused by Flue Gases and Their Deposits," *Trans. ASME*, July 1971, pp. 307–312.

23. D. N. French, "Ash Corrosion Under Reducing Conditions," Presented at Corrosion '80, NACE, Chicago, Illinois, March 3–7, 1980, Paper No. 59.

24. W. D. Niles and H. R. Sanders, "Reactions of Magnesium with Inorganic Constituents of Heavy Fuel Oil and Characteristics of Compounds Formed," *Trans. ASME*, April 1962, pp. 178–186.

25. S. H. Stoldt, R. P. Bennett, and D. C. Meier, "Chemical Control of Heavy Fuel Oil Ash in a Utility Boiler," Presented at Corrosion '79, NACE, Atlanta, Georgia, March 12–16, 1979, Paper No. 76.

26. B. Less, "Magnesium Based Additives Reduce High Temperature Problems in Large Oil-Fired Boilers in Europe," Presented at Winter Annual Meeting ASME, NY, Nov. 26–30, 1972, Paper No. 72-WA/CD-4.

27. D. N. French, "Corrosion of Superheaters and Reheaters in Fossil Fired Boilers," Proceedings of 2nd Annual Conference on Environmental Degradation of Materials, Virginia Polytechnic Institute, Blacksburg, Virginia, Sept., 1981.

28. K. L. Atwood and C. L. Hale, "A Method for Determining Need for Chemical Cleaning of High Pressure Boilers," Presented at American Power Conference, April 1971.

29. D3483-83 Standard Test Methods for Accumulated Deposition in a Steam Generator Tube, *1990 Annual Book of ASTM Standards, Section 11, Water and Environmental Technology*, ASTM, Philadelphia, Pennsylvania, 1990.

30. J. A. Alice, J. A. Janiszewski, and D. N. French, "Liquid Ash Corrosion, Remaining Life Estimation, and Superheater/Reheater Replacement Strategy in Coal Fired Boilers," Presented at The Joint Power Generation Conference, Milwaukee, Wisconsin, October 1985.

31. D. N. and S. M. French, "Improvement of Superheater Life Through Chemical Cleaning," Presented at NACE Corrosion '91, Cincinnati, Ohio, March 1991.

32. J. J. Dillon and R. D. Port, "Stress-Assisted Corrosion in Boilers," Presented at NACE Corrosion '90, Las Vegas, Nevada, April 1990.

33. *Principles of Industrial Water Treatment*, Drew Chemical Corp., Boonton, New Jersey, 1979.

34. F. N. Keemer, *Water: The Universal Solvent*, Nalco Chemical Co., Oak Brook, Illinois, 1979.

35. *Drew Principles of Industrial Water Treatment*, Drew Chemical Corp., Boonton, New Jersey.

36. T. C. Breske, J. C. Bovankovich, J. D. DeRuyter, and A. N. Jackson, "Boiler Tube Failure from Localized Chelant Attack," Presented at NACE Corrosion '88, St. Louis, Missouri, March 1988, Paper No. 332.

37. R. D. Port and H. M. Herro, *The NALCO Guide to Boiler Failure Analysis*, McGraw-Hill, New York, 1991.

CHAPTER SEVEN

WELD FAILURES

No discussion of metallurgical failures in a boiler would be complete without some mention, however brief, of weld failures. Utility-size units may contain more than 50,000 welds, about 60% done in the shop and 40% at the construction site. For the most part, welds are given a suitable nondestructive examination by ultrasonic or radiographic means before the unit operates. All welds are given a hydrostatic pressure test at 1½ times the design pressure as part of the ASME Code requirements. Even after all of this, a few leaks will be found during the first few months of unit operation.

Everyone has his or her favorite example of a poor weld that gives successful, trouble-free service for many years. Slag inclusion, porosity, lack of fusion, or incomplete penetration do cause leaks after start-up, even though radiographic examination has indicated these welds fit for service. Usually, the first few months of the shakedown of a boiler consist of a severe set of cycling runs until all systems of the complex structure that is the complete power plant are functioning smoothly. Small defects can, and do, propagate and cause leaks.

However, rather than offer examples of the usual and the unique weld failures, since these have been covered in detail elsewhere,[1] a couple of other topics should be covered.

HEAT-AFFECTED ZONE

The temperature profile from the centerline of the weld pool is presented, schematically, in Fig. 7.1. At the edge of the fusion zone in the base metal, the temperature is the melting point of the alloy and decreases at distances

HEAT-AFFECTED ZONE

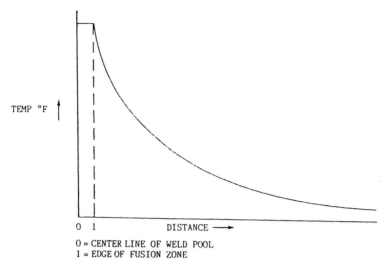

FIGURE 7.1. Schematic presentation of temperature profile during welding.

away from the weld. At some position remote from the weld, no temperature change can be measured. When welding is complete, heat flows out of the molten weld metal into the base metal and the weld solidifies. The cooling rate depends on the distance from the centerline of the weld. The cooling rate, also schematically, is presented in Fig. 7.2. The centerline of the weld pool has a slightly lower cooling rate than the edge of the fusion zone. Two factors govern the behavior of the metal adjacent to the weld pool: peak temperature and cooling rate.

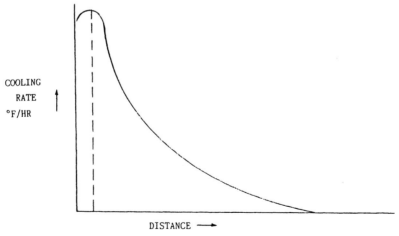

NOTE: CENTERLINE OF WELD POOL HAS A SLIGHTLY SLOWER RATE THAN THE EDGE OF THE FUSION ZONE.

FIGURE 7.2. Schematic presentation of the cooling rate following welding.

424　CHAPTER 7　WELD FAILURES

The heat-affected zone (HAZ) may be defined as that region near the edge of the fusion zone or weld metal that during welding has a peak temperature high enough to alter the microstructure. For austenitic alloys similar to 304 stainless steel, these changes would be a grain-size increase or a recrystallization of a cold-worked microstructure. For ferritic alloys similar to SA-210 or SA-213 T-22, these changes would include a transformation from normal low-temperature ferrite and pearlite to ferrite and austenite or all austenite. Schematically these are presented in Fig. 7.3. The region from 1 to 2 has peak temperatures above the upper critical transformation temperature and, for a moment, has been transformed completely to austenite. The region between 2 and 3 has peak temperatures above the lower critical transformation temperature and has only partially transformed to austenite. At distances greater than position 3, the base metal has been essentially unaffected by the heating and cooling during welding. No spheroidization of the carbide phase in pearlite is expected, because the time at temperature is too short. The heat-affected zone is defined as that distance from the edge of the fusion

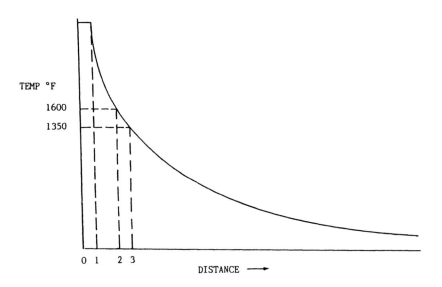

0 = CENTERLINE OF WELD
1 = EDGE OF FUSION ZONE
2 = 1600°F POSITION
    BETWEEN 1-2 MICROSTRUCTURE IS ALL AUSTENITE
    DURING WELDING
3 = 1350°F POSITION
    BETWEEN 2-3 MICROSTRUCTURE IS MIXTURE OF
    AUSTENITE AND FERRITE
POSITIONS 1-3 ARE HEAT AFFECTED ZONE

FIGURE 7.3. Definition of the heat-affected zone in ferritic steels.

zone to position 3. In plain low-carbon steels, there is a third region immediately adjacent to the fusion zone where delta ferrite forms. Reference to the iron–carbon diagram on page 86 shows a transformation of face-centered-cubic austenite to body-centered-cubic ferrite at temperatures above about 2540°F.

The microstructure of the heat-affected zone of ferritic steels depends on peak temperature, cooling rate, and alloy content. Rapid cooling or quenching of austenite can form martensite, a hard, brittle material. Slower cooling forms mixtures of ferrite and softer bainite or pearlite. Factors that determine the microstructure in the heat-affected zone may be understood from the changes that occur in a plain-carbon steel similar to SA-192.

Figure 7.4 shows a schematic of a time–temperature–transformation diagram for a low-carbon steel, similar to SA-192. Three cooling curves are superimposed to indicate the cooling rates needed to form three types of HAZ microstructures. Cooling rate a, the slowest, forms ferrite and pearlite. From a practical viewpoint these HAZ structures are seldom seen. Cooling rate b leaves a mixture of ferrite and a quenched structure from the austenite, often a Widmanstätten structure. This intermediate cooling rate is most often found. Cooling rate c forms a martensitic HAZ and is to be avoided, especially in the higher-chromium alloys.

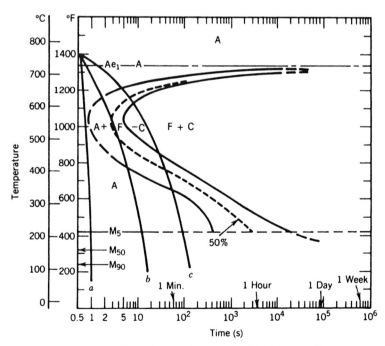

FIGURE 7.4. T-T-T curve for a low-carbon steel with three cooling rates superimposed.

The cooling rate from the all-austenite region determines the final microstructure. Under equilibrium conditions, cooling rate c and slower, ferrite begins to appear at the upper critical transformation temperature. The relative amounts of austenite and pearlite change, as does the composition of austenite. At the lower critical transformation temperature, the austenite transforms to pearlite. At a more rapid cooling rate, b, the transformation of austenite is suppressed and ferrite only begins to form at temperatures below the upper critical transformation temperature. The dashed line in Fig. 7.4 indicates the formation of ferrite, which forms preferentially along the austenite grain boundaries and along particular crystallographic planes in the austenite. The cooling rate is now fast enough to prevent the formation of pearlite, and the austenite transforms to bainite or, perhaps, martensite. Extremely rapid cooling rates, a, prevent formation of ferrite, pearlite, *and* bainite, and the final microstructure is martensite. In low- and medium-carbon steels (SA-192 and SA-210 A-1), cooling rates are seldom quick enough to form martensite, and the usual HAZ structure is ferrite and bainite.

Figure 7.5 shows the expansion and contraction characteristics when heated to and cooled from about 1800°F. The body-centered-cubic arrangement of atoms in ferrite is not as densely packed as the face-centered-cubic arrangement of atoms in austenite. Thus, at about 1350°F, the lower critical transformation temperature, as ferrite transforms to austenite, the denser atomic packing appears as a contraction of the length of a sample. The lower critical transformation temperature depends on the alloy content. Table XV in the appendix gives the approximate temperatures for the more common

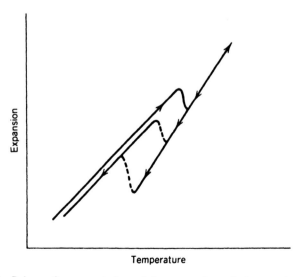

FIGURE 7.5. Schematic presentation of the expansion of a low-carbon steel during heating and cooling.

ferritic boiler steels. At about 1550°F, the upper critical transformation temperature, the ferrite-to-austenite transformation is complete. The slope of the heating curve below 1350°F is in the thermal-expansion coefficient of ferrite and pearlite; the slope of the heating curve above 1550°F is the thermal-expansion coefficient of austenite. During welding the base-metal heat-affected zone will follow a heating curve similar to that in Fig. 7.5. The slope for ferrite and pearlite is not as steep as the slope for austenite; that is, the coefficient of expansion for austenite is greater than that of ferrite and pearlite.

When welding is complete, the HAZ begins to cool. When the cooling rate is very slow, the transformation back to ferrite and iron carbide or pearlite follows the heating curve, and the final transformation product in the HAZ will be ferrite and pearlite. For more rapid cooling rates, the transformation to ferrite and carbide is suppressed, and martensite begins to form at a temperature of about 400°F; see Fig. 7.4. For carbon steels, the cooling rate to prevent the formation of ferrite and iron carbide is very rapid, and, for the most part, the heat-affected zones show structures that are free of martensite.

Another consideration in carbon steels is the cooling rate through the two-phase region of ferrite and austenite. Again, if the cooling rate is slow enough to allow ferrite to form, the initiation sites for ferrite formation are along the austenite grain boundaries and along particular crystallographic planes in the austenite. Under these cooling rates, the microstructure just prior to austenite transformation is a mixture of austenite and ferrite with each austenite grain nearly surrounded by ferrite. The final microstructure will be a Widmanstätten structure.

The second effect of the cooling of the heat-affected zone is the residual stress that remains following the austenite transformation. The change from BCC ferrite to FCC austenite involves a rearrangement of the iron atoms into a more densely packed structure. The discontinuous change in length at 1350°F in Fig. 7.5 reflects the closer spacing in the atomic alignment. The change in volume of the HAZ must be accommodated by the surrounding metal. Stresses, or more accurately strains, within this material leave both the HAZ and adjacent material with locked-in residual stresses. Such strains lead to distortion or, perhaps, cracking.

Figure 7.5 shows that when the transformation occurs at an elevated temperature, the volume change (shown in Fig. 7.5 as a change in length) is smaller than when the transformation occurs at a low temperature. Extrapolation of the cooling of the austenite line to 500°F, for example, leads to a bigger change in length because the coefficient of expansion of austenite is larger than that of ferrite. When the transformation of austenite occurs at a high temperature, the surrounding material is both more ductile and weaker than at low temperatures; see Appendix Table VIII for example. When the transformation of austenite occurs at a low temperature, the volume change is greater and the surrounding material is stronger and less ductile. The net result is a greater strain and residual stress within the heat-affected zone when the transformation occurs to martensite rather than to ferrite and iron carbide.

The effect of alloying elements on the T-T-T curve of Fig. 7.4 is to shift the nose of the pearlite transformation to the right. Slower cooling rates in SA-213 T-11 or SA-213 T-22 than in SA-192 will form martensite, for example. The transformation to ferrite and carbide is suppressed, and martensite will form at slower cooling rates. Cooling rates for the formation of satisfactory heat-affected zones in low-carbon steel will form unsatisfactory, martensitic, heat-affected zones in alloy steels. Thus, for low-alloy, ferritic steels, cooling rates need to be reduced to prevent excessive residual stresses. The easiest way to accomplish this is to preheat the area surrounding the weld. Depending on the alloy content, higher preheat temperatures will be needed to retard HAZ cooling rates enough to prevent damage.

How much preheat depends on the composition. One such concept is the carbon equivalent. The carbon equivalent essentially determines the effect of alloying elements on the cooling rate needed to avoid martensitic transformations. One carbon equivalent, (CE), equation is[6]

$$CE = \%C + \frac{\%Mn}{6} + \frac{\%Ni}{15} + \frac{\%Cr}{5} + \frac{\%Cu}{13} + \frac{\%Mo}{4}$$

and it applies for compositions of ferrite steels. For

%C	<0.5	%Cr	<1.0
%Mn	<1.5	%Cu	<1.0
%Ni	<3.5	%Mo	<0.5

This equation was developed for low-alloy steels and does not actually cover the chromium–molybdenum range for SA-213 T-11 and SA-213 T-22. The amount of preheat required depends on the carbon equivalent as the following table suggests:

CE, %	Preheat Temperature, °F
<0.45	Optional
0.45–0.60	200–400°
>0.060	400–700°

The *ASME Boiler and Pressure Vessel Code* presents the preheat requirements differently. Alloys are grouped by P number—in effect, a recognition that higher alloys require different welding parameters.

In Appendix A-100 of Section I of the *Boiler and Pressure Vessel Code* are given nonmandatory preheat recommendations.

Material	Preheat, °F
Carbon steel (<1 in and <0.30% C)	50
Carbon steel (>1 in and >0.30% C)	175
Carbon + ½Mo (>⅝ in)	175
Carbon + ½Mo (<⅝ in)	50
1¼Cr–½Mo (>½ in)	250
1¼Cr–½Mo (<½ in)	50
2¼Cr–1Mo	300

The heat-affected zones of austenitic stainless steels are different in the sense that there is no transformation between ferrite and austenite on heating and cooling during welding. The heat-affected zones of the austenitic stainless steels reflect only the peak temperature during welding. As temperature is increased, residual-stress relief, recrystallization of cold-worked structures, and grain growth occur. In those regions immediately adjacent to the centerline of the weld, there will be fairly large austenite grains, and the grain size will decrease until the cold-worked structures are not altered by the heating and cooling cycle of welding.

HEAT-AFFECTED ZONE MICROSTRUCTURES

For plain-carbon steels, the only difference expected in HAZ microstructures is caused by differences in carbon content. Figures 7.6–7.8 present the base metal and two HAZ microstructures. In Fig. 7.7 the peak temperature was between the lower and upper critical transformation temperatures, 1340–1550°F. Some austenite formed and on cooling was transformed to very fine pearlite or bainite. In Fig. 7.8, the peak temperature was well above 2000°F, note the size of the bainite islands. The cooling rate is as b in Fig. 7.4. These three structures are from an SA-106B header.

Similar structures are found in SA-178A or SA-192 materials. With a lower carbon content, the relative amounts of ferrite and pearlite differ. Figures 7.9–7.11 present microstructures from SA-178A waterwall tubes. The HAZ structures are taken from a corrosion-resistant weld overlay.

Figures 7.12–7.14 present microstructures from SA-213 T-22. The microstructure at the edge of the fusion zone, Fig. 7.14, is a mixture of bainite and ferrite with a hardness of about Rockwell B 100. In the base metal, Fig. 7.12, the hardness is R_B 81.

Heat-affected zone microstructures will transform during normal boiler operations at temperatures high enough to allow such changes. The final microstructures will be similar to those seen in wrought products. However,

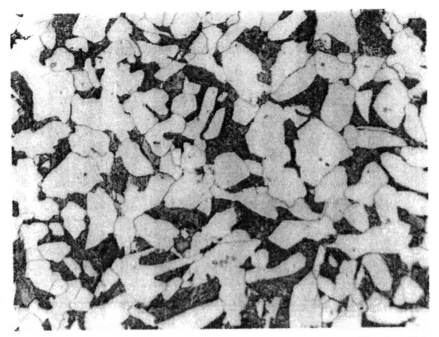

FIGURE 7.6. Base metal, ferrite and pearlite, SA-106 B (Magnification 500×, etched).

FIGURE 7.7. HAZ with peak metal temperature between lower and upper critical transformation temperature (Magnification 500×, etched).

FIGURE 7.8. HAZ with peak metal temperature above about 2000°F (Magnification 500×, etched).

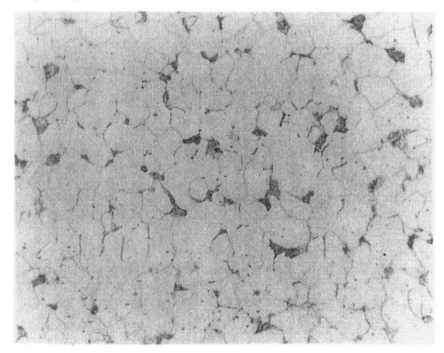

FIGURE 7.9. Base metal, ferrite and pearlite, SA-178A (Magnification 500×, etched).

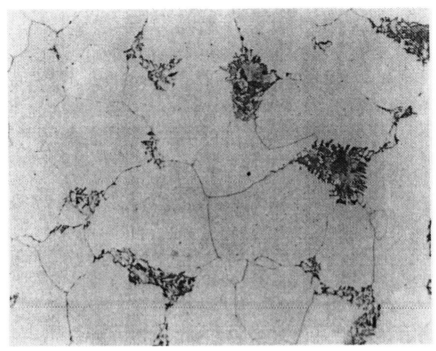

FIGURE 7.10. HAZ with peak metal temperature between lower and upper critical transformation temperature (Magnification 500×, etched).

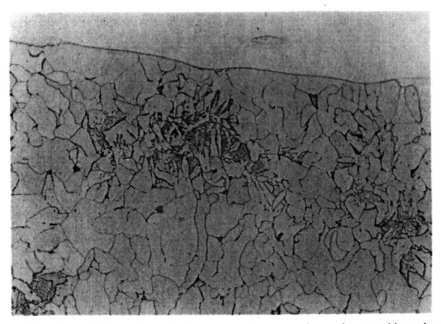

FIGURE 7.11. HAZ at edge of fusion zone under a corrosion-resistant weld overlay (Magnification 500×, etched).

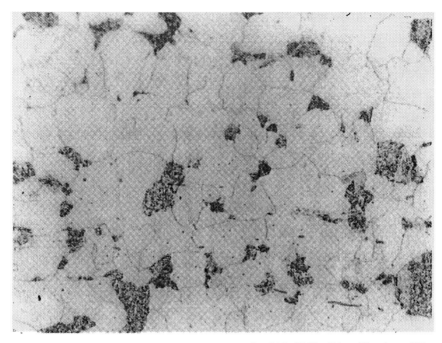

FIGURE 7.12. Base-metal microstructure, SA-213 T-22 (Magnification 500×, etched).

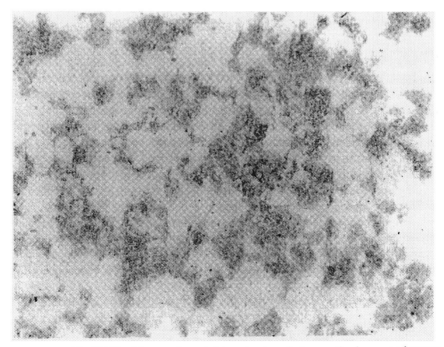

FIGURE 7.13. HAZ with a peak metal temperature between the lower and upper critical transformation temperatures (Magnification 500×, etched).

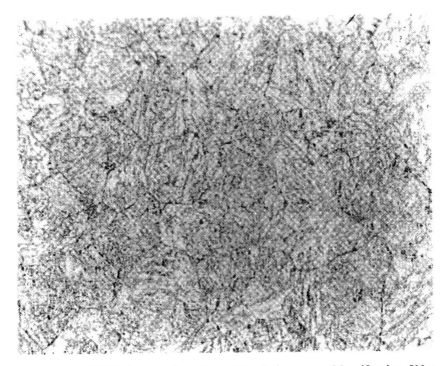

FIGURE 7.14. HAZ close to the edge of the fusion zone (Magnification 500×, etched).

the quenched structures will transform sooner than ferrite and pearlite, so a HAZ will show the effects of elevated temperatures first. For attachment welds, higher localized stresses will also promote more rapid transformations. Figure 7.15 presents a graphitized HAZ from an SA-209 T-1 primary-superheater tube. No graphitization was seen in the base metal; see Fig. 7.16. Since alloys with greater than 0.5% chromium will not graphitize, T-11 and T-22 HAZ structures simply spheroidize as do the tube structures; see Fig. 7.17.

In stainless steel, the HAZ structures will reflect the extent of cold work in the base metal. For annealed tubes, the HAZ will display only grain growth in the material adjacent to the weld metal; see Fig. 7.18. For heavily-cold-worked material, the HAZ will recrystallize the cold-worked grains. The grain size of the recrystallized austenite will increase as the peak temperature during welding increases. Figure 7.19 displays a HAZ in a cold-worked 304 stainless steel.

Figures 7.20–7.22 present the 304 HAZ from the edge where the cold-worked structure is still evident, to the edge of the fusion zone where grain growth has occurred. Compare the grain size in Figs. 7.21 and 7.22 to see this effect. Figure 7.20 shows the very early stages of recrystallization, since the austenite grain size within the cold-worked grains is quite small.

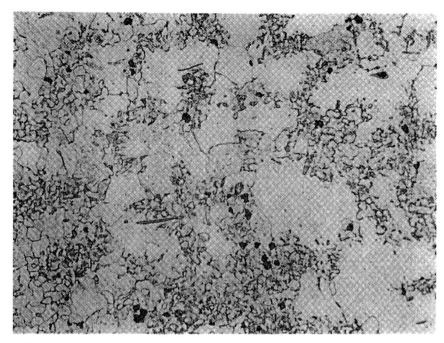

FIGURE 7.15. HAZ from an SA-209 T-1 tube, after more than 20 years service, has graphitized (Magnification 500×, etched).

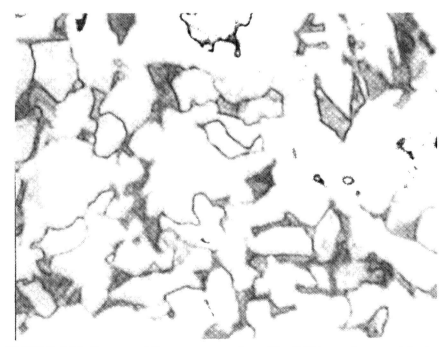

FIGURE 7.16. Base metal from same sample as Fig. 7.15 is nearly normal ferrite and pearlite (Magnification 500×, etched).

FIGURE 7.17. HAZ from a T-22 weld after several years service is fully spheroidized (Magnification 500×, etched).

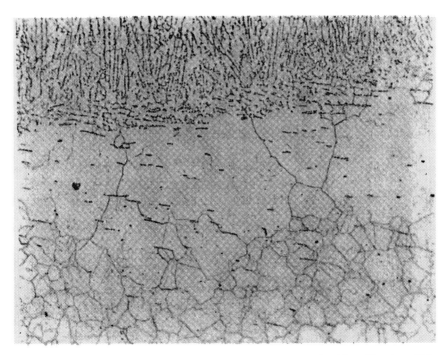

FIGURE 7.18. Annealed 304 HAZ shows grain growth (Magnification 200×, etched).

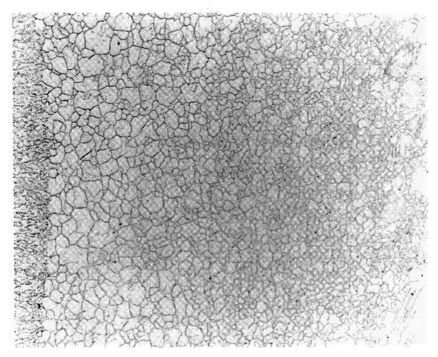

FIGURE 7.19. Cold-worked 304 HAZ (Magnification 100×, etched).

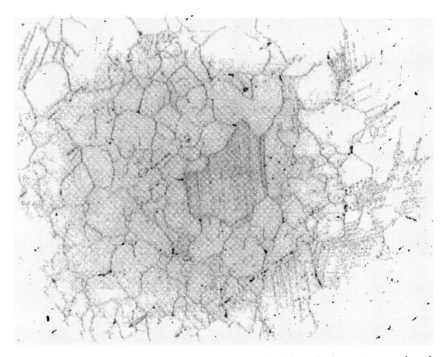

FIGURE 7.20. Edge of HAZ shows remnants of cold-worked structure and early stages of recrystallization (Magnification 500×, etched).

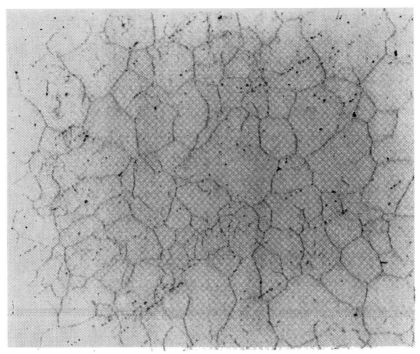

FIGURE 7.21. Fully recrystallized austenite (Magnification 500×, etched).

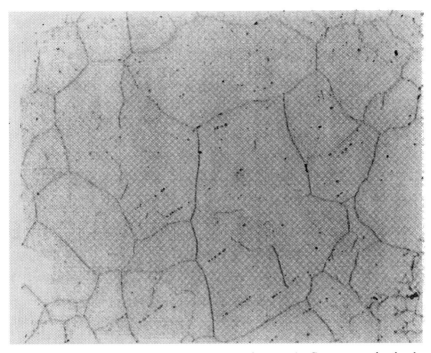

FIGURE 7.22. Edge of fusion zone shows grain growth. Compare grain size here with Fig. 7.21 (Magnification 500×, etched).

INTERNAL CONFIGURATION

Backing rings are no longer used in water-wetted circuits, because the built-in crack between the ring and the ring seat is a convenient place for chemical hide-out; and smoother ID surfaces can be manufactured with automatic or semiautomatic welding processes. A protrusion on the ID may, under certain fluid-flow conditions, cause a disturbance downstream of the weld, similar to an orifice. The flow speeds up through the constriction, and leaves a lower-pressure zone downstream.

Figure 7.23 schematically presents the problem. Downstream, of the weld, eddies may form that give rise to a lower-pressure zone. Feedwater chemicals may drop out of solution and cause corrosive attack locally that may lead to hydrogen damage and/or failure. Figure 7.24 shows such a region of localized corrosion at the upset from a flash weld. Within a superheater or reheater, similar low-pressure zones form; but, in this case, the eddies create a region of inadequate cooling. Hot spots then form, and premature failure occurs. Figure 7.25 shows a tube failure at the weld, steam flow is from left to right in the figure. Microstructural analysis shows, as can be seen in Figs. 7.26 and 7.27, a spheroidized structure at the failure, and normal ferrite and pearlite about 12 in upstream of the failure. Both tubes were installed at the same time and were of SA-213 T-22 material.

Figure 7.28 illustrates another problem with backing-ring joints; this was from an economizer tube that had been chemically cleaned several times. Acid accumulated in the crevice between the ring and the tube, and was not adequately neutralized for some time after cleaning. The acid corroded the tube until a leak occurred, and the sample was submitted for analysis.

These examples illustrate the need for caution in designing joint geometries. Aside from the usual and needed requirements for soundness and cleanliness, the welding process used should leave the ID smooth in order to ensure undisturbed fluid flow under all conditions.

DISSIMILAR WELDS

Two types of welds between ferritic steel, usually SA-213 T-22, and austenitic stainless steel need to be addressed: tube butt welds and tube support welds. Tube-to-tube butt welds, welds between pressure parts, have re-

FIGURE 7.23. Schematic presentation of the fluid-flow disturbance caused by an ID protuberance.

440 CHAPTER 7 WELD FAILURES

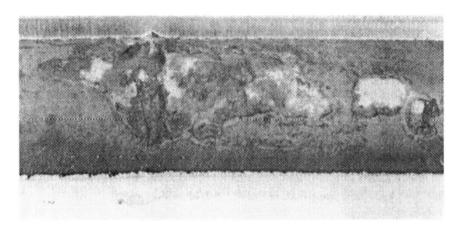

FIGURE 7.24. Corrosion at the upset caused by flash weld.

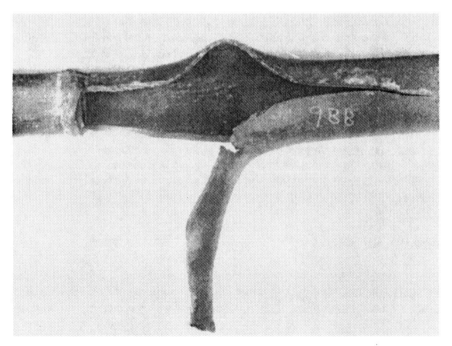

FIGURE 7.25. Superheater-tube failure downstream from a weld at the hot spot caused by the weld.

DISSIMILAR WELDS 441

FIGURE 7.26. Microstructure through the failure; note spheroidized carbides indicative of a long-time condition of overheating (Magnification 500×, nital etch).

FIGURE 7.27. Microstructure 12 in upstream from the weld shows normal ferrite and pearlite (Magnification 500×, nital etch).

FIGURE 7.28. Acid corrosion at the backing ring from an economizer tube.

ceived considerable attention; a bibliography on the subject runs to 160 entries.[2] Of equal importance, but less discussed, are the tube-to-tube tie or support-clip welds. Since the support gets less cooling from the steam-cooled tube, higher alloys are required, and high alloy means the use of austenitic materials.

Austenitic–ferritic alloy welds were originally made with a stainless-steel alloy weld. Most failures, industry wide, have been in these welds. Briefly, the problems arise from:[3]

1. Differences of coefficients of expansion between austenite and ferrite put large temperature-induced stresses at the weld interface. Table 7.1 gives these expansion coefficients for the common boiler steels.

TABLE 7.1. Linear Thermal Expansion from 70°F (21°C) to Indicated Temperature (in/ft)

	400°F (204°C)	600°F (316°C)	800°F (427°C)	1000°F (538°C)	1200°F (649°C)
Carbon steel, SA-210	0.027	0.046	0.067	0.089	0.111
Carbon + ½ Mo SA-209	0.027	0.046	0.067	0.089	0.111
1¼ Cr–½ Mo SA-213 T-11	0.027	0.046	0.066	0.089	0.111
2¼ Cr–1 Mo SA-213 T-22	0.027	0.046	0.066	0.089	0.111
18Cr–8Ni	0.038	0.062	0.088	0.115	0.142
18Cr–8Ni + Ti	0.038	0.062	0.088	0.115	0.142
Inconel 601®	0.032	0.052	0.073	0.095	0.125
Inconel 625®	0.029	0.047	0.067	0.087	0.111

2. Carbon diffusion from the ferrite to austenite at the weld interface reduces the creep strength of the ferrite so that creep voids develop, and failure occurs by creep along the weld interface in the weakened ferritic alloy.
3. Differences in oxidation resistance between the two give rise to an oxide wedge that forms in the ferritic alloy adjacent to the weld deposit.

At one time or another, welds have been made with stainless-steel Types 309 or 310 rods, with ferritic-alloy rods similar in composition to T-22, and high-nickel-alloy rods; also, flash or forge welds have been made with no filler metal added. None of the above choices is perfect, but the best appears to be the high-nickel-alloy welding materials. Austenitic alloys with substantial amounts of nickel have coefficients of expansion closer to ferrite than do stainless steels; see Table 7.1. Thus, the temperature-induced stresses are less, as noted in item 1. Nickel does not form a carbide, while iron and chromium do; as a consequence, carbon diffusion from the ferritic alloy, T-22, is less of a problem. Less carbon loss at the weld interface means that grade T-22 steel will maintain its creep strength better; hence, problem 2 becomes less likely a failure mechanism. Item 3 is still a concern, because all of the nickel alloys are more oxidation-resistant than low-alloy ferritic steels.

Figures 7.29 and 7.30 show the weld interfaces for a grade T-22 Incoweld A®–321H stainless-steel weldment. Note the formation of an oxide wedge on the T-22 side. The example is from a superheater that had been in service nearly 15 years without a single weld failure in the T-22–321H joints.[4,5]

Figure 7.31 shows a failure at a tube-to-tube tie weld. The tube is grade SA-213 T-22 steel, the lug is Type 347 stainless steel, and the weld is Type 310 stainless steel. Note that the failure is in the pressure part and forced the unit off the line. Figure 7.32 shows the oxide wedge at the grade T-22–310 stainless-steel interface; Fig. 7.33 displays the weld–grade T-22 steel interfaces away from the surface. Note the carbon migration into the stainless that forms a dark band in the stainless steel. Figure 7.34 shows creep voids that form in the T-22 adjacent to the weld interface. Loss of carbon from the T-22 reduces the creep strength, and ultimately fracture occurs when these voids link up. Figure 7.34 is unetched, and these creep voids are clearly visible.

In the case of tube-to-tube tie or support-clip welds, the same three problems exist; however, the location of the failure can be transferred from the tube to the support. While this allows the superheater or reheater to fall out of proper alignment, it does not force the unit off the line with a tube leak. By using a welding alloy that closely matches the ferritic-alloy composition, the weld problems are thus transferred to the alloy support material; and failure, when and if it occurs, will be away from the pressure part.

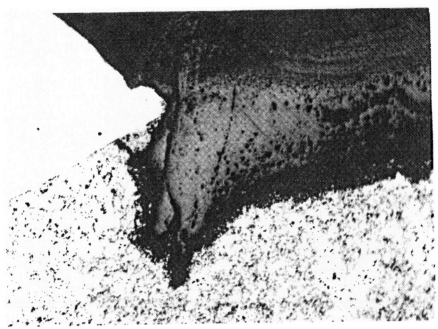

FIGURE 7.29. Oxide wedge on the OD surface at the interface between SA-213 T-22 steel and Incoweld A® alloy weld deposit. The weld is to the upper left and is unaffected by the etch (Magnification 100×, nital etch).

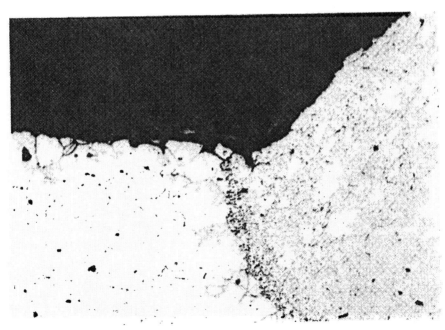

FIGURE 7.30. Weld interface on the OD surface between Incoweld A® alloy weld deposit and 321H stainless steel. The weld is to the right and is etched by the electrolytic etch (Magnification 100×, electrolytic etch).

FIGURE 7.31. Tube-tie failure at the tube-to-weld interface.

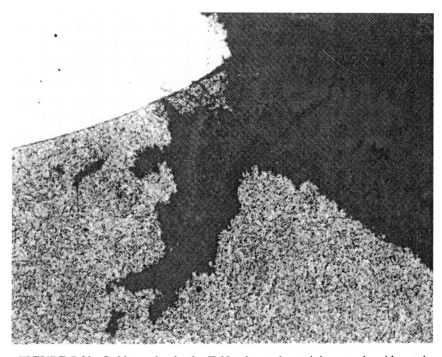

FIGURE 7.32. Oxide wedge in the T-22 tube at the stainless-steel weld metal.

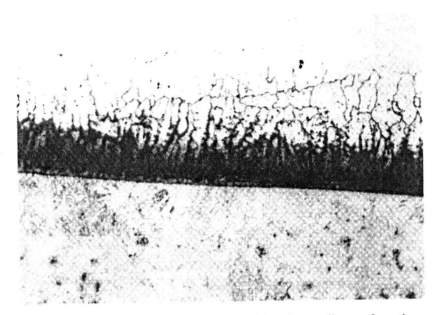

FIGURE 7.33. Grade T-22 to 310 stainless-steel interface well away from the surface; note dark band in the stainless indicative of carbon diffusion from the T-22 steel.

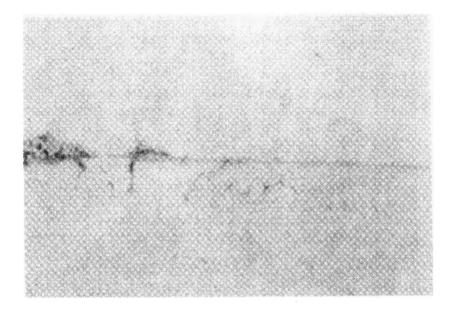

FIGURE 7.34. T-22 to 310 stainless-steel interface close to the surface; note the creep voids in the T-22 adjacent to the weld. Stainless steel is above, T-22 is below (Magnification 500×, unetched).

FIGURE 7.35. Circumferential crack in a superheater tube at the toe of a fillet weld caused by improper blending of the weld to the tube.

Tube-tie or support-clip welds should be made with a smooth blending of the fillet to prevent the formation of a stress-concentration notch that may cause failure. Figure 7.35 shows a circumferential crack in a tube at the toe of the tube-tie weld caused by improper fillet radius.

While this is anything but an exhaustive failure list, these topics are the ones most frequently seen over the years. The ID configuration, dissimilar-metal welds between tubes and support members, and sharp discontinuities at fillet welds are the three types of weld-related failures submitted for study. Almost without exception, weld leaks during the first months of operation are caused by poor welding techniques and not by design-related defects.

DISSIMILAR-METAL WELD FAILURE PREVENTION

Extensive studies of dissimilar-metal weld (DMW) failures have been sponsored by EPRI.[7] All of these studies indicate that substantial improvement in the expected life can be achieved by following four guidelines:

1. Use nickel-based welding alloys for all DMWs, both tube-to-tube butt welds and the attachment of stainless-steel support or alignment attachments.
2. Locate the DMW at as low a temperature as possible. This means the weld at the outlet end of a superheater or reheater should be in the penthouse or header enclosure. The metal temperature here will reflect the steam temperature only. No heat transfer between flue gas and

steam occurs at this location, so the metal temperature cannot be greater than the steam. The low-temperature weld toward the inlet end should be made at a metal temperature below 1000°F if possible, or at as low a metal temperature on the inlet as can be arranged.

3. Make the welds in vertical rather than horizontal runs of tubing. The vertical leg will have only the weight below and will minimize any bending moments that may contribute to the overall stress.

4. Make the DMWs in a shop and use semiautomatic or automatic-welding processes to keep the geometric discontinuities to a minimum. The weld should be smoothly blended to the base metal with no undercut or other abrupt section change that would act as a stress raiser.

On final comment, design DMWs so that postweld heat treatment (PWHT) is avoided. Austenitic-alloy tubes should be welded to T-22 stub tubes, not directly to P-22 headers. At temperatures and times used for PWHT of chromium–molybdenum alloy headers, carbon will diffuse out of the ferrite, form a plane of carbides at the weld fusion line, and leave a low-carbon HAZ. Without any carbides, the creep strength of this region will be reduced and premature failures likely, even when nickel-based welding materials are used for the DMW.

By following these recommendations, the life expectancy of a DMW can be improved by a factor of nearly $2\frac{1}{2}$ over those made with stainless-steel filler metal.

Figures 7.36–7.45 present a catalog of failures in DMWs made with stainless-steel electrodes and pressure welds made without filler metal. In all cases, the failure is in the HAZ of the ferritic grade. In Fig. 7.45, failure has occurred, and a remnant of the T-22 tube is still attached to the stainless steel. Note also that the sensitization in the stainless steel is greatest at the interface, which supports the concept of carbon loss from the T-22 HAZ. However, since the failure is 0.008 in from the fusion line, carbon loss is probably not a significant factor in premature failures.

Figure 7.46 is taken from a new, unused DMW made with a nickel-based alloy. There is a greater amount of ferrite visible just at the edge of the fusion zone. The diamond pyramid hardness (DPH) at the fusion line in the T-22 is 195. In the HAZ away from the edge, the DPH rises to 260.

STRESS-ASSISTED CORROSION

DMWs made with stainless-steel electrodes may fail in the T-22 HAZ even at temperatures below the creep range. Differences in coefficients of thermal expansion (see Table 7.1) lead to the formation of large temperature stresses. Under severe ash-corrosion conditions, corrosion rates in the HAZ can be rapid. Figure 7.47 presents such a stress-assisted corrosion attack. Estimated metal temperature from steam-side scale-thickness measurements

FIGURE 7.36. Representative failure of a DMW made with stainless-steel filler metal. Fracture is on the T-22 side (Magnification 1.8×).

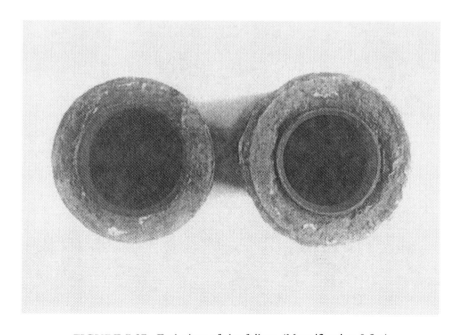

FIGURE 7.37. End view of the failure (Magnification 0.8×).

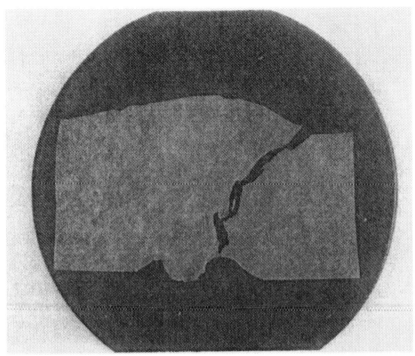

FIGURE 7.38. Cross section through failure (Magnification 3×).

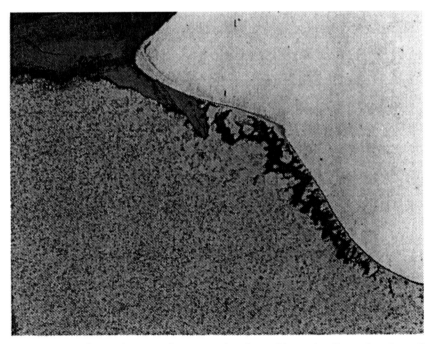

FIGURE 7.39. Creep damage often precedes the oxide-wedge formation from the OD (Magnification 25×, etched).

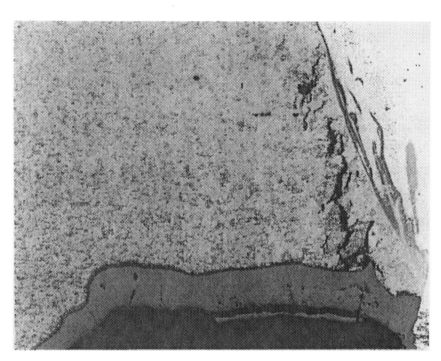

FIGURE 7.40. Creep damage can also initiate at the ID surface (Magnification 25×, etched).

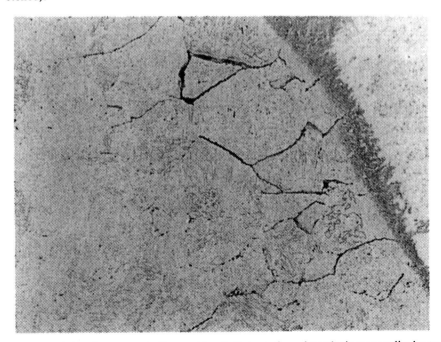

FIGURE 7.41. Creep cracks form midwall. Orientation of cracks is perpendicular to the edge of the fusion zone and also perpendicular to the temperature stress that parallels the weld.

FIGURE 7.42. Pressure weld with no filler metal added. Note severe wastage and oxidation to the T-22 (Magnification 18¾×, etched).

FIGURE 7.43. ID oxide-wedge growth in a pressure weld. Note the crack is not exactly along the weld line (Magnification 50×, etched).

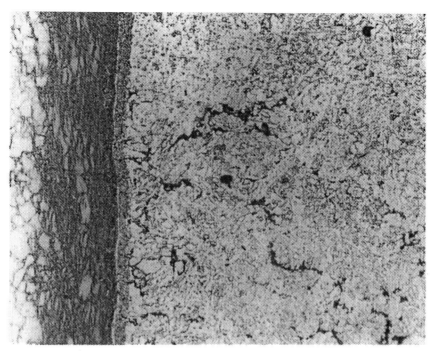

FIGURE 7.44. Creep damage midwall in the T-22 (Magnification 150×, etched).

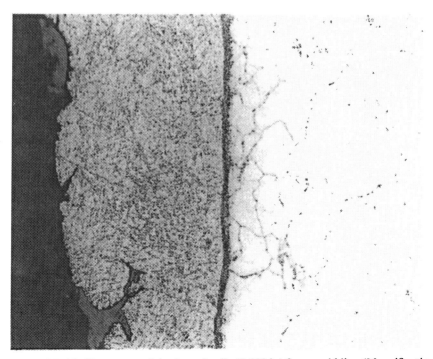

FIGURE 7.45. Fracture path is about 8 mils (0.008 in) from weld line (Magnification 200×, etched).

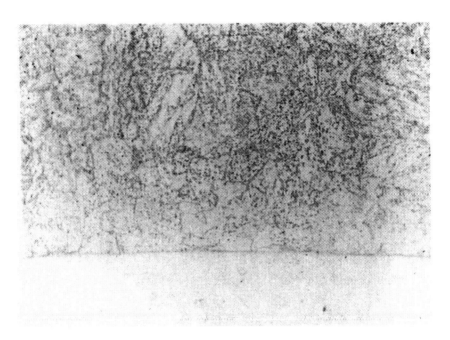

FIGURE 7.46. New DMW made with nickel-based welding alloy does show some loss of carbon in the T-22. DPH hardness is 195 (R_B 90) at the edge of the weld in the T-22 and 260 (R_B 100) a short distance away (Magnification 500×, etched).

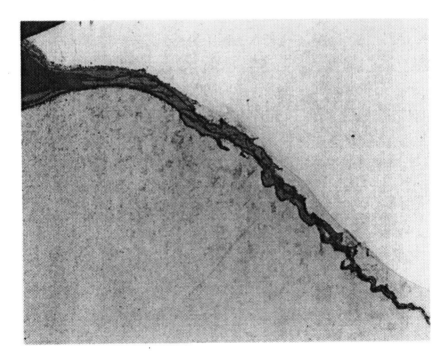

FIGURE 7.47. Stress-assisted corrosion along a DMW HAZ in a coal-ash corrosion environment (Magnification 25×, etched).

was less than 975°F. No creep damage precedes the crack. Since temperature stresses are lower in nickel-based weldments, oxide-wedge formation will also be less, as shown in Fig. 7.40.

MICROSTRUCTURAL EFFECTS ON CORROSION

Heat-affected zones of welds have microstructures very different from the base metal. Under some corrosion conditions, these rapidly cooled structures will be attacked more severely than the normalized structures in the base metal.[8] Figures 7.48 and 7.49 present examples of preferential HAZ corrosion in carbon steel.

REHEAT CRACKING

Low-alloy steels containing chromium and molybdenum, similar to T-11 and T-22, are susceptible to reheat cracking.[9] Reheat cracking occurs in the heat-affected zone during postweld heat treatment. In boiler operation, similar cracking can occur during normal operation at superheater and reheater temperatures. It is more likely to occur in thick sections where the residual

FIGURE 7.48. HAZ corrosion in carbon steel (Magnification 1.8×).

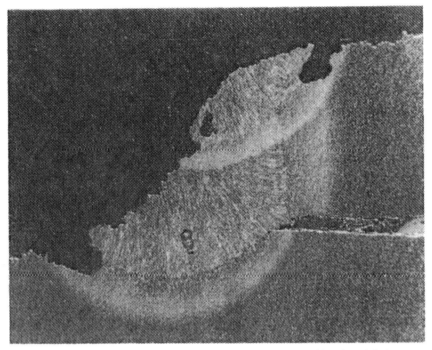

FIGURE 7.49. HAZ corrosion in carbon steel (Magnification 6.6×).

stresses may be larger, but it can develop in thin sections if the restraint is sufficient. The driving force is a tensile stress which can be a residual stress from welding, an applied stress from the system, or a temperature stress from differences in expansion.

The explanation for this phenomenon is that chromium and molybdenum carbides form during the postweld heating and form mainly within the grains. The carbide precipitation makes the grain interior relatively stronger than the grain boundary. The relaxation of the residual stresses in the thin HAZ is achieved by the formation of cracks along the grain boundaries.

Figure 7.50 shows a reheat crack in the heat-affected zone of a lug-attachment weld in a T-11 superheater tube. The weld was made without any preheat, and the restraint on the weld by the lug attachment prevented the stress relief without crack formation. The reheat crack formed during the normal operation of the boiler as the superheater was put back in service. As the superheater temperature increased during start-up, cracking occurred. In effect, the normal superheater operation was a postweld heat treatment, and the system stresses were large enough to develop a reheat crack. The crack was not discovered until growth through the tube wall by a thermal-fatigue mechanism led to a steam leak.

Figure 7.51 shows the similar heat-affected zone cracking in a P-22 stub-tube-to-header weld. In this case, proper welding preheat was used, but to

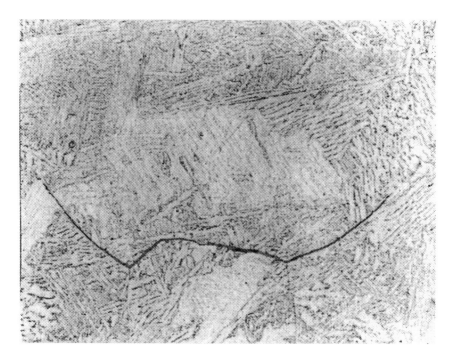

FIGURE 7.50. Reheat crack in a T-11 lug weld made without any preheat. Crack follows prior austenite grain boundaries (Magnification 500×, etched).

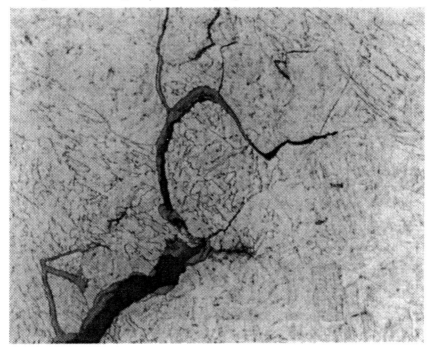

FIGURE 7.51. Reheat crack in a P-22 stub-tube socket weld. Crack was caused by rigid alignment during postweld heat treatment (Magnification 500×, etched).

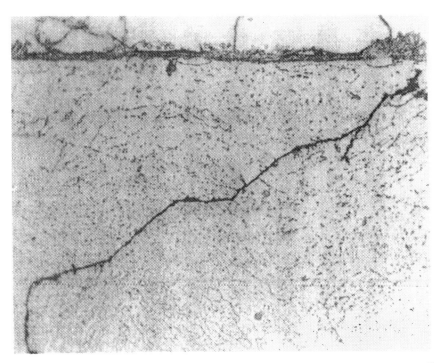

FIGURE 7.52. Reheat crack in T-22 HAZ of a DMW made without preheat (Magnification 500×, etched).

prevent distortion during the stress-relief anneal, the header and tube connectors were held rigidly in position.

On rare occasions, reheat cracks can occur in dissimilar-metal welds if the preheat temperature during welding is inadequate. Figure 7.52 shows such a reheat crack within the T-22 heat-affected zone adjacent to a dissimilar-metal weld.

Reheat-cracking tendencies can be prevented by reducing the constraint during welding and postweld heat treatment, and the use of adequate preheat.

COPPER PENETRATION

Waterwall tubes are often pad-welded to restore correct wall thickness in wasted tubes, in the vicinity of sootblowers for example. Pad welds have been known to be used to repair waterwall leaks caused by hydrogen damage, creep, or oxygen pitting. There is a potential danger in this type of repair when copper deposits exist on the water side of the tube. During welding, the ID surface temperature can rise above 2080°F, the melting point of copper. When this occurs, the copper melts and penetrates the ferrite grain boundaries and leads to further steam leaks.[10]

Figure 7.53 shows a pad weld along the OD surface of a waterwall tube that had failed by hydrogen damage. Figure 7.54 shows the copper penetration close to the ID surface. Even at some distance away from the ID surface, in the immediate vicinity of the edge of the fusion zone, molten copper had penetrated the grain boundaries, Fig. 7.55.

Similar problems can occur during a proper repair when a dutchman is installed. To prevent copper penetration, the ID surface should be carefully machined or ground to a distance of $\frac{1}{4}$ to $\frac{1}{2}$ in away from the edge of the weld prep. All copper deposits will then be removed, and no copper penetration problems will develop. When pad welds are installed, it is impossible to clean the ID surface of the copper deposits, and the risk of copper penetration and future leaks is always present.

Similar liquid–metal embrittlement, but caused by lead, has been reported in superheaters and reheaters.[11] When shotguns are used to deslag the convection pass, lead pellets can be imbedded in the tube. Lead melts and penetrates the ferrite grain boundaries with the same steam leak and forced-outage result as copper penetration.

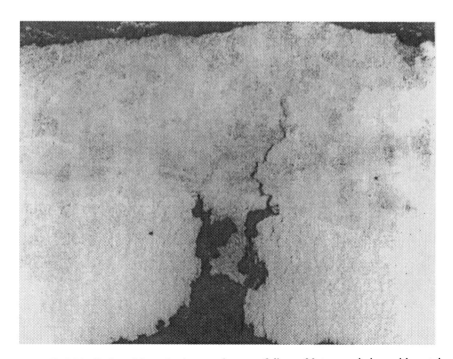

FIGURE 7.53. Pad weld on hydrogen-damage failure. Note crack in weld metal from propagation of existing crack (Magnification $18\frac{3}{4}\times$, etched).

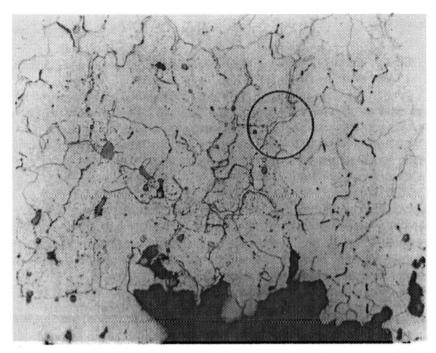

FIGURE 7.54. Copper penetration along ferrite-grain boundaries next to tube ID under pad weld (Magnification 500×, etched).

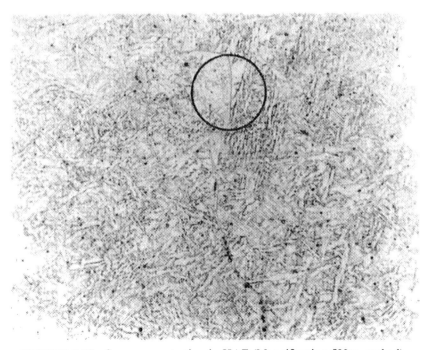

FIGURE 7.55. Copper penetration in HAZ (Magnification 500×, etched).

REFERENCES

1. H. Thielsch, *Defects and Failures in Pressure Vessels and Piping*, Reinhold, New York, 1965.
2. C. D. Lundin, "Dissimilar Metal Welds-Transition Joints Literature Review," *Welding Journal*, Feb., 1982, pp. 595–635.
3. A. Schaefer, "Dissimilar Metal Weld Failure Problems in Large Steam Generators," *Power*, Dec. 1979, pp. 68–69.
4. D. N. French, "High-Nickel Joints Unite Dissimilar Steels," *Welding Design & Fabrication*, May 1981, pp. 92–93.
5. D. N. French, "Welding Austenitic Stainless Steel (321H) to Ferritic Steel (T-22)—Fifteen Years of Successful Service," presented at ASM Conference, Philadelphia, Pennsylvania, December, 1980.
6. G. E. Linnert, *Welding Metallurgy*, Vol. 2, 3rd ed., American Welding Society, Miami, 1967.
7. G. A. Lamping and R. M. Arrowood, Jr., *Manual for Investigation and Correction of Boiler Tube Failures*, EPRI, April 1985.
8. ASM Committee on Corrosion of Weldments, *Metals Handbook*, 9th ed., Vol. 13, "Corrosion of Weldments," ASM International, Metals Park, Ohio, 1987.
9. S. Kou, *Welding Metallurgy*, Wiley, New York, 1987.
10. R. E. Hargrave, "Unusual Failures Involving Copper Deposition in Boiler Tubing," NACE Corrosion '90 Paper No. 187, Las Vegas, Nevada, April 1990.
11. R. D. Port, "Two Unusual Case Histories of Brittle Ruptures in Fossil Fuel Boiler Tubes," NACE Corrosion '87, Paper No. 167, San Francisco, March 1987.

CHAPTER EIGHT

FAILURE PREVENTION

The preceding seven chapters have discussed the causes of metallurgical failures. What can be learned from these examples to prevent or reduce the occurrence of failures in the future? For the most part, the common denominator of the majority of these cases has been a tube-metal temperature higher than expected in the original design. Tube-metal temperatures may increase slowly over many years or rapidly over a few hours. Internal oxide scale or deposit formation usually results in long-term overheating that gradually increases the temperature. Loss of boiler feedwater or the formation of a steam blanket usually results in rapid overheating and often rapid failure. Internal scale formation causes premature loss of expected life due to external corrosion, oxidation and tube wastage, microstructural degradation, and creep failures. Sudden high-temperature failures occur when the tube-metal temperature is raised to 1400°F (760°C) or so and are usually caused by loss of internal steam or water flow. Such failures are generally caused by operational upset.

For purposes of discussion, the problems affecting the expected life of a boiler, and especially a superheater or reheater, may be divided into broad categories of design, fuels, and operation. The furnace proper, or waterwalls, is characterized by high-heat-transfer rates, high gas temperatures, and a mixture of water and steam on the inside of the tubing. Superheaters and reheaters are characterized by much lower heat-flux rates, lower gas temperatures, but higher metal temperatures due to higher fluid temperatures and lower film conductances, and nearly pure steam on the inside of the tubing.

DESIGN OF SUPERHEATERS AND REHEATERS

Superheaters and reheaters have especially difficult operating conditions, particularly during start-up when steam flow has not been fully established. The highest metal temperatures within the boiler are in the finishing legs of superheaters and reheaters. Metal temperatures can be in the neighborhood of 1100°F (595°C) or higher for 1005°F (540°C) steam temperature. Since metal temperatures are the highest, the fire-side oxidation and corrosion potentials are the greatest. It is for these reasons that the actual life of this tubing is about half that of the rest of the unit. Typically superheaters and reheaters last 15 years or so.

The problems inherent to the operation of superheaters and reheaters are well known to the design engineer. At the time of the initial design, the intent is to furnish superheaters and reheaters that will last the life of the unit, upward of 30 years. High steam velocity raises the inside steam-film conductance and, thus, the tube-metal temperature more nearly approaches the local steam temperature. However, a high steam velocity requires a large pressure drop, which causes cycle degradation, especially within reheaters, due to additional pumping pressure and loss of expansion. Proper header design and sizing are necessary to ensure even steam distribution to the individual tubes. To ensure uniform steam distribution within the tube, tube-flow studies are undertaken, leading to selective tube orificing and/or tube shortening. The intent of all these considerations is to have a uniform steam flow tube to tube, whose final temperature meets the specifications.

The hottest tubes will be the first to fail. As a hot tube develops, its pressure drop increases, which causes less steam flow, less coolant, and higher metal temperatures. If the specified finishing steam temperature is 1005°F (540°C) across the steam manifold, individual tube steam temperature can vary by ±40°F (23°C) from the average. Thus, the hottest tube will have a steam temperature of 1045° (565°C) and a metal temperature of perhaps 1065° (575°C). While the coolest tube will be nearly 80°F (45°C) lower. The amount of unbalance is related to the overall pressure drop through the superheater or reheater. At the time boiler specifications are written, if a higher pressure drop is allowed, this temperature unbalance can be reduced. The penalty, of course, is a slightly lower overall net cycle efficiency as the pumping power requirements increase. The obvious benefits are a more uniform steam temperature, lower tube-metal temperatures, and longer life.

Internal scale acts as an insulating layer to the transfer of heat from the flue gas to the steam and raises the tube-metal temperature; see Chapter 2. As the tube-metal temperature increases, so does the rate of internal scale formation. As the scale thickness increases, so does the tube-metal temperature, and the cycle continues progressively becoming higher each year. The published literature on the scale formation of low-chromium-molybdenum alloys are for constant temperature.[1-3] As has been discussed, superheater and reheater tubes operate at a continually increasing temperature, and an estimate must be made of scale thickness as a function of time and temperature and not taken directly from published values at constant temperature. In order to make this estimate, we use the data of Rehn et al.,[4] who have correlated scale thickness with the Larson–Miller parameter. Figure 8.1 shows the data of Rehn et al. The data of Fig. 8.1 may be approximated by the equation

$$\log X = 0.00022P - 7.25 \tag{8.1}$$

where X is scale thickness and P is the Larson–Miller parameter:

$$P = T(20 + \log t) \tag{8.2}$$

T is absolute temperature (°F + 460), and t is time in hours. From Fig. 6.43, the metal temperature increase, ΔT, over the scale-free conditions may be

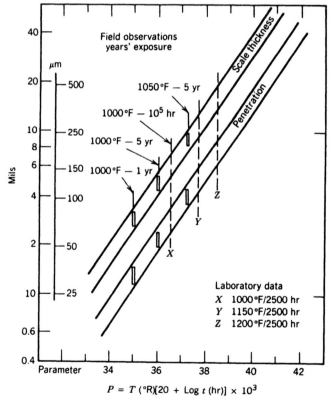

FIGURE 8.1. Steam-side scale formation for ferritic steels of 1–3% chromium correlated with the Larson–Miller parameter. From *Materials Performance*, Vol. 20, No. 6, June, 1981.

approximated for:

1. A typical superheater, $Q/A = 18,000$ Btu/hr $\cdot$ ft^2, by

$$\Delta T = 2.4X \tag{8.3a}$$

2. A typical reheater, $Q/A = 8000$ Btu/hr $\cdot$ ft^2, by

$$\Delta T = 1.1X \tag{8.3b}$$

An iterative process is used to calculate the scale thickness as both temperature and time increase. Time intervals of 1000 hr are used; and for each interval of time, the temperature is assumed to be constant. Further scale formation is assumed to occur at a constant temperature for each 1000-hr increment.

The general shape of the scale-thickness-versus-time curve should be parabolic; as scale thickness increases, the incremental increase over each succeeding 1000-hr interval should decrease. Starting with the design temperature T_0, Eqs. (8.1) and (8.2) are used to calculate a scale thickness X_1. At the end of 1000 hr, a temperature increase is calculated from X_1 and Eq. (8.3). The ΔT is added to T_0 to give a new temperature T_1 of the tube as a result of the internal scale. The higher temperature T_1 is then used to calculate the incremental thickness increase from 1000 to 2000 hr from Eqs. (8.1) and (8.2). For 2000 hr and T_1, P is calculated [Eq. (8.2)] and X is found from Eq. (8.1). For 1000 hr and T_1, P is calculated [Eq. (8.2)] and X is found from Eq. (8.1). Subtracting one from the other produces the incremental scale formation from 1000 to 2000 hr, which is added to X_1 to give X_2. From X_2 and Eq. (8.3) a new ΔT is found, and, thus, a higher T_2 is found. The process is repeated up to 100,000 hr, about 12 years.

Figure 8.2 plots the scale thickness for design temperatures of 1050°, 1075°, and 1125°F (566°, 580°, and 607°C). For all conditions except for the superheater example with a design temperature of 1125°F (607°C), the curves have the general shape of a parabola. The unusual shape of the superheater example with a starting temperature of 1125°F (607°C) is related to the poor oxidation resistance above about 1175°F (635°C). Once the tube temperature reaches this value, the increase in scale thickness becomes so great that more rapid metal-temperature increases occur. Based on the type of frequency of superheater and reheater failures, and the measured scale thickness noted, the qualitative shape of these curves is correct. In the early 1970s, the acceptable oxidation limit for 2¼ Cr–1 Mo SA-213 T-22 material was lowered from 1125 to 1075°F (607 to 580°C). Dramatic improvements in the oxidation resistance have been noted, and the type of superheater failures has changed in the last 15 years. Further improvements in oxidation resistance may be achieved by reducing the limit to 1050°F (565°C).

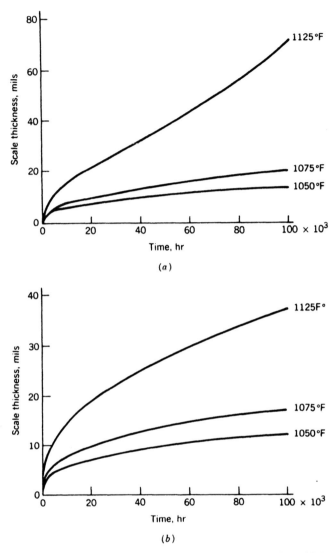

FIGURE 8.2. Calculated scale thickness as a function of time for initial tube-metal temperatures of 1050, 1075, and 1125°F (566, 580, and 607°C) for a typical (*a*) superheater and (*b*) reheater.

At the time the boiler is designed or specified, improved thermal life may be achieved by increasing the amount of high-alloy steel used. That is, reduce the oxidation limits for ferritic steels so that more austenitic stainless-steel materials would be used. Immediate benefits would be lower oxidation rates, as the tubes would be operating at temperatures much lower than the oxidation limit; and, thus, internal scale would form at a much slower rate. This causes both increased life of the tubing and decreased exfoliation. Exfoliation has been shown to contribute to turbine-blade erosion. The economic impact would be higher superheater or reheater initial cost, but an improvement in life expectancy would result.

Regardless of the oxidation limits established for materials of construction, internal scale inevitably forms. Scale removal during the normal life by suitable chemical-cleaning methods will return tube-metal temperatures to the design value. Superheaters and reheaters should be designed to be drainable to facilitate chemical cleaning. Part of the present reluctance to clean superheaters chemically is the widespread use of nondrainable superheaters and reheaters. It is difficult but possible to ensure that all of the scale-removal chemicals used are circulated uniformly through all circuits, completely neutralized, and adequately flushed when the chemical cleaning is finished. Drainable superheaters are more expensive initially, but the design would aid in extending the life of the tubing.

The damage done to turbine blades by erosion caused by exfoliation of internal scale particles is a serious problem. Since most of the oxide particles spall from the ID of these high-temperature steam lines during rapid temperature changes either during cool-down or start-up, the use of a bypass arrangement to route the initial steam generated during start-up around the turbine will help. The oxide particles will be flushed out and harmlessly bypassed around the turbine. The first steam to the reheater from the turbine during start-up may be as cold as 300°F (150°C), and a significant quench will occur from tube-metal temperatures around 1100°F (595°C). The 1100°F (595°C) temperature is caused by zero steam flow through the reheater during a cold start as the reheater metal approaches local flue-gas temperature. This quench causes thermal fatigue and releases oxide particles from the reheater tubing. Thermal shock of the superheater, and especially the reheater, may be avoided during start-up by using drum steam through these portions of the unit. Again, the initial cost for these bypass arrangements is higher, but the overall life cost may be substantially reduced by preventing erosion damage to the turbine. Steam-bypass systems for both superheater and reheaters for cycling-duty boilers are particularly attractive and are in wide use in Europe.

The suggestions for improvement in useful service life of superheaters and reheaters center on reducing steam-side scale formation and tube-metal temperatures. Not all suggestions are equally effective. At the time of purchase, serious consideration is required of the total cost, including both capital and operating and maintenance costs of the unit over its entire life. If oxidation limits are reduced, metal temperatures kept at design levels, and superheater and reheaters are chemically cleaned at regular intervals, then exfoliation and turbine erosion may be controlled.

No steam-bypass arrangements would be necessary for exfoliation control, but still might be helpful for turbine-temperature matching. For reheater, and certainly in double reheat units, steam flow should be established at start-up by routing steam from the drum before steam flow is fully established through the turbine. The design considerations in order of preference are:

1. For ease of chemical cleaning, superheaters and reheaters should be of the drainable design rather than the pendant nondrainable style. Periodically chemically cleaning the superheater minimizes the effects of internal scale deposit on tube-metal temperature. Periodic chemical cleaning of the superheater and reheater as well as the furnace will keep heat-transfer coefficients at the design level and keep tube-metal temperatures under control.

2. Changes in metal selection for the superheater and reheater should be made with an eye toward the use of higher alloys. In effect, this lowers the oxidation limit by about 25°F (15°C). The substitution of SA-213 T-11 for SA-210 A-1, T-22 for T-11, and 304 or 321 stainless steel for T-22 should be made, especially if the unit is expected to see cycling duty.

3. Specifications should allow a greater pressure drop through the superheater and reheater to improve the balance of steam flow from tube to tube. Presently, a ±40°F (23°C) unbalance is generally being recorded on the steam temperature across units, and an increase in pressure drop during the design would cause the unbalance to be lowered. If the design temperature at the outlet averages 1005°F (540°C) for the entire superheater or reheater, the hottest tube will be 1045°F (565°C) steam temperature, with a metal temperature of perhaps 1060°F (570°C). Increasing the pressure drop reduces the unbalance, and the hottest tube-metal temperature would decrease.

4. Since the reheater sees no steam flow until flow has been established through the superheater to the turbine and back to the reheater, the reheater is the most vulnerable to high-temperature excursions. Bypass steam from the drum can be used to cool the reheater and keep metal temperatures at design limits during start-up operations.

5. The erosion of turbine blades by particles of scale entrained in superheater or reheat steam can be a serious problem. Since exfoliation is greatest during transients, steam may be bypassed around the turbine so that during start-up any particles of scale will be harmlessly bypassed around the turbine, and no erosion will occur. Periodic chemical cleaning of superheaters and reheaters will keep the scale formation to a minimum and reduce the opportunity for turbine-blade erosion by entrained scale.

DESIGN OF THE FURNACE

As in the case of the superheater and reheater, the design considerations for waterwalls are similar. High fluid velocities entering the tubes and high subcooling are desirable for cooling for the same reasons that high steam velocities are desirable in superheaters. Proper header design and sizing are also necessary to ensure even water distribution to the bottom of the furnace so that each waterwall tube has adequate flow. The highest-heat-flux areas within the furnace are in the burner zones. Inevitably, tube failures occur first in these regions, since they are high-heat-transfer regions. At the time the boiler is designed, special attention is necessary in these critical areas. Tubing with internal spiral fins or rifling may be used in these high-heat-flux areas to promote better heat transfer. Better mixing between the steam–water mixture on the unheated side will prevent hot spots from forming. Higher-alloyed tubing, for instance SA-209 T-1, carbon–$\frac{1}{2}$ Mo or SA-213 T-2, $\frac{1}{2}$ Cr–$\frac{1}{2}$ Mo, rather than SA-210 A-1 may be used. The higher-alloyed tubing allows a higher metal temperature.

The purchase specification should include adequate and thorough feedwater-treatment facilities to ensure the best quality of the boiler feedwater that can be attained at all operating conditions. In the higher-pressure units it is becoming more popular to use a zero-solids water treatment or an all-volatile treatment to prevent dissolved solids from precipitating on heating surfaces. Careful monitoring of the boiler-water quality is necessary to correct any preboiler leaks that may develop. Condensers and boiler-feedwater heaters are sources of trouble, and leaks in these systems will deposit copper and other corrosion products on the inside of the boiler tubes reducing overall heat transfer. It is quite possible to have extensive hydrogen damage as a result of corrosion in the furnace walls after as little as six or eight months of operation if the boiler-feedwater chemistry control is not properly monitored.

It has become common practice nowadays to clean the furnace portion of a steam generator chemically at regular intervals to remove scale and deposits. It is not unusual to remove several hundred pounds of copper and up to a few thousand pounds of iron oxide during such a chemical cleaning.

As might be expected, all of these added components will increase the initial cost of the steam generator, but will result in a much smoother, trouble-free, operational life. Increased availability and, thus, return on investment must be balanced against capitalization costs.

OPERATIONAL PROBLEMS

The manner in which a boiler is operated can affect the life expectancy of the high-temperature portions of the superheater and reheater. For the most part, the expected life at the time the boiler is purchased is approximately 30 years. Actual tubing life may be nearer to 15 years owing to premature tube failures. Some of the operational problems that have contributed to premature tube-metal failures are:

1. Units have been run with small steam or water leaks for an appreciable time after the leak was detected. This puts an added burden on the

feedwater-treatment equipment because the amount of makeup required is higher than expected. Escaping steam can cause erosion of adjacent tubing and lead to failure. Also, escaping steam or water causes maldistribution in the remaining circuits as the fluid selectively goes to the lower-pressure regions and can cause other failures.

2. The unit-load change or ramp rates are higher than the designer expected. Changes in loading from both a temperature and pressure point of view put additional stresses on virtually all components of the boiler that may or may not have been allowed for in the initial design. Rapid ramp rates may increase superheater and reheater metal temperature above the safe-operating temperature, and these operational spikes lead to a more rapid oxidation rate and scale buildup than operating strictly at design levels. With more rapid pressure and temperature swings than designed for, thermal fatigue of some components may become a problem.

3. Less-than-ideal water treatment or boiler-feedwater quality can lead to substantial corrosion failures, especially in the high-heat-release area of the lower furnace and around the burners. Less-than-proper water chemistry may be related to operation with leaks in the preboiler circuitry, the boiler-feedwater heaters and condensers, for example. Any condition that increases the amount of solids or impurities in the water beyond the expected level will lead to increased corrosion in the hotter portions of the furnace. Direct chemical attack can result in an all-volatile water-treatment method if the condensate demineralizers are not handled properly.

4. Many older units have changed from base-loaded operation to cycling duty. Unless there has been an upgrade or change in the boiler design, the cycling duty puts thermal strains on many portions of the steam generator not anticipated by the designer for a base-loaded unit.

5. In spite of all the precautions that operators may take, the number of units that operate without water because of human error is surprisingly high. The amount of damage that can be done in a very short time by operating with a low water level can run into the millions of dollars.

FUEL

In the normal course of several decades of operation, the changes that may occur in the type of fuel used can have a dramatic effect on the boiler performance. Changes in coal from that which was specified to off-specification coal can affect the boiler operation in the following ways:

1. The change from a low-fouling to a high-fouling coal, for example, directly affects the superheater and reheater. At the time of initial

design, the type of coal expected will determine the superheater or reheater bundle spacing to prevent pluggage by the coal ash. If the expected coal is a low-fouling coal, bundle spacing can be closer and the number of sootblowers fewer as the passages are less likely to close off. The frequency of sootblower use will be less, and, therefore, the erosion that these sootblowers may cause is diminished. Within the furnace, a change from a low- to a high-slagging coal decreases furnace heat-absorption and increases furnace-exit-gas temperature. A high-slagging coal would require more frequent use of wall blowers. An increase in the furnace-exit-gas temperature will increase the heat available for the superheater and reheater and require greater use of superheat or reheat spray to maintain temperature. In extreme conditions, the amount of superheat and reheat spray available may be too little to maintain steam temperatures, and superheater and reheater surface adjustments may need to be made. Conversely, a change from a high-slagging coal to a low-slagging coal will give greater heat absorption in the furnace and a lower furnace-exit-gas temperature, and, thus, a lower superheater and reheater steam temperature.

2. A change from a low- to a high-Btu coal is more easily accomplished than the change from a high- to a low-Btu coal. The heat content per pound of fuel dictates the amount of coal required to make the proper steam flow, and changes may have adverse consequences. An increase in coal flow above the design causes problems in the coal-crushing and -grinding equipment, the rate of erosion in coal pipes, and problems in the ash-handling systems. A decrease in heat content often means an increase in inert ingredients. Increases in ash content over the design coal may lead to greater erosion of superheater and reheater because the frequency of sootblowing will increase. The composition of the ash, especially its sodium and sulfur level, can have a dramatic effect on the corrosiveness of the ash deposits.

3. Changes in fuel-oil analysis or the method of shipment of the oil affect the boiler performance, especially as it relates to the potential corrosion in the high-temperature portions of the unit. Changes in the sodium, sulfur, and vanadium contents can alter dramatically the potential for liquid-ash corrosion. The source of the fuel oil may change such that a portion of its transportation is done by ocean tankers or barges, which may lead to increased chlorine content from seawater entrainment. The addition of even a small amount of sodium chloride from seawater can increase the corrosion rate as low-melting-point iron chloride may form in the superheater and reheater. Mixtures of chlorides, sulfates, and oxides may form mixtures of such low melting points that liquid-ash attack is inevitable.

The preceding discussion has touched briefly on some of the changes in operation that can lead to diminished life of portions of the boiler. Some of these problems can be reduced or prevented by careful specification at the time the boiler is purchased. Some of the design changes that may be contemplated are:

1. Drainable superheaters and reheaters and routine chemical cleaning.
2. Use alloy steel and reduce oxidation limits.
3. Increase allowable pressure drop to give more uniform steam flow and less unbalance in steam temperature.
4. Route steam from the drum through the reheater to prevent excessive metal temperature during start-up.
5. Route the first steam from the turbine to the condenser rather than to the reheater. This scheme is only possible when the reheater is cooled by steam flow from the drum during start-up.
6. Route the initial steam from the superheater to the condenser to reduce turbine-blade erosion from exfoliation of the superheater.
7. Design for the best possible water treatment and use the best possible boiler feedwater, and furnace corrosion problems will be reduced.

REMAINING-LIFE ASSESSMENT

In the normal service life of a superheater, failure usually occurs by a creep or stress-rupture mechanism. Regardless of the design temperature, steam-side scale or oxide forms by the reaction of steam with steel.

$$3Fe + 4H_2O = Fe_3O_4 + 4H_2 \tag{8.4}$$

The oxidation is parabolic with time; that is, the oxide-scale thickness X is proportional to the square root of time.

$$X = kt^{1/2} \tag{8.5}$$

The proportionality constant k increases with temperature.

The thermal conductivity of the oxide is much less than the thermal conductivity of the steel tube. The net effect is to raise tube-metal temperature. See Chapter 2. The thicker the scale is, the larger the temperature increase of the tube metal is. The sequence of aging within a superheater tube may be summarized as follows:

1. Steam reacts with steel to form iron oxide along the inside of the tube. The thermal conductivity of the scale is approximately 5% that of the steel and insulates the tube from cooling by the steam.

2. Oxidation or corrosion along the fire side of the tube occurs by contact of the steel with the flue gas or fuel ash. The effect of this wastage is to reduce tube-wall thickness and increase the hoop stress within the tube wall.
3. The effects of increased metal temperature and increased stress are to increase the rate of steam-side oxide formation and flue-gas-side corrosion. Steps 1 and 2 proceed at ever-increasing rates as the tube temperature continuously increases.
4. The increase in stress and temperature decreases the creep life or the time to failure.
5. The superheater tube fails by a creep or stress-rupture mechanism.

The prediction of remaining life is based on published data of stress-rupture curves and log stress vs. Larson–Miller parameter (LMP) curves. Figure 8.3 presents representative data for SA-213 T-22 at 1100°F.[5] Note that at any stress, failure will occur over a wide range of times, which in turn leads to the concept of "minimum," "maximum," and "mean" failure times. Table 8.1 presents stress-rupture data for T-22 at 1000, 1100, and 1200°F.[5]

A more convenient way to present stress-rupture data is to combine time and temperature into the Larson–Miller parameter.

$$P = (T + 460)(20 + \text{Log } t) \qquad (8.6)$$

where P = Larson–Miller parameter
T = temperature, °F
t = time to rupture, hr
20 = empirical constant

Since there are minimum, maximum, and mean failure times on the stress-rupture data, there are minimum, maximum, and mean failure curves for Larson–Miller parameter data. Figure 8.4 plots the data of Table 8.1 for T-22 material. What these curves imply is that for a given stress, the first failure will occur when a combination of time and temperature is reached that falls on the minimum failure curve. For example, at 5000 psi, the LMP for the first or minimum failure is 39,000; any combination of T and t that satisfies the LMP relationship will lead to the first creep failures. All tubes will have failed at 5000 psi when an LMP of 40,600 is reached. These two values of LMP give widely differing results of expected life at a given temperature.

To illustrate,

$$39,000 = (T + 460)(20 + \text{Log } t)$$

At 1075°F, $t = 255,000$ hr, thus the first failures are expected at 255,000 hr.

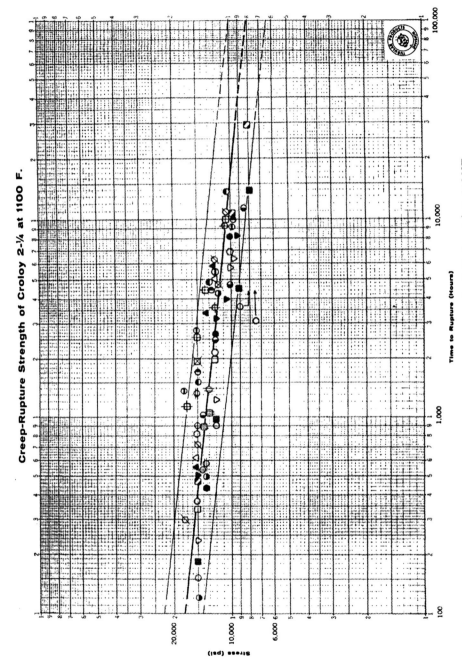

FIGURE 8.3. Creep-rupture strength of Croloy 2-¼ at 1100°F.

TABLE 8.1. Comparison of Statistically Determined Creep-Rupture Strength Properties—Least-Squares Averages and 95% Confidence Limits for Linear and Quadratic Treatments

	100 hr		1000 hr		10,000 hr		100,000 hr[a]	
	Linear	Quadratic	Linear	Quadratic	Linear	Quadratic	Linear	Quadratic
			1000°F					
UCL[b]	35,408	38,189	26,797	26,764	20,280	20,331	(15,348)	(16,741)
Average	28,515	30,701	21,519	21,517	16,238	16,345	(12,253)	(13,458)
LCL[c]	22,938	24,682	17,281	17,298	13,002	13,140	(9,783)	(10,820)
			1100°F					
UCL[a]	22,530	17,763	17,163	17,468	13,074	11,968	(9,960)	(5,713)
Average	17,857	14,492	13,623	14,252	10,394	9,765	(7,929)	(4,661)
LCL[c]	14,153	11,824	10,814	11,628	8,262	7,967	(6,313)	(3,803)
			1200°F					
UCL[a]	16,789	11,721	10,168	10,157	6,158	4,771	(3,729)	(1,215)
Average	12,988	9,783	7,885	8,428	4,787	3,982	(2,906)	(1,014)
LCL[c]	10,048	8,165	6,115	7,076	3,721	3,324	(2,264)	(846)

[a] Values in parentheses not statistically valid. Shown for information only.
[b] UCL = upper confidence limit.
[c] LCL = lower confidence limit.

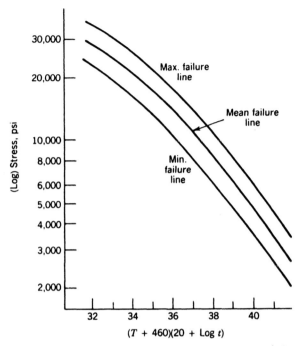

FIGURE 8.4. Larson–Miller parameter, SA-213 T-22, $2\frac{1}{4}$ Cr–1 Mo.

The last failure will occur when

$$40{,}600 = (T + 460)(20 + \text{Log } t)$$

or time is 2,800,000, more than a factor of 10 greater!

This scatter to the published data for the material property is one factor that makes life prediction difficult. Figures 8.5–8.10 present the variation of the Larson–Miller parameter with stress for carbon steel, C–$\frac{1}{2}$ Mo (T-1) $\frac{1}{2}$ Cr–$\frac{1}{2}$ Mo (T-2), 1 Cr–$\frac{1}{2}$ Mo (T-12), $1\frac{1}{4}$ Cr–$\frac{1}{2}$ Mo (T-11), and $2\frac{1}{4}$ Cr–1 Mo (T-22).

However, a reasonable estimate of the condition of a superheater (or reheater) may be obtained from metallographic examinations of 10–12 samples. A ring section is cut from each sample, and OD, ID, wall thicknesses, and ID scale thicknesses are measured. Since the ID is unaffected by ash corrosion and is nearly circular, it is better to calculate stress from

$$S = \frac{P(D_m)}{2W} = \frac{P(R_m)}{W} \quad \text{where } R_m \text{ is mean radius} \tag{8.7}$$

and

$$R_m = R_I + \frac{W}{2}$$

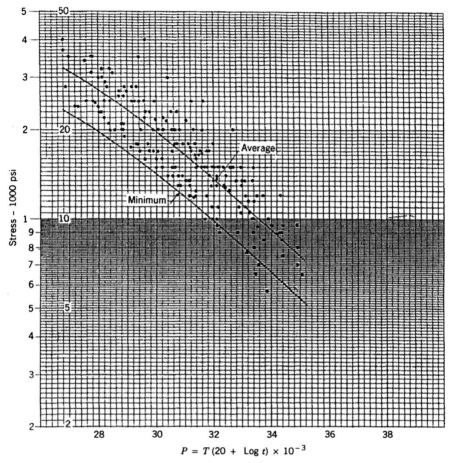

FIGURE 8.5. Variation of Larson–Miller parameter with stress for rupture of carbon-steel pipe and tube. Copyright ASTM. Reprinted with permission.

where R_I is inside radius and S, P, and W have their usual meanings. Thus,

$$S = \frac{P(R_I + W/2)}{W} \qquad (8.8)$$

The stress on the cylinder is calculated at the thinnest point.

The calculation of stress within a tube is relatively simple. More difficult to assess is the average operating temperature of a superheater tube. Several equations have been published that relate the steam-side scale thickness for the chromium–molybdenum steels, similar to T-11 and T-22, and operating time. One such equation is

$$\text{Log } X = 0.00022(T + 460)(20 + \text{Log } t) - 7.25 \qquad (8.9)$$

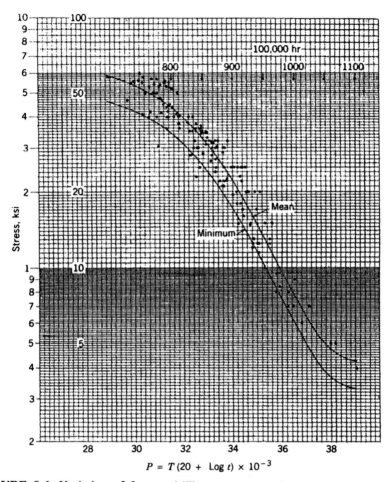

FIGURE 8.6. Variation of Larson–Miller parameter with stress for rupture of C–Mo steel. Copyright ASTM. Reprinted with permission.

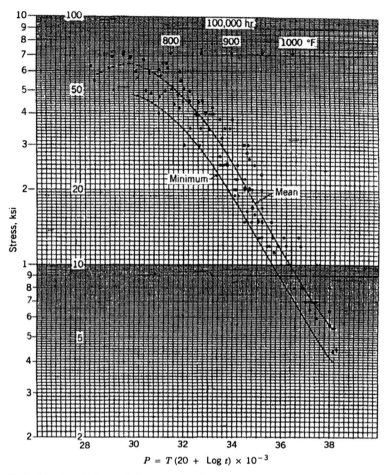

FIGURE 8.7. Variation of Larson–Miller rupture parameter with stress for ½ Cr–½ Mo steel. Copyright ASTM. Reprinted with permission.

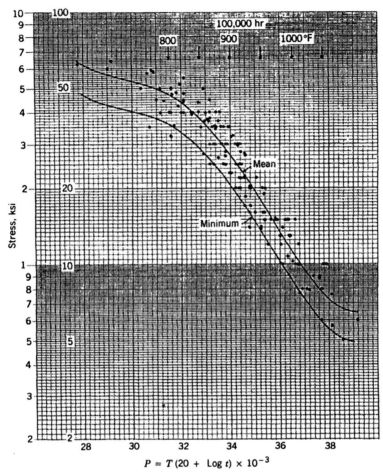

FIGURE 8.8. Variation of Larson–Miller rupture parameter with stress for 1 Cr-½ Mo steel. Copyright ASTM. Reprinted with permission.

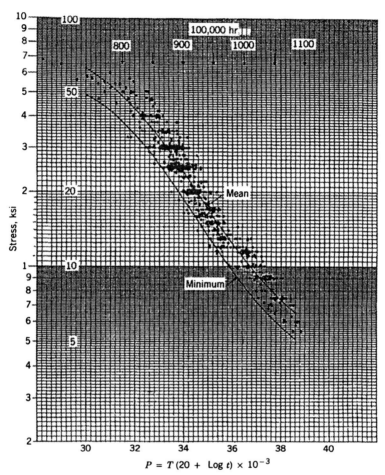

FIGURE 8.9. Variation of Larson–Miller rupture parameter with stress for wrought 1¼ Cr–½ Mo steel. Copyright ASTM. Reprinted with permission.

482 CHAPTER 8 FAILURE PREVENTION

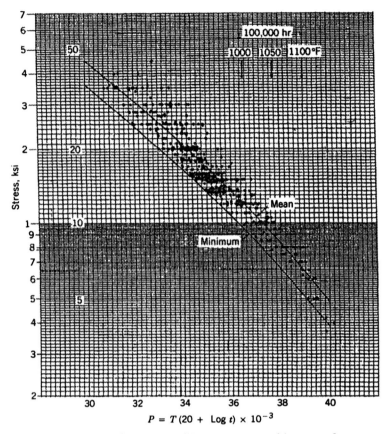

FIGURE 8.10. Variation of Larson–Miller parameter with stress for rupture of annealed material. $2\frac{1}{4}$ Cr–1 Mo steel. Copyright ASTM. Reprinted with permission.

TABLE 8.2.

Tube No.	OD × MWT Specified, in	OD × MWT Actual (min), in	Stress, psi (actual)	ID scale, mils	T_A, °F	Larson–Miller Parameter
1[a]	1½ × 0.284	1.33 × 0.200	5500	34	1125	39,900
2	1½ × 0.284	1.345 × 0.275	3800	29.4	1115	39,600
3	1½ × 0.284	1.320 × 0.260	3980	28.3	1110	39,500
4[a]	1½ × 0.284	1.410 × 0.194	6080	41.4	1140	40,200
5	1½ × 0.360	1.395 × 0.340	3030	15.3	1065	38,400
6	1½ × 0.284	1.450 × 0.275	4200	16.4	1070	38,500
7	1½ × 0.360	1.440 × 0.350	3040	16.4	1075	38,500
8	1½ × 0.360	1.410 × 0.340	3100	16.4	1070	38,500
9	1½ × 0.360	1.445 × 0.355	3000	18.5	1080	38,700

[a] Failed tubes

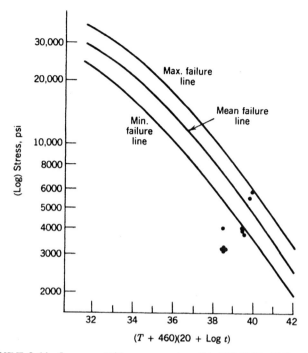

FIGURE 8.11. Larson–Miller parameter, SA-213 T-22, 2¼Cr–1 Mo.

where X = steam-side scale thickness in mils, and T and t are defined in Eq. (8.2).

For the 10–12 samples, stress is calculated from Eq. (8.8), and the average temperature, T, is calculated from Eq. (8.9). Service hours are obtained from plant records or estimated from years since start-up and availability percent. With a value for t, the Larson–Miller parameter is calculated from Eq. (8.2). All the data are then plotted on a log stress vs. LMP plot. Table 8.2 presents a set of data for nine samples from a superheater which are plotted on Fig. 8.11. If possible, it is better to have at least one data point that is a failure. This superheater is in fair condition. While several points are near the minimum failure line, failure will not occur until corrosion wastage reduces the wall thickness to about 0.200 in. Thus, the recommendation would be to measure wall thickness during each annual outage and replace all tubes with a measured wall less than 0.220 in.

COMPUTER SIMULATIONS

Computer programs have been designed to simulate the aging of a superheater tube in service and used to predict failure times.[10] The program does a series of calculations until failure occurs. Built into the computer program's data base are equations that relate the log stress to the Larson–Miller parameter for minimum and mean values of the failure curve, similar to the data shown in Fig. 8.9 for T-11 and Fig. 8.10 for T-22 material. Calculations are performed until the failure line is reached. The logic that the computer program uses may be summarized by the following sequence:

1. Calculates the average operating metal temperature T from steam-side scale thickness and operating time by using Eq. (8.9):

$$\text{Log } X = 0.00022(T + 460)(20 + \text{Log } t) - 7.25 \qquad (8.9)$$

2. Calculates the scale-growth constant k from Eq. (8.5).

$$X = kt^{1/2} \qquad (8.5)$$

3. The heat-flow analysis of Chapter 2, Figs. 6.43 and 6.44, and Eq. (8.3) show that the increase in tube-metal temperature, δT, is essentially linear with steam-side scale thickness X and varies with heat flux. For a given heat flux, Q/A, the temperature increase due to the steam-side scale is of the form

$$\delta T = CX \qquad (8.10)$$

4. The constant C is a function of the heat flux and tube dimensions. Appropriate values for C may be calculated, but C varies from less than 1 to nearly 7°F/mil.
5. Calculates the hoop stress, S psi, from an equation similar to Eq. (8.7):

$$S = \frac{P(D_m)}{2W} = \frac{P(R_m)}{W} \qquad \text{where } R_m \text{ is mean radius} \qquad (8.7)$$

6. The wall thickness decreases due to oxidation and corrosion. The form of the wastage is usually assumed to be linear with time. The wall thickness at anytime is

$$W = W_0 - C_R t \qquad (8.11)$$

where W_0 = initial wall thickness, in
 C_R = corrosion rate, in/hr

Subroutines may be inserted to allow for changes in corrosion rate, for example, when liquid-ash corrosion begins. Figures 6.50 and 6.51 present corrosion data which indicate that the wastage as a function of metal temperature is shaped like a hockey stick. The more rapid corrosion rate corresponds to the melting point of the liquid-ash species.

7. Calculates the OD temperature from the thermal gradient across the tube wall:

$$T_{OD} = \frac{(Q/A)R_0 \ln(r_o/r_i)}{D_{T22}} + T \qquad (8.12)$$

8. Calculates the midwall effective temperature from T (the ID effective temperature) and T_{OD}:

$$T_{eff} = \frac{T + T_{OD}}{2} \qquad (8.13)$$

9. The temperature increase, δT, due to steam-side scale is a combination of Eqs. (8.5) and (8.10):

$$\delta T = kCt^{1/2} \qquad (8.14)$$

10. An estimate of the initial tube-metal temperature, T_i, may be obtained from the average temperature, T, found from Eq. (8.9) minus the temperature increase for the average time of operation. δT is calculated from Eq. (8.14) with t equal to half the total service time; thus T_i is

$$T_i = T - kCt_A^{1/2} \qquad (8.15)$$

where t_A = half the operating time used in Eq. (8.6).

11. Calculates P from Eq. (8.6).

$$P = (T + 460)(20 + \text{Log } t) \qquad (8.6)$$

The tacit assumption is that failure occurs by a creep or stress-rupture mechanism. The input data needed are tube dimensions (OD and wall thickness in inches) operating steam pressure (psi), operating time (hours), steam-side scale thickness (mils), oxidation or corrosion rate (in/3000 hr), and finally some estimate of the heat flux in Btu/(hr · ft²). The heat flux in a superheater varies from location to location, with a range of a factor of $\pm 2\frac{1}{2}$ from the average given by the designer. If the average is 18,000 Btu/(hr · ft²), the range is likely to be 7000–45,000 Btu/(hr · ft²).

The program is an iterative series of calculations at 3000-hr intervals and continues until failure is predicted. The sequence is as follows:

1. Calculates the average operating metal temperature from Eq. (8.1).
2. Calculates the scale growth constant, k, in Eq. (8.5).

3. Calculates the tube-scale thickness at half the operating time from the value of k calculated in step 2, and Eq. (8.5).
4. Calculates δT with the scale thickness just calculated and Eq. (8.10) and an appropriate value of C; that is,

$$\delta T = kCt^{1/2} \qquad (8.14)$$

To find the initial tube-metal temperature, it subtracts the δT found in Eq. (8.14) from the average metal temperature found in Eq. (8.1) at step 1. This is the metal temperature in the absence of any steam-side scale.

5. Calculates the hoop stress from Eq. (8.8).
6. Calculates the midwall temperature from the initial ID temperature (step 1) and the thermal gradient across the tube wall, as shown in Eqs. (8.12) and (8.13).
7. With the midwall metal temperature (step 6) and a time interval of 3000 hr, calculates P from Eq. (8.6).
8. Searches its data base and determines an appropriate P for the stress (step 5) that would cause failure at the initial operating temperature calculated in step 4.
9. Solves Eq. (8.6) with the P at failure (step 7) and the initial metal temperature (step 1) for the time to cause failure.
10. The life fraction consumed in this first step is then 3000 hr divided by the failure time found in step 8. To illustrate: at a stress of 5000 psi, the first failure occurs at a P of 39,000. At an initial temperature of, say, 1050°F, Eq. (8.6) gives the time to failure of 673,000 hr, and the life fraction consumed is

$$L_f = \frac{3000}{673,000} = 0.00446 \qquad (8.16)$$

11. The second interval has a wall thickness reduced by the oxidation rate (Eq. 8.11), so the hoop stress in Eq. (8.8) increases. Steam-side scale has increased as in Eq. (8.5), and so has the metal temperature from Eq. (8.10). A new life fraction is calculated and summed to the previous one, and the process continues step by step until a life fraction of 1 is found. A life fraction equal to 1 is taken as failure. In effect, the programs track, on the log stress vs. Larson–Miller parameter, a curve similar to that in Fig. 8.12.
12. Changes in corrosion rate may be inserted when the calculated OD surface temperature reaches the melting point of species that cause more rapid liquid-ash corrosion.

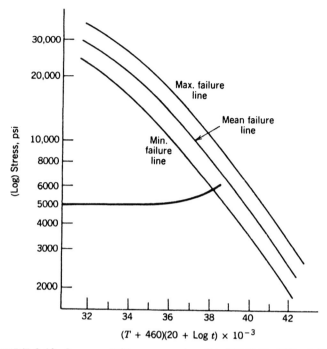

FIGURE 8.12. Larson–Miller parameter, SA-213 T-22, 2¼Cr–1 Mo.

Different organizations that perform remaining-life assessments often give very different recommendations on the remaining life: the computer simulations are similar, but the results are not. Another potential problem in the use of computer programs to predict the behavior and remaining life of a superheater or reheater tube is the choice of correlation used between steam-side scale thickness, operating time, and metal temperature. Several correlations have been published and are summarized by Viswanathan et al.[11] These correlations predict slightly different metal temperatures based on the steam-side scale thickness.

To illustrate the effects that slightly different tube-metal temperatures have on the remaining life, we compare two equations for calculating the metal temperature in the preceding computer program. Equation (8.17), developed by Aptech Engineering Services Co. under contract to EPRI,[5] will be one equation; the other equation, (8.9), is based on Foster–Wheeler data. The Aptech Engineering Services Co. correlation is

$$\text{Log } X = C_1 + C_2 T + C_3 T \log t \qquad (8.17)$$

where X = steam-side scale thickness, mils
 T = metal temperature, °R
 t = operating time, hr

$C_1 = -6.839869$
$C_2 = 0.003860$
$C_3 = 0.000283$

Equation (8.17) may be rearranged in a more familiar form:

$$\text{Log } X = 0.000283[(T + 460)(13.6396 + \log t)] - 6.839869 \quad (8.17a)$$

where the term in brackets is a form of Larson–Miller parameter with a constant of 13.6396 rather than 20; see Eq. (8.6).

Equation (8.17) predicts a slightly different metal temperature than Eq. (8.9). For example, in 100,000 hr of service, a scale thickness of 23.6 mils was measured on a T-22 superheater tube. Equation (8.17) predicts a temperature of 1097°F, while Eq. (8.9) predicts a temperature of 1108°F.

There are other similar correlations between steam-side scale thickness and operating time and temperature. They predict metal temperatures either higher or lower, depending on the equation used. In 100,000 hr, and a measured scale thickness of 23.6 mils, the estimated metal temperatures range from 1067°F to 1121°F.[5] For illustrative purposes, the preceding computer program was run with exactly the same data input, except that two different equations for metal temperature were used, Eqs. (8.17) and (8.9).

The input data are:

Tube OD	1.5420 in
Tube wall	0.4040 in
Hours of service	123,000
ID scale thickness	21.0
Stage 1 corrosion rate	0.00090/3500 hr
No liquid-ash corrosion	

The results are shown in Table 8.3.

When an assumed liquid-ash melting temperature of 1075°F is added with a stage-2 corrosion rate of 0.00180 in/3500 hr, the results are similar; see Table 8.4.

TABLE 8.3. Calculated Metal Temperatures

	Metal Temp. Eq. (8.17)	Metal Temp. Eq. (8.6)
Initial temp.	1044°F	1057°F
Failure temp.	1108°F	1114°F
Failure time	199,500 hr	157,500 hr
Wall thickness at failure	0.3527 in	0.3635 in

TABLE 8.4. Calculated Metal Temperatures

	Metal Temp. Eq. (8.17)	Metal Temp. (DNF) Eq. (8.6)
Initial temp.	1044°F	1057°F
Time at start of liquid- ash corrosion,	45,500 hr	17,500 hr
Failure temp.	1106°F	1112°F
Failure time,	185,500 hr	147,000
Wall thickness at failure	0.3212 in	0.3329 in

Assuming 7000 hr actual operating time per year, a small temperature difference of only 13°F between the two equations leads to a difference in the time-to-failure of six years! When other such equations are used that predict metal temperatures either higher or lower, the differences in time-to-failure will be correspondingly greater.

Since, even with a computer program, there are unknowns (heat flux for example), predictions on remaining life can be further improved when data on a failed tube can be used. The computer model can be forced to predict the known failure, both time at failure and wall thickness at failure. The wall thickness is measured at the ends of the opening and used as the wall thickness at failure. The number of hours at failure is also known. Thus, different heat-flux values can be used in the model until the "correct" value is found that "predicts" the actual failure. With this better value of heat flux, the remaining life or life expectancy can be predicted.

Without a failure to work with, the model predicts a wall thickness at failure. Annual wall-thickness measurements and replacement at predicted wall thickness at failure, plus a margin for error, will improve overall reliability of any superheater or reheater by reducing forced outages due to tube failures.

REFERENCES

1. H. L. Solberg, G. A. Hawkins, and A. A. Potter, "Corrosion of Unstressed Steel Specimens and Various Metals by High Temperature Steam," *Trans ASME*, May 1942, pp. 303–316.
2. F. B. Eberle and J. H. Kitterman, "Scale Formation on Superheater Alloys Exposed to High Temperature Steam," in *Behavior of Superheater Alloys in High Temperature, High Pressure Steam*, ASME, New York, 1968.
3. F. B. Eberle, J. W. Seifert, and J. H. Kitterman, "Scaling of Ferritic Superheater Steel during 36,000 Hours' Exposure in 980/1030°F Steam of 2350 psi, with Particular Respect to Scale Exfoliation Tendency," Vol. XXVI, *Proceed-*

ings of the American Power Conference, American Power Conference, Inc., 1964, pp. 501–510.

4. I. M. Rehn, W. R. Apblett, Jr., and John Stringer, "Controlling Steamside Oxide Exfoliation in Utility Boiler Superheaters and Reheaters," *Materials Performance*, June 1981, pp. 27–31.

5. *The Creep Properties of Croloy* $2\frac{1}{4}$. Babcock and Wilcox Co. Tubular Products Division, Beaver Falls, Pennsylvania.

6. G. V. Smith, *An Evaluation of the Elevated Tensile and Creep-Rupture Properties of Wrought Carbon Steel DS11S1*, ASTM, Philadelphia, Pennsylvania, 1970.

7. G. V. Smith, *Evaluation of the Elevated Temperature Tensile and Creep-Rupture Properties of C-Mo, Mn-Mo, and Mn–Mo–Ni Steels DS47*, ASTM, Philadelphia, Pennsylvania, 1971.

8. G. V. Smith, *Evaluation of the Elevated Temperature Tensile and Creep-Rupture Properties of* $\frac{1}{2}$*Cr–*$\frac{1}{2}$*Mo, 1 Cr–*$\frac{1}{2}$*Mo, and 1*$\frac{1}{4}$*Cr–*$\frac{1}{2}$*Mo Steels*, ASTM, Philadelphia, Pennsylvania, 1973.

9. G. V. Smith, *Supplemental Report on the Elevated Temperature Properties of Cr–Mo Steels (An Evaluation of* $2\frac{1}{4}$*Cr–1 Mo Steel) DS6S2*, ASTM, Philadelphia, Pennsylvania, 1971.

10. J. A. Alice, J. A. Janiszewski, and D. N. French, "Liquid Ash Corrosion, Remaining Life Estimation and Superheater/Reheater Replacement Strategy in Coal Fired Boilers," Presented at the JT ASME/IEEE Power Generation Conference, Milwaukee, Wisconsin, October 1985, Paper No. 85-JPGC-PWR-3.

11. R. Viswanathan, J. R. Foulds, and D. I. Roberts, "Methods for Estimating the Temperature of Reheater and Superheater Tubes in Fossil Boilers," Presented at Conference on Boiler Tube Failures in Fossil Plants, EPRI Research Project 1890, Atlanta, Georgia, Nov. 1987.

APPENDIX

TABLE I	Typical Coal Analysis
TABLE II	Range of Analyses of Fuel Oils
TABLE III	Typical Percentage Compositions of Oil Ashes
TABLE IV	Bark Ash Analysis
TABLE V	Alloys Used in Boilers
TABLE VI	Oxidation Limits
TABLE VII	Allowable Stresses (psi), ASME Code
TABLE VIII	Short-Time High-Temperature Tensile Strength
TABLE IX	Composition and Properties of Some Stainless Steels
TABLE X	ASME Code Heat Treatment for Austenitic Stainless Steels
TABLE XI	High-Alloy Austenitic Alloys
TABLE XII	Trace Elements in Some Crude Oils
TABLE XIII	Strength of Boiler Steels
TABLE XIV	Effects of Sigma-Phase on Mechanical Properties of 304H Stainless Steel
TABLE XV	Approximate Lower Critical Temperatures

TABLE I Typical Coal Analysis

	Eastern Bituminous	Midwest Bituminous	Sub-bituminous C	North Texas Lignite	Dakota Lignite
Total moisture (%)	5.0	15.4	30.0	31.0	39.6
Ash (%)	10.3	15.0	5.8	10.4	6.3
Volatile matter (%)	31.6	33.1	32.6	31.7	27.5
Fixed carbon (%)	53.1	36.5	31.6	26.9	26.6
Total (%)	100.0	100.0	100.0	100.0	100.0
Btu/lb, as fired	13,240	10,500	8,125	7,590	6,523
Btu/lb, moisture and ash free	15,640	15,100	12,650	12,940	12,050
Ash analysis (%)					
SiO_2	40.0	46.4	29.5	46.1	23.1
Al_2O_3	24.0	16.2	16.0	15.2	11.2
Fe_2O_3	16.8	20.0	4.1	3.7	8.4
CaO	5.8	7.1	26.5	16.6	23.7
MgO	2.0	0.8	4.2	3.2	5.8
Na_2O	0.8	0.7	1.4	0.4	7.3
K_2O	2.4	1.5	0.5	0.6	0.7
TiO_2	1.3	1.0	1.3	1.2	0.4
P_2O_5	0.1	0.1	1.1	0.1	—
SO_3	5.3	6.0	14.8	12.7	17.6
Not accounted for	1.5	0.2	0.6	0.2	1.2
Sulfur (%)	1.8	3.2	0.34	0.6	0.7
Lb moisture/million Btu	3.8	14.7	36.9	40.8	60.7
Lb ash/million Btu	7.8	14.3	7.1	13.7	9.7
Lb sulfur/million Btu	1.36	3.05	0.42	0.79	1.1

Source: H. E. Burboch and H. Bogot, "Design Considerations for Coal Fired Steam Generators," Presented at Annual Association of Rural Electric Generating Cooperatives, Wichita, Kansas, June 13–16, 1976. Reprinted with permission of Combustion Engineering, Inc.

TABLE II Range of Analyses of Fuel Oils

	Grade of Fuel Oil				
	No. 1	No. 2	No. 4	No. 5	No. 6
Weight (%)					
Sulfur	0.01–0.5	0.05–1.0	0.2–2.0	0.5–3.0	0.7–3.5
Hydrogen	13.3–14.1	11.8–13.9	(10.6–13.0)[a]	(10.5–12.0)[a]	(9.5–12.0)[a]
Carbon	85.9–86.7	86.1–88.2	(86.5–89.2)[a]	(86.5–89.2)[a]	(86.5–90.2)[a]
Nitrogen	Nil–0.1	Nil–0.1	—	—	—
Oxygen	—	—	—	—	—
Ash	—	—	0–0.1	0–0.1	0.01–0.5
Heating value					
Btu/lb, gross (calculated)	19,670–19,860	19,170–19,750	18,280–19,400	18,100–19,020	17,410–18,990

[a] Estimated

Source: Steam/Its Generation and Use, 39th ed., Babcock and Wilcox, Co., 1978. Courtesy of Babcock and Wilcox.

TABLE III Typical Percentage Compositions of Oil Ashes

Constituent	California	Midcontinent	Texas	Pennsylvania	Iran
SiO_2	38.8	31.7	1.6	0.8	12.1
Fe_2O_3					
Al_2O_3	17.3	31.8	8.9	97.5	18.1
TiO_2					
CaO	8.7	12.6	5.3	0.7	12.7
MgO	1.8	4.2	2.5	0.2	0.2
MnO	0.3	0.4	0.3	0.2	Trace
V_2O_5	5.1	Trace	1.4	—	38.5
NiO	4.4	0.5	1.5	—	10.7
Na_2O	9.5	6.9	30.8	0.1	—
K_2O	—	—	1.0	—	—
SO_3	15.0	10.8	42.1	0.9	7.0
Chloride	—	—	4.6	—	—

Source: W. T. Reid, *External Corrosion and Deposits: Boilers and Gas Turbines*, Elsevier, New York, 1971. Courtesy of William T. Reid.

TABLE IV Bark Ash Analysis

	Pine	Oak	Spruce[a]	Redwood[a]
Ash content (dry) (%)	2.9	5.3	3.8	0.4
Sulfur content (dry) (%)	0.1	0.1	0.1	0.1
Ash constituents				
Silicon dioxide, SiO_2 (%)	39.0	11.1	32.0	14.3
Aluminum oxide, Al_2O_3 (%)	14.0	0.1	11.0	4.0
Iron oxide, Fe_2O_3 (%)	3.0	3.3	6.4	3.5
Calcium oxide, CaO (%)	25.5	64.5	25.3	6.0
Magnesium oxide, MgO (%)	6.5	1.2	4.1	6.6
Sodium oxide, Na_2O (%)	1.3	8.9	8.0	18.0
Potassium oxide, K_2O (%)	6.0	0.2	2.4	10.6
Titanium dioxide, TiO_2 (%)	0.2	0.1	0.8	0.3
Manganese oxide, MnO (%)	Trace	Trace	1.5	0.1
Sulfite, SO_3 (%)	0.3	2.0	2.1	7.4
Chloride, Cl (%)	Trace	Trace	Trace	18.4
Other	4.2	8.6	6.4	10.8

[a] Logs stored in saltwater

Source: R. Schwieger, "Power from Wood," *Power*, Feb. 1980. Courtesy of Babcock & Wilcox.

TABLE V Alloys Used in Boilers

Product Form	ASME Specification Number	Minimum Tensile Strength (psi)	Minimum Yield Strength (psi)	Carbon (Min.–Max.)	Manganese (Min.–Max.)	Silicon (Min.–Max.)	Nickel (Min.–Max.)	Chromium (Min.–Max.)	Molybdenum (Min.–Max.)	Other
Tubes	SA-178A			0.06–0.18	0.27–0.63					
	SA-192	(47,000)	(26,000)	0.06–0.18	0.27–0.63	0.25				
	SA-210A1	60,000	37,000	0.27	0.93	0.10				
	SA-178C	60,000	37,000	0.35	0.80					
Pipe	SA-106B	60,000	35,000	0.30	0.29–1.06	0.10				
Plate	SA-515-70	70,000–90,000	38,000	0.35	0.90	0.15–0.30				
	SA-516-70	70,000–90,000	38,000	0.31	0.85–1.20	0.15–0.30				
Tubes	SA-209-T1	55,000	30,000	0.10–0.20	0.30–0.80	0.10–0.50			0.44–0.65	
	SA-213-T11	60,000	30,000	0.15	0.30–0.60	0.50–1.00		1.00–1.50	0.44–0.65	
	SA-213-T22	60,000	30,000	0.15	0.30–0.60	0.50		1.90–2.60	0.87–1.13	
	SA-213-TP304H	75,000	30,000	0.04–0.10	2.00	0.75	8.00–11.00	18.00–20.00		
	SA-213-TP321H	75,000	30,000	0.04–0.10	2.00	0.75	9.00–13.00	17.00–20.00		Ti 0.60 Max.
	SA-213-TP347H	75,000	30,000	0.04–0.10	2.00	0.75	9.00–13.00	17.00–20.00		Ta 1.00 Max.
	SA-213-TP316H	75,000	30,000	0.04–0.10	2.00	0.75	11.00–14.00	16.00–18.00	2.00–3.00	
Plate	SA-240-304	75,000	30,000	0.08	2.00	1.00	8.00–10.50	18.00–20.00		
	SA-240-321	75,000	30,000	0.08	2.00	1.00	9.00–12.00	17.00–19.00		Ti 0.70 Max.
	SA-240-347	75,000	30,000	0.08	2.00	1.00	9.00–13.00	17.00–19.00		Ta + Cb 1.10 Max
	SA-240-316	75,000	30,000	0.08	2.00	1.00	10.00–14.00	16.00–18.00	2.00–3.00	

Source: Reprinted, with permission, from *Annual Book of ASTM Standards*, Part #1. Copyright, American Society for Testing and Materials, 1916 Race Street, Philadelphia, PA 19103.

TABLE VI Oxidation Limits

Material	ASME Specification	Temperature
Carbon steel	SA-178, SA-210, SA-192, etc.	850°F (454°C)
Carbon + ½Mo	SA-209-T1	900°F (482°C)
1¼Cr–½Mo	SA-213-T11	1025°F (552°C)
2¼Cr–1Mo	SA-213-T22	1075°F (579°C)
18Cr–10Ni	SA-213-321H	1500°F (816°C)

Source: Riley Stoker Corp., Metallurgy Dept., Standard. Courtesy of Riley Stoker Corporation.

TABLE VII Allowable Stresses (psi) ASME Code

ASME Specification Number	For Metal Temperatures Not Exceeding								
	400°F (204°C)	500°F (260°C)	600°F (316°C)	700°F (371°C)	800°F (427°C)	900°F (482°C)	1000°F (538°C)	1100°F (593°C)	1200°F (649°C)
SA-178A	11,800	11,800	11,800	11,500	7,700	(4,300)	(1,300)		
SA-192	11,800	11,800	11,800	11,500	9,000	(5,000)	(1,500)		
SA-210-A1	15,000	15,000	15,000	14,400	10,800	(5,000)	(1,500)		
SA-178-C	15,000	15,000	15,000	14,400	9,200	(4,300)	(1,300)		
SA-106-B	15,000	15,000	15,000	14,400	10,800	(5,000)	(1,500)		
SA-515-70	17,500	17,500	17,500	16,600	12,000	(5,000)	(1,500)		
SA-516-70	17,500	17,500	17,500	16,600	12,000	(7,300)			
SA-209-T1	13,800	13,800	13,800	13,800	13,500	12,700	(4,800)		
SA-213-T11	15,000	15,000	15,000	15,000	14,400	13,600	6,300	2,800	(1,200)
SA-213-T22	15,000	15,000	15,000	15,000	15,000	13,100	7,800	4,200	(1,600)
SA-213 T-91	21,200	21,100	20,800	20,000	18,700	16,700	14,300	10,300	4,300
SA-213-TP304H	13,000	12,200	11,400	11,100	10,600	10,200	9,800	8,900	6,100
SA-213-TP321H	17,100	17,100	16,400	15,800	15,500	15,300	14,000	9,100	5,400
SA-213-TP347H	15,500	14,900	14,700	14,700	14,700	14,700	14,000	9,100	4,400
SA-213-TP316H	13,400	12,500	11,800	11,300	11,000	10,800	10,600	10,300	7,400
SA-240-304	13,000	12,200	11,400	11,100	10,600	10,200	9,800	8,900	6,100
SA-240-321	12,900	12,000	11,400	11,000	10,800	10,600	10,400	6,900	3,600
SA-240-347	15,500	14,900	14,700	14,700	14,700	14,700	14,000	9,100	4,400
SA-240-316	13,400	12,500	11,800	11,300	11,000	15,600	15,300	12,400	7,400

Source: ASME Boiler and Pressure Vessel Code, Section I, Power Boilers, ASME, New York, 1989.

TABLE VIII Short-Time High-Temperature Tensile Strength

ASME Specification Number	Test Temperature °F (°C)	0.2% Offset Yield Strength (psi)	Tensile Strength (psi)	Elongation (% in 2 in.)	Reduction of Area (%)
SA-178A	80 (27)	35,000	55,000	35	64
SA-192	300 (149)	28,000	59,000	27	61
	500 (260)	23,500	59,500	25	60
	700 (371)	20,000	52,600	33	68
	900 (482)	16,000	41,000	42	76
	1100 (593)	11,200	20,000	55	87
	1300 (704)	5,600	9,900	72	94
	1500 (816)	—	5,600	90	97
SA-210	80 (27)	36,000	64,000	37	60
SA-178C	300 (149)	30,200	64,000	25	56
SA-106B	500 (260)	27,800	63,800	28	55
SA-515	700 (371)	25,400	57,000	35	63
SA-516	900 (482)	21,500	44,000	42	65
	1100 (593)	16,300	25,200	50	65
	1300 (704)	7,700	9,000	71	65
SA-209-T1	80 (27)	40,000	66,200	39	70
	300 (149)	34,900	69,000	30	68
	500 (260)	30,700	70,500	27	69
	700 (371)	27,600	69,500	26	71
	900 (482)	25,400	59,000	38	75
	1100 (593)	22,000	38,000	45	82
	1300 (704)	9,400	18,000	62	90
SA-213-T11	80 (27)	35,500	72,000	38	72
	300 (149)	31,700	71,000	31	68
	500 (260)	30,400	70,000	28	65
	700 (371)	29,200	68,000	30	67
	900 (482)	27,200	64,000	29	70
	1100 (593)	22,200	44,000	34	80
	1300 (704)	11,000	18,000	54	93
SA-213-T22	80 (27)	39,500	72,000	33	65
	300 (149)	35,800	70,000	26	60
	500 (260)	34,500	67,000	21	55
	700 (371)	34,000	64,200	20	50
	900 (482)	28,000	60,000	22	60
	1100 (593)	16,000	41,000	35	65
	1300 (784)	—	22,000	60	85

TABLE VIII *(Continued)*

ASME Specification Number	Test Temperature °F (°C)	0.2% Offset Yield Strength (psi)	Tensile Strength (psi)	Elongation (% in 2 in.)	Reduction of Area (%)
SA-213-321H	80 (27)	32,400	84,000	60	72
	300 (149)	27,500	68,000	52	70
	500 (260)	24,700	62,500	45	69
	700 (371)	22,800	60,000	40	68
	900 (482)	20,900	56,400	37	68
	1100 (593)	18,800	49,300	37	68
	1300 (704)	16,400	38,000	48	71
	1500 (816)	13,400	23,000	70	75
	1700 (927)	—	12,500	60	80
	1900 (1038)	—	6,700	65	78

Source: U. S. Steel Corp., *Steels for Elevated Temperature Service*, 1972. Courtesy of United States Steel Corporation.

TABLE IX Composition and Properties of Some Stainless Steels

Grade	Cr (%)	Ni (%)	C (%) (Max.)	Other	Typical Mechanical Properties			
					Tensile (psi)	Yield (psi)	Elongation (% in 2 in.)	BHN
Ferritic								
405	11.5/14.5		0.08		60,000	25,000	20	183
410	11.5/13.5		0.15		50,000	30,000	20	217
430	16/18		0.12		50,000	30,000	30	183
Austenitic								
304	18/20	8/10.5	0.08		75,000	30,000	40	183
304L	18/20	8/12	0.03		70,000	25,000	40	183
309	22/24	12/15	0.08		75,000	30,000	40	217
310	24/26	19/22	0.08		75,000	30,000	40	217
316	16/18	10/14	0.08	Mo 2.0/3.0	75,000	30,000	40	217
321	17/19	9/12	0.08	Ti 5 × C, Min.	75,000	30,000	40	183
347	17/19	9/13	0.08	Cb + Ta 10 × C, Min.	75,000	30,000	40	183

Source: Reprinted, with permission, from *Annual Book of ASTM Standards*, Part #1. Copyright, American Society for Testing and Materials, 1916 Race Street, Philadelphia, PA 19103.

TABLE X ASME Code Heat Treatment for Austenitic Stainless Steels

All austenitic tubes shall be furnished in the heat-treated condition. The heat-treatment procedure, except for the H grades, shall consist of heating the material to a minimum temperature of 1900°F (1038°C) and quenching in water or rapidly cooling by other means.

The heat treatment of cold-worked Grade TP-321H shall be at a minimum temperature of 2000°F (1093°C). As evidence that the material has received this treatment, it shall exhibit a grain size of No. 7 or coarser as determined in accordance with the Methods of Estimating the Average Grain Size of Metals (ASTM Designation: E 112).

All H grades shall be furnished in the solution-treated condition. If cold working is involved in processing, the minimum solution treating temperature for Grades TP-347H and TP-348H shall be 2000°F (1093°C) and for Grades TP-304H and TP-316H, 1900°F (1038°C). If the H grade is hot rolled, the minimum solution treatment for Grades TP-321H, TP-347H, and TP-348H shall be 1925°F (1052°C), and for Grades TP-304H and TP-316H, 1900°F (1038°C).

Source: Reprinted, with permission, from *Annual Book of ASTM Standards*, Part #1. Copyright, American Society for Testing and Materials, 1916 Race Street, Philadelphia, PA 19103.

TABLE XI High-Alloy Austenitic Alloys

Alloy	Cr (%)	Ni (%)	C (%) (Max.)	Other	Typical Mechanical Properties		
					Tensile (psi)	Yield (psi)	Elongation (% in 2 in.)
Inconel 601[a]	21/25	58/63	0.10	Al 1.0/1.7	90,000	40,000	50
Inconel 617[a]	22	52	0.07	Co 12.5, Mo 9.0, Al 1.0	105,000	52,000	60
Incoloy 800[a]	30/35	19/23	0.10	Al 0.15/0.60, Ti 0.15/0.60	90,000	40,000	40
Incoloy 825[a]	19.5/23.5	38/46	0.05	Mo 2.5/3.5, Ti 0.15/0.60	85,000	35,000	50
Hastelloy C-276[b]	14.5/16.5	Bal	0.02	Fe 4/7, Co 2.50 Max.	110,000	50,000	60
Hastelloy X[b]	20.5/23.0	Bal	0.15	Fe 17/20, Co 2.50 Max.	114,000	53,000	41
RA 330[c]	17/20	34/37	0.08		85,000	42,000	41
RA 333[c]	24/27	44/47	0.05	W 2.5/4.0, Mo 2.5/4.0 Co 2.5/4.0	100,000	50,000	50

[a] International Nickel Co.
[b] Cabot Corp.
[c] Rolled Alloys Corp.

TABLE XII Trace Elements in Some Crude Oils

Element	Alberta, Canada (Devonian)	Athabasca Tar Sand Crude, Canada	Boscan, Venezuela	Iranian Light Crude	Nigerian Crude	California (Tertiary) Heavy Crude
			Percentage			
S	0.44	3.9	0 002	1.28	0.35	0.98
			Parts per Million			
Ni	5	72	117	12	5.9	94
V	0.02	178	1,120	53	9.5	7.5
Na	0.04	21	25	0.6	1.1	11
Fe	0.1	254	0.33	1.4	37	73
			Parts per Billion			
Co	6.8	2,000	11	300	198	13,000
Zn	—	—	41	324	2,600	9,300
Cu	—	—	17	32	210	930
Cr	5	1,700	19	17	380	634
Se	3	520	39	58	370	360
As	1.5	320	4.8	95	1,200	660
Sb	0.01	28	0.1	0.1	273	52
Cs	0.9	69	—	—	—	—
Rb	10	720	—	—	—	—
Sc	0.1	190	—	0.07	4.4	8.8
Eu	0.2	23	0.3	0.21	—	—
Br	8.0	104	91	16	—	—
Cl	3,800	8,000	8,700	2,270	—	—
Hg	20	20	5.6	20	139	21,000

Source: R. H. Filby, K. R. Shah, and F. Yaghmaie, "The Nature of Metals in Petroleum Fuels and Coal-Derived Synfuels," in *Ash Deposits and Corrosion Due to Impurities in Combustion Gases*, R. W. Bryers, ed., Hemisphere Publishing Corp., 1978, pp. 51–64.

TABLE XIII Strength of Boiler Steels

Material	TS	YS	EL	R_B
178A	—	—	—	58.9 (48/67)
192	—	—	—	67.0 (60/76)
210 A-1	67.51	46.9	53.2	74.8
	65.2/72.3	44.4/51.1	40/63.8	72/78
SA-209	68.8	49.2	58.5	76.7
T-1A	67.2/70.4	47.1/51.3	51.5/67.1	74/79
T-5	76.8	54.0	53.1	79.8
	74.8/78.7	50.3/57.8	52/55	78/81
T-11	71.2	50.9	55.1	81.6
	67.1/77.6	33.6/63.0	47.0/64.4	76/85
T-22	74.3	49.8	51.5	80.1
	67.7/87.0	33.7/64.7	30/63.3	74/85
T-9	84.2	51.3	48.5	84.0
	83.7/84.6	46.6/56.0	47/50	83/85
T-91	119.0	98.8	21	101[a]
	109.0/129.0	83.5/113.0	20/23	98/103
304H	88.7	40.7	75.5	78.0
	82/102	36/49.6	57/84	72/81
321H	86.9	39.7	55.8	79.1
	77.1/93.4	30.9/49.8	49/62	74.5/84
347H	101.5	47.9	52.8	80.9
	91.4/108.8	40.1/52.9	49/58	73/87

[a] R_B 100 is approximately R_C 20.

TABLE XIV Effects of Sigma-Phase on Mechanical Properties of 304H Stainless Steel

	UTS		YS			
% Sigma	Sigma	S.A.	Sigma	S.A.	Sigma	S.A.
2–3	88.9	82.7	34.8	35.9	46	64
3–5	100.0	91.0	39.0	38.1	43	60
5–7	103.0	89.0	40.4	37.6	32	58
2–3	81.3	78.0	35.0	32.5	24	64
3–5	82.2	82.0	37.3	32.1	20	68

Charpy Tests, ft · lb

2–3%	17	83
3–5%	20	76
5–7%	10	91
2–3%	9	60
3–5%	23	78

UTS and YS in ksi.
S.A. = solution annealed 1950°F for 4 hr and water quenched.
Materials Performance, Dec. 1985. Data of Victor E. May.

TABLE XV Approximate Lower Critical Temperatures

Carbon steel (SA-192, SA-210)	1340°F
Carbon + ½Mo (SA-209)	1350°F
1¼Cr–1 Mo (T-11)	1430°F
2¼Cr–1 Mo (T-22)	1480°F
5 Cr–1 Mo (T-5)	1505°F

ASME Code for Pressure Piping, Power Piping ASME B31.1, 1989 edition.

INDEX

Additives:
  coal, 350-351
  oil, 367-368
Alligator hide, ash corrosion:
  cause, 348
  corrosion fatigue, 357
Allowable stress, Table, 497
Alloying elements, effect of, 122-123
  carbon, 85
  chromium, 123, 286, 242-243
  manganese, 122
  molybdenum, 122
  nickel, 242-243
  silicon, 122
Alloys of construction, 1, 16, 495
Annealing, 122. *See also* Solution anneal
Anneals, *see* Solution heat treatment
Anode, 325
  reactions, 325
Ash analysis:
  bark, 494
  coal, 346, 356
  oil, 494
  refuse, 365
  superheater, 346
  waterwall, 356
Ash corrosion, 110
  alligator hide, 348
  circumferential grooving, 348, 352-361

  control, 350-351
  effect of carbon on, 357, 367
  iron sulfide in, 344
  liquid species, 358
  reducing conditions, 357-358
  reheaters, 344, 350
  shields, 290, 350
  slag shedding, 357, 358
  superheaters, 344, 350
  trisulfates in, 344
  variable stress in, 358
  waterwall, 348, 352-361
  weld overlay, 359
ASME Code, 4
  allowable stresses, 497
  heat treatment, 501
  pressure calculation, 157, 159
  stress calculation, 25, 50-51
Atomic structure:
  austenite, 85
  body-centered cubic, 76
  face-centered cubic, 76
  ferrite, 85
  martensite, 88
Austenitic alloys:
  composition, 500, 502. *See also* Microstructures
  structures, 75
Austenitic stainless steel, 242

INDEX

Backing rings, 439
  chemical cleaning effect, 439, 442
  corrosion at, 439–440
  hydrogen damage at, 439
  local temperature at, 439–440
Backside (cold side), definition, 124
Bainite, 91
  formation, 91
  hardness, 193
  microstructure, 114
Bark (wood) analysis, 494
Bend, return, cross section, 28
Boiler:
  design, 7–10, 12–15, 21
  feedwater, control of, 381–384
  fire tune, 21
  operation, 22–23
  water tube, 21
Carbon equivalent, 428
  equation for, 428
  preheat for crack prevention, 428
Carburization, 286–290
Carburization corrosion:
  carbon content of, 287, 322
  etchant to reveal, 320
  of stainless steel, 307–309, 319–323
  temperature of, 287
Cathode, 324
  reactions, 324
Caustic attack, 340
  embrittlement, 335
  grooving, 340
Cell, concentration, 325
  differential aeration, 325
  differential temperature, 325
  dissimilar electrode, 325
Cementite, 87
  decomposition to ferrite and graphite, 87
Charpy curves, 107
Chelants, 406
  corrosion, example, 406–410
Chemical cleaning:
  decisions, 372–375
  effect on life, 377–378
  need for, 372–373
  sample location, 372
  superheater and reheater, 377–378
Chlorine:
  in coal, 366
  in refuse, 364
Chordal thermocouples, 36, 37, 357
  use, 39–40
Chromium carbide formation, 249

Clock positions, definition, 124–125
Coal analysis Table, 492
Coal-ash corrosion:
  superheater example, 393–394
  waterwall, 348, 352–361
Coefficients, heat transfer, 56, 70
  thermal conductivity:
    deposits, 59
    steam, 73
    steels, 37
  thermal expansion, 442
Cold work, *see also* Microstructures
  effect on corrosion, 326, 328
  failure by, 183–186
  Inconel 601®, 271
Congruent control, 383
Control of ash corrosion, 350–351
Convective heat, transfer, 35
Coordinated phosphate, control, 383
Copper, *see also* Microstructures
  deposits, 340, 342
  location, 459
  penetration, 458–460
  source, 342
Corrosion:
  ash:
    coal, 342–361
      superheater/reheater, 344–350
      waterwalls, 348, 352–361
    oil, 361–364
    refuse, 364–366
  cell, 325
    concentration, 325
    differential aeration, 325
    dissimilar metal, 325
  dew-point, 378
  during water washing, 404–405
    prevention, 404
  electrochemical, 324–325
  fatigue, example, 189–192, 332–333, 390–393
  fretting, 330, 331
  intergranular, 332
  low-temperature, 378
  pitting, 327
    pitting factor, 327
  stress assisted, 378–381
    cause, 379–381
    location, 378–379
  uniform attack, 326
Creep, 99–106
  ASME Code allowable, 100
  causes, 149

damage, 81, 272–276
  grain boundary, 81–82
  steam-side scale, 103–105
data, T-22, 474–475
definition, 100
expansion, 100, 294
grain size, 100
T-22 graph, 58

Decarburization, 284
Departure from nucleate boiling (DNB), 24, 148
  locations, 24
Deposit:
  definition, 125
  effect on tube temperature, 370–372
  thickness variation, 373–375
Design:
  alterations, 467–468
  furnace, 468–469
    effect of fuel, 47–49
  superheater and reheater, 40–41, 463
    effect on life, 463
Dew-point corrosion, 378
Differential-temperature cell, 325–326
Diffusion:
  carbon, 284
    in welds, 443–446
  iron, 285
  oxygen, 285
Diffusion, carbon sensitization, 249
Dissimilar-electrode cell, 325
Dissimilar-metal welds (DMW), 439, 442–447
  carbon diffusion in, 443, 446
  creep in, 443, 446
  failures in, 442–443, 448–453
    prevention, 447–448
  oxide wedges in, 443–445
  PWHT of, 448
  reheat cracking in, 458
  stress-assisted corrosion in, 454–455
  stress raiser in, 447
Ditched structure, 260
DMW, see Dissimilar-metal welds
DNB, see Departure from nucleate boiling

Elastic limit, 94
Electric-resistance weld (ERW), see Microstructures
Endurance limit, definition, 108
Energy balances, 43–45
Equilibrium microstructures, 120–121

Erosion, 290. See also Exfoliation
  corrosion, 330–332
  damage, 229–230, 291
  microstructure, 292
  operation, effect on, 330–332
  solid particle, 230
  turbine blade, 467
ERW, see Electric-resistance weld
Etch, 124
  carburization corrosion, 320
Eutectoid temperature, 89
Exfoliation, 104, 279–283
  cause, 281
  microstructure, 282–283
  and turbine blade erosion, 467
Expansion:
  coefficients, 442
  lateral, Charpy test, 107

Failure:
  ash corrosion of superheater, 344, 350
  carburization, 286–290, 307–309, 319–323
  chelant corrosion, 406–410
  cold work, 183–186
  corrosion fatigue, 108–110, 189–192, 224, 332–333, 390–393
  creep, 99–106, 210–216
  decarburization, 304–307
  erosion, 290–293
  furnace wall:
    ID, 395–396
    OD, 348, 352–361
  graphitization, 115–120, 161–167, 203–213
  high-temperature:
    long-term, 159–161
    reheater, 51–59
    SA-178, 149–153, 181–183, 231–235
    SA-209 T-1, 153–156
    SA-210 A-1, 156–159
    SA-213 T-11, 170–172
    SA-213 T-22, 172–178, 178–181
    SA-213 TP304, 267–270
    SA-213 TP321H, 254–259, 272–276
  hydrogen damage, 384–385, 410–414
  improper material, 167–169
  low-water upset, 192–199
  oil ash:
    reheater, 400–402
    superheater, 402–404
  oxygen pitting, 386–390
  prevention, 462–489
  sigma phase:
    309 stainless, 259–260

INDEX **509**

Failure, sigma phase (*Continued*)
  321H stainless, 261-264, 272-276
  statistics, 2-3
    corrosion, 2
    location, 3
    materials, 3
    mechanical, 2
  thermal fatigue, 108-110, 221-222, 225-229
    SA-178A, 183-186
    SA-210 A-1, 199-203
    SA-213 TP347H, 265-267
  U-bend, 170-172
  at wall-thickness transition, 216-221
  waterwalls:
    creep, 210-216
    ID deposit, 395-396, 397-400
    OD deposit, 351-361
  water washing, 404-405
  welds, 18
    dissimilar metals, 439-447
Fatigue, 107-108
  corrosion fatigue, 108-110
  thermal fatigue, 108-110
Ferrite, 85
  carbon solubility in, 85
  delta, 85
  microstructure, 78
  pro-eutectoid, 90
    definition, 93
  structure, 85
Ferritic stainless steels, 242
Fire side:
  corrosion, 269-342
  definition, 124
Flame impingement, 148
Flow resistance, effect on tube temperature, 47
Fluid flow, 41-47
  restriction, 148
Fly-ash erosion, 230, 292
Fracture appearance transition temperature (FATT), 107
Fretting corrosion, 330, 331
Friction factor, 45
Fuel:
  bark analysis, Table, 494
  coal analysis, Table, 493
  effects of changes in, 470-472
    on furnace size, 47-49
  heat values:
    coal, 47
    gas, 47
    oil, 47
  oil analysis, Table, 494
Fuels, 21
  bagasse, 11
  by-product gas, 11
  coal, 6
  coke, 15
  miscellaneous, 16
  municipal refuse, 11
  natural gas, 6
  oil, 6
  wood, 11
Furnace design, 468-469
Furnace wall corrosion, ID example, 395-396

Grain:
  boundary, 77
    oxidation, 79, 80, 287
    growth, 122, 268, 269, 304
    and decarburization, 284
  shape, 81-82
  size, 79
    ASME Code requirement, 321H, 79
Graphitization, 111, 115-120
  effect of temperature, 115
  effect of time, 117
  microstructures, 113, 116-119
  proof of, 120

Hardness:
  bainite, 193
  Brinell, 96
  cold-worked SA-178C, 184
  Rockwell, 98
  and tensile strength, 98
HAZ, *see* Heat-affected zone
Heat absorption, furnace, 37-38
Heat-affected zone (HAZ), 251, 422-429
  changes in, 424-426
    austenitic steels, 424
    ferritic steels, 424-426
  corrosion of, 442, 448, 454-455
  definition, 424
  effect of cooling rate on, 425-426
  grain growth in, 424, 429, 431
  and iron-carbon diagram, 425
  martensite in, 425
  microstructures, 429-438
    effect of service on:
      SA-209 T-1, 434-435
      SA-213 T-22, 434, 436
  recrystallization in, 424, 431

510   INDEX

SA-106B, 430–431
SA-178A, 431–432
SA-213 T-22, 433–434
SA-213 TP304, 436–437
residual stress in, 428
thermal expansion of, 426
Widmanstätten structure, 427
Heat-flow analysis, 31–41, 51–54
  of blisters, 60–74
Heat flux, 32, 36
Heat transfer:
  coefficient, 32, 43, 56
  conduction, 31–35
  convection, 35
  effect of ID scale, 51–59
  plane area, 31–32
  radiation, 35
  tubes (cylinders), 32–35
Heat treatment, 120, 122
  annealing, 122
  grain growth, 122, 256, 268
  normalization, 122
  quenching, 122
  solution, 122, 246, 248
  spheroidization, 122
  stabilization, 250
  stress-relieving, 122
  tempering, 122
Hematite, 285
Hideout, 283
High-temperature:
  failure, 142
  tensile strength, 98, 498, 499
Hooke's law, 93
Hoop stress:
  ASME Code equation for, 25
  calculation, 238, 240, 397, 476, 477
    thin-wall cylinder, 25–26, 30
      comparison, 26–27
    thick-wall cylinder, 26
Hot spots, reheater tubes, example, 312–318
Hydrazine, 382
  oxygen scavenger, 382
Hydrogen damage, 335–340, 410–419
  appeaance, 337
  boiler locations, 335
  example, 384–386
  grain boundary, 81
  tests for, 384–385
  and under-deposit corrosion, 335
Hydrogen sulfide, in, refuse ash, 366

Impact properties, 106–107

Inconel:
  601®, 252, 271
  625®, 252
  stress-corrosion cracking in, 405–406
  weld overlays of, 369
Indicator deposits, 340
In-situ spheroidization, 152
  definition, 111
Intergranular attack, 80
  corrosion, 81, 332
  304H, 251–321
  321H, 263, 275
  347H, 80
  oxidation, 79–80
  T-22, 302
Ions, 324
Iron–carbon phase diagram, 85–86
  –chromium phase diagram, 243
  –graphite phase diagram, 87
  –oxygen phase diagram, 286

Larson-Miller parameter (LMP), 98, 99,
  114, 300, 464, 473
  and scale formation, 177–178, 464–465
  and stress curves, 473, 477–483, 487
  use of, 473, 476
Lateral expansion, Charpy test, 107
Lay-up procedures, 326, 388–390
Lead penetration, 459
Lever rule, 85–87, 90
  peak temperature from, 114–115
Life assessment, commputer simulations,
  483–489
Liquid-ash corrosion:
  coal ash, 343, 351–361
  oil ash, 361–364
  refuse ash, 364–366
LMP, *see* Larson-Miller parameter
Lower critical transformation temperature,
  Table, 505
Low-temperature corrosion, 378
Low-water upset, example, 192

Magnesium hydroxide, oil additive, 367
Magnesium oxide, oil additive, 367
Magnetite formation, 277–279, 284–285,
  464–466
Martensite, 88–89. *See also* Microstructures
  microstructure, 89
  start temperature, Mg, 91
Mechanical properties, 16, 93
Metals, boiler use, 1, 16, 495
Methane, in hydrogen damage, 335–340

INDEX **511**

Microstructures:
  annealed Inconel 601®, 271
  austenite, 78
  bainite, 91, 114, 137, 138, 140, 141, 143, 144, 155, 158, 214, 233
  banded, 143, 144, 147, 198, 233
  carbide precipitation, from ferrite, 120–121
  carburized, 288–289, 309
  caustic attack, 341
  cold-worked, 185, 245
    Inconel 601®, 271
  copper deposits, 342, 459–460
  corrosion fatigue, 109, 110, 191, 223–225
  creep damage, 68, 82, 101–106, 212, 213, 214, 218, 273, 296, 315–316
  decarburized, 179, 180, 234, 284, 306
  deformation, 95, 97
    twinning, 96, 97
  ditched, 260
  effect on corrosion, 455–456
  electric-resistance weld (ERW):
    failure, 231–235, 235–240
    fusion line, 97, 188, 239
    lack of fusion, 236–237
  erosion, 229–230
  exfoliation, 282, 283
  ferrite, 78
  ferrite, pearlite and bainite, 142
  flow instability, 228–229
  grain growth, 269
  graphitized, 68, 163, 165–166, 168, 169, 213, 229
  hydrogen-damaged, 337, 339, 340, 385, 386
  Inconel 601®, 253, 254
  Inconel 625®, 253
  In-situ spheroidization, 152
  martensite, 89
  nickel-based alloys, 253, 254, 271
  oxide wedge, 444, 445, 450, 452
  oxygen penetration, 80, 287
  oxygen pitting, 327, 328, 329, 330, 387, 388, 389, 411, 412
  pearlite, 90, 111, 126, 128, 130, 145, 146, 147, 152, 156, 158, 166, 197, 198, 215, 233, 239
  plastic deformation, 95
  recrystallized, 246, 247
  renormalized, 197
  scale, 103–105, 279–281, 317–318, 319
  sensitized, 250, 251
  sigma phase, 248, 264, 274, 275
  slip, 244, 271
  spheroidized carbides, 83, 212
  steels:
    SA-106B, 145
    SA-106C, 145
    SA-178A, 63, 65, 67, 68, 126–127, 160–161, 239
    SA-178C, 151–152, 182–185
    SA-192, 117
    SA-209 T-1, 91, 111–114, 116, 117, 119, 128, 129, 154–156, 158
    SA-210 A-1, 118, 163, 194–198, 201–202
    SA-213 T-5, 135, 295–297
    SA-213 T-11, 130, 131, 171
    SA-213 T-22, 53, 80, 132, 133, 174–177, 179, 180, 218, 298, 303
    SA-213 T-9, 135, 136
    SA-213 T-91, 134
    SA-213 TP304H, 244, 247, 251, 256–258
    SA-213 TP309, 260
    SA-213 TP321H, 244, 250, 263, 264
    SA-213 TP347H, 80, 266, 267
    SA-515-70, 90, 146
    tool steel, 36
  subgrains, 140
  thermal fatigue, 109, 187–188, 192, 201, 222, 225
  from sootblower, 226–227
  twins, 244
  weld:
    carbon steel, 456
    304, 436
    Incoweld A®, 444
    310, 446
  Widmanstätten, 92, 139

Nickel-based alloys, 252–254
  composition, 502
  microstructure, 253, 254, 271
Nil ductility transition temperature (NDTT), 107
Normalizing, 122
Nusselt number, 36

Oil:
  analysis, 493
  ash:
    analysis, 402, 403
    Table, 494
    x-ray diffraction of, 403
  corrosion, 361
    effects of carbon on, 402–404
      reheater tube, example, 400–402
      superheater tube, example, 402–404

temperature effect, 363
Operational problems, 469–470
Orientation, clock positions, definition, 124–125
Overall heat transfer, coefficient, 34, 56
Oxidation, 284–286, 324, 325, 463–466
 limit, 123, 370, 496
 parabolic rate, 285
Oxide wedge, 108, 333, 443, 444, 445, 450, 452
Oxygen pitting, 326–330. *See also* Microstructures, oxygen pitting
 superheater example, 386–390
Oxygen scavenger:
 hydrazine, 382
 sodium sulfite, 382

Parabolic rate law, 277, 285, 465, 472
 graph, 278
Pearlite, 87
 morphology, 90
Phase:
 Cr-Fe (sigma), 248–249, 260, 274
   effect on strength, Table, 504
 Cu-Ni, 85
 definition, 83
Phase Diagrams, 83–86
 Cr-Fe, 243
 Cu-Ni, 84
 Pb-Sn, 84
 Fe-C, 86
 Fe-O, 286
 Na_2O-V_2O_5, 362
 V_2O_5-Na_2SO_4, 363
 $Na_3FE(SO_4)_3$-$K_3Fe(SO_4)_3$, 372
Pitting:
 effects of manufacturing, 326, 328
  storage, 386
 location, 327
 oxygen, 326–330
 ratio, 326
 in superheater tube, 386–390
Plastic deformation, 94
 microstructure, 95
Prandtl number, 36
Preheat, ASME Code, 429
Pressure drop, 45
Pyrosulfates, 351

Quenching, 122

Radiation heat transfer, 35. *See also* Heat transfer
Rankine temperature, 98

Recrystallization, 122
 in HAZ, 424, 429, 434
  microstructures, 436, 437
 effect of temperature, 434, 437
Reducing atmosphere, 364
Reduction, 324
Refuse ash analysis, 365
 corrosion, 364–366
 effects of chlorine, 364
 erosion, 364
 reducing atmosphere, 364
Reheat cracking, 455–458
 cause, 455–456
 in DMW, 458
 explanation, 456
 in P-22, 457
 in T-11, 457
 susceptible alloys, 455
Reheater tube:
 design calculation, example, 50–51
 failure, example, 159–161
Remaining-life calculations, 472–489
Reynold's number, 36
Rifled tubing, 38–39
 DNB in, 39
Rust, 326

Sample preparation, metallographic, 123
Scale, *see also* Steam-side scale
 definition, 125
 effects on OD temperature, 34–35, 51–59, 370–371
 formation, 277–279, 472
  parabolic rate law, 277–278
 thickness on T-22, 279
 thickness on 304, 279
Sensitization, 249–251
 and carburization corrosion, 322
Shields, 290
Sigma phase, 248–249. *See also* Phase
 effect on strength, 504
 ferrite effect on, 251–252
 microstructure, 248, 264, 274, 275
 temperature of formation, 259
Slip, 95
Sodium sulfite, 382
 oxygen scavenger, 382
Solid particle erosion, 230
Solid solutions, 77
 interstitial, 77
 substitutional, 77
Solution heat treatment (solution anneal), 122, 246, 248
Spectator deposits, 340

Spheroidization, 110–111, 122
Stainless steel:
  austenitic, 242–243
  composition of, Table, 500
  definition of, 77, 242
  ferritic, 242
  low-carbon, 249
  heat treatment of, 252, 501
  sensitization, 249–251
  sigma phase, 248–249
  solution anneal, 246
Steam:
  blanket, 151, 340
  cycle, 21, 23–24
  film coefficient, 36, 56
  superheating, 41
  temperature profile, 42
  thermal conductivity, 73
  -washing, 292, 309–310
Steam-side scale, 277
  effects of creep on, 100, 103–105
  effects on tube temperature, 278, 299–300, 465–466
  microstructures, 103–105, 279–281
  temperature calculation, 177–178, 464–465
  thickness, T-22 and 321 comparison, 279
Strength:
  ASME Code, 497
  Charpy, 106–107
  SA-515-70, 107
  fatigue, 107–108
  high-temperature, 98, 498–499
  impact, 106–107
  S-N curve, 108
  steels, 500, 502, 504
  stress-rupture, 98–99, 474–475, 477–482
  tensile, 93–94
    elevated-temperature, 498
    room-temperature, 504
  yield, 94
Stress-assisted corrosion, 378–381, 448, 454–455
  in liquid-ash corrosion, 454–455
Stress calculations:
  hoop stress, 25–31
    cylinder, 25–26
    at failure, 157–158
    and remaining life, 476, 477, 482, 484
    torus, 27–28
Stress-corrosion cracking, 333–335
  ferritic steels, 334
  Inconel 625®, example, 405–406
  stainless steels, 336

Stress relieving, 122
  effect on hardness, 186
Subgrains, 140
Sulfates, in ash corrosion, 344, 351, 356
Sulfuric acid:
  dew-point corrosion, 378–379
Sulfur print, 344–345, 346, 356, 395, 403
Sulfur trioxide, 343–344, 356
  equilibrium with SO_2, 343
Supercritical pressure, 24
Superheaters and reheaters:
  corrosion in, 375–377
  effect of temperature on corrosion in, 376–377
  life expectancy, 371

Tarnishing of silver, 326
Temperature:
  calculation, 177–178, 477
    effect of assumptions on, 487–489
  design limit, 303
  drop:
    gas-side, 35, 55
    deposit-scale, 35, 55
    steam-side, 35, 55
    tube-wall, 35, 55
  effect of ID scale on, 278, 299–300, 465–466
  Rankine, 98
Tempering, 122
Tensile strength, see Strength, tensile
Thermal conductance, 32
Thermal conductivity:
  ID scale-deposits, Table, 59
  steam, 73
  steels, 37
Thermal fatigue, 186–189, 201, 265–266
Thermal resistance, 32
  gas-side, 34, 35
  ID scale-deposit, 34, 55
  steam-side, 34, 55
  tube-wall, 34, 55
Thermocouples:
  chordal, 36, 37, 357
  use, 39–40
T-T-T curves, 88–93
  effects of alloying, 88
  graph, 88, 425
Titanium nitride, in Inconel 601®, 271
Trace elements, crude oils, Table, 503
Trisodium phosphate, 383
Trisulfates:
  decomposition temperature, 350
  formation, 344

melting point, 372
Tube-wall thinning, 29–30
  safe limit, 30–31
Tubing, rifled, 38–39
Twinning:
  austenite, 95, 96, 244, 245
  ferrite, 96–97
  microstructure, 96

Vanadium pentoxide, 362
  phase diagram:
    with sodium oxide, 362
    with sodium sulfate, 363
Volatile treatment, 383

Water:
  treatment, 381–383
    congruent control, 383
    coordinated-phosphate control, 383
    hydrazine, 382–383
    sodium-sulfite, 382
    volatile, 383
    zero-solids control, 383

washing, 404–405
Waterwall, see also Failures, waterwall
  design calculation, 50–51
  wall-thickness calculation, 50
Welding:
  dissimilar-metal, 439–448
  effects of notch on, 444–445
  heat-affected zone, 422–438
  internal configuration, 439–440
  stainless steel, 251–252, 434, 436–438
Weld overlays:
  Inconel 625®, 369
  stainless steel, 359–360
Widmanstätten structure, 92–93
  formation, 93
Work hardening, 96, 243–245
Wüstite, 285

Yield point, 94
Young's modulus, 93

Zeolite softening, 382
Zero-solids control, 383

Printed in the United States
114459LV00001B/4/A